T0250067

CHEMICAL AND
B■ OCHEMICAL TECHNOLOGY

Materials, Processing, and Reliability

CHEMICAL AND
BIOCHEMICAL TECHNOLOGY
Materials, Processing, and Reliability

Edited by
Sergei D. Varfolomeev, DSc

A. K. Haghi, PhD, and G. E. Zaikov, DSc
Reviewers and Advisory Board Members

Apple Academic Press

TORONTO   NEW JERSEY

Apple Academic Press Inc. | Apple Academic Press Inc.
3333 Mistwell Crescent | 9 Spinnaker Way
Oakville, ON L6L 0A2 | Waretown, NJ 08758
Canada | USA

©2015 by Apple Academic Press, Inc.

First issued in paperback 2021

Exclusive worldwide distribution by CRC Press, a member of Taylor & Francis Group
No claim to original U.S. Government works

ISBN 13: 978-1-77463-360-1 (pbk)
ISBN 13: 978-1-77188-044-2 (hbk)

Library of Congress Control Number: 2014952503

Library and Archives Canada Cataloguing in Publication

Chemical and biochemical technology : materials, processing, and reliability/edited by Sergei D. Varfolomeev, DSc; A.K. Haghi, PhD, and G.E. Zaikov, DSc, Reviewers and Advisory Board Members.

Includes bibliographical references and index.
ISBN 978-1-77188-044-2 (bound)
. Chemistry, Technical. 2. Biochemistry. 3. Chemical engineering. I. Varfolomeev, Serge™i Dmitrievich, editor

TP145.C54 2014   660   C2014-906825-5

Apple Academic Press also publishes its books in a variety of electronic formats. Some content that appears in print may not be available in electronic format. For information about Apple Academic Press products, visit our website at **www.appleacademicpress.com** and the CRC Press website at **www.crcpress.com**

ABOUT THE EDITOR

Sergei D. Varfolomeev, DSc

Sergei D. Varfolomeev, DSc, is Professor and Director of N. M. Emanuel Institute of Biochemical Physics and Head of Division in Chemical Faculty of M. V. Lomonosov Moscow State University, Moscow, Russia. He is a specialist in the field of biochemical kinetics, biochemical physics, and chemistry and physics of biopolymers and composites. He is the contributor of 1000 original papers and reviews as well as 100 books.

REVIEWERS AND ADVISORY BOARD MEMBERS

A. K. Haghi, PhD

A. K. Haghi, PhD, holds a BSc in urban and environmental engineering from University of North Carolina (USA); a MSc in mechanical engineering from North Carolina A&T State University (USA); a DEA in applied mechanics, acoustics and materials from Université de Technologie de Compiègne (France); and a PhD in engineering sciences from Université de Franche-Comté (France). He is the author and editor of 65 books as well as 1000 published papers in various journals and conference proceedings. Dr. Haghi has received several grants, consulted for a number of major corporations, and is a frequent speaker to national and international audiences. Since 1983, he served as a professor at several universities. He is currently Editor-in-Chief of the *International Journal of Chemoinformatics and Chemical Engineering* and *Polymers Research Journal* and on the editorial boards of many international journals. He is a member of the Canadian Research and Development Center of Sciences and Cultures (CRDCSC), Montreal, Quebec, Canada.

Gennady E. Zaikov, DSc

Gennady E. Zaikov, DSc, is Head of the Polymer Division at the N. M. Emanuel Institute of Biochemical Physics, Russian Academy of Sciences, Moscow, Russia, and Professor at Moscow State Academy of Fine Chemical Technology, Russia, as well as Professor at Kazan National Research Technological University, Kazan, Russia. He is also a prolific author, researcher, and lecturer. He has received several awards for his work, including the Russian Federation Scholarship for Outstanding Scientists. He has been a member of many professional organizations and on the editorial boards of many international science journals.

CONTENTS

LIST OF CONTRIBUTORS

E. A. Mamin Eldar Alievich
Emanuel Institute of Biochemical Physics of Russian Academy of Sciences

I. Baranov
Vernadsky Institute of Geochemistry and Analytical Chemistry, Russian Academy of Sciences, ul. Kosygina 19, Moscow, 119991 Russia, E-mail: baranov_50@mail.ru

S. A. Bekuzarova
Gorsky State Agrarian University, 362040, Republic of North Ossetia – Alania, Vladikavkaz, Kirov st., 37, Vladikavkaz, Russia

A. A. Belov
University of Chemical Technology of Russia (RChTU) him. D.I.Mendeleev, ch. Biotechnologies, Research Institute of Textile Materials, Moscow, Russian Federation E-mail: ABelov2004@ yandex.ru

V. A. Belyayeva
Institute for Biomedical Research of and Government of the Republic North Ossetia – Alania, (362019, North Ossetia Alania, Vladikavkaz, Pushkinskaya st., 40), Vladikavkaz, Russia

A. Ja Bome
Tyumen basing point of the All-Russia research institute of plant growing of N.I. Vavilov, Bol'shaya morskaya str., 190000, St. Petersburg, Russia

N. A. Bome
Tyumen State University, Semakova str., 1, 625003, Tyumen, Russia. Tel: +7(3452) 46-40-61, 46-81-69. E-mail: bomena@mail.ru

V. V. Chernova
Bashkir State University, 450074, Russia, Republic of Bashkortostan, Ufa, Zaki Validi st., 32

Jakub Czakaj
AIB Ślączka, Szpura, Dytko spółka jawna, Knurów

G. S. Dmitrieva
Lomonosov Moscow State University of Fine Chemical Technologies, Moscow, Russia

A. V. Doronin
National Scientific Centre "Institute of Agrarian Economics" of Ukraine 10, Heroiv Oborony Str., Kyiv, Ukraine

I. A. Dudareva (Cherkashina)
Tobolsk Complex Scientific Station, Ural Division of the Russian Academy of Sciences

M. R. El-Aassar
Polymer Materials Research Department, Institute of Advanced Technology and New Material, City of Scientific Research and Technology Applications, New Borg El-Arab City 21934, Alexandria, Egypt

M. S. Mohy Eldin
Polymer materials research Department, Institute of Advanced Technology and New Material, City of Scientific Research and Technology Applications, New Borg El-Arab City 21934, Alexandria, Egypt

L. S. Ghishkayeva
Chechen State University (364907, Chechen Republic, Grozny, Sheripovst., 32; Tel: +7(8712)21-20-04), Grozny, Russia

M. D. Goldfein
Saratov State University named after N.G. Chernyshevskiy

M. A. Goldshtrakh
N.N. Semenov Institute of Chemical physics Russian Academy of Sciences 119991 Moscow, street Kosygina, 4

S. V. Gorbachev
M.V. Lomonosov Moscow State University of Fine Chemical Technologies, 119571 Moscow, pr. Vernadskogo, 86

A. S. Gorshkova
M.V. Lomonosov Moscow State University of Fine Chemical Technologies, 119571 Moscow, pr. Vernadskogo, 86

G. A. Gromova
M.V. Lomonosov Moscow State University of Fine Chemical Technologies, Moscow, Russia

E. A. Hassan
Department of Chemistry, Faculty of Science, Al-Azhar University, Cairo/Egypt

A. M. Ionov
Institute of Solid State Physics RAS, 142432 Chernogolovka, 2 Academician Ossipyan str., Russia; E-mail: ionov@issp.ac.ru

A. V. Ivanov
Saratov State University named after Chernyshevskiy

I. M. Khaniyeva
V.M. Kokov's Kabardino-Balkarsky State Agrarian University, 360000, Nalchik, Lenin Avenue, 1v; Tel: +7(8662) 47-41-77, Nalchik, Russia

S. D. Khizhnyak
Tver State University, Russia 170100 Tver, Zhelyabova, 33; E-mail: sveta_khizhnyak@mail.ru

N. N. Kolokolova
Tyumen State University, Semakova str., 1, 625003, Tyumen, Russia

E. V. Kopylova
M.V. Lomonosov Moscow State University of Fine Chemical Technologies, 119571 Moscow, pr. Vernadskogo, 86

A. I. Korotaeva
University of Chemical Technology of Russia (RChTU) him. D.I.Mendeleev, ch. Biotechnologies, Research Institute of Textile Materials, Moscow, Russian Federation

N. V. Kozhevnikov
Saratov State University named after N.G. Chernyshevskiy

B. E. Krisyuk
The Institute of Problems of Chemical Physics of the Russian Academy of Sciences (IPCP RAS)

L. Z. Kravtsova
the "NTC BIO", LLC, 309292 Russia, Belgorod region, Shebekino town, Dokuchayev str., 2

Nataliya M. Kucher
National dendrological park "Sofiyivka" of NAS of Ukraine, 12-aKyivska Str., Uman, Cherkassy region, Ukraine 20300

E. I. Kulish
Bashkir State University, 450074, Russia, Republic of Bashkortostan, Ufa, Zaki Validi st., 32. Tel./Fax : +7 (347) 229 97 24

N. G. Lebedev
Volgograd State University, Volgograd, Russia

C. A. Liman
the "Agroakademia", LLC, 309290 Russia, Belgorod region, Shebekino town, A. Matrosov str. 2A

V. I. Lobyshev
Faculty of Physics, M.V. Lomonosov Moscow State University, Moscow, 119991 Russia, E-mail: lobyshev@yandex.ru

A. V. Maiorov
Emanuel Institute of Biochemical Physics of Russian Academy of Sciences

K. Majewski
Military Institute of Chemistry and Radiometry, 105 Allea of General A. Chrusciela, 00-910 Warsaw, Poland

A. F. Meysurova
Tver State University, Russia 170100 Tver, Zhelyabova, 33

V. Mikhailov
Vernadsky Institute of Geochemistry and Analytical Chemistry. Russian Academy of Sciences, ul. Kosygina 19, Moscow, 119991 Russia

R. N. Mozhchil
National Nuclear Research University of Moscow Engineering Physics Institute, 115409, Moscow, Kashirskoe shosse, 31

Nenka Maksym Mykolayovych
Post-Graduate Student Institute of Bioenergy Crops and Sugar Beet of National Academy of Agrarian Science of Ukraine 44 Lenina Str., Dmytrushky village, Uman district, Cherkasy region

G. Nyszko
Military Institute of Chemistry and Radiometry, 105 Allea of General A. Chrusciela, 00-910 Warsaw, Poland, E-mail: Grzegorz.Nyszko@wichir.waw.pl

Korneeva Myroslava Oleksandrivna
Institute of Bioenergy Crops and Sugar Beet of National Academy of Agrarian Science of Ukraine

A. A. Olkhov
N.N. Semenov Institute of Chemical physics Russian Academy of Sciences 119991 Moscow, street Kosygina, 4, E-mail: aolkhov72@yandex.ru

Anatoliy Iv. Opalko
Uman National University of Horticulture, 1 Instytutska Str., Uman, Cherkassy region, Ukraine 20305

Olga A. Opalko
National dendrological park "Sofiyivka" of NAS of Ukraine, 12-aKyivska Str., Uman, Cherkassy region, Ukraine 20300; E-mail: opalko_a@ukr.net

P. M. Pakhomov
Tver State University, Russia 170100 Tver, Zhelyabova, 33

D. S. Pavlov
1 - A.N. Severtsov Institute of Ecology and Evolution, Russian Academy of Sciences, 119071 Russia, Moscow, Leninskij prosp., 33

J. Pielichowski
Cracow University of Technology, Department of Polymer Science and Technology, Warszawska str., 31-155 Krakow, Poland

S. V. Ponomarev
the "Bioaquapark" Innovation Centre– the Scientific Centre of the Aqua-Culture at the ASTU, 414025, Astrakhan, Tatischev str., 16

A. A. Popov
Emanuel Institute of Biochemical Physics of Russian Academy of Sciences, E-mail: popov@sky.chph.ras.ru

N. I. Poteshnaya
Vernadsky Institute of Geochemistry and Analytical Chemistry. Russian Academy of Sciences, ul. Kosygina 19, Moscow, 119991 Russia

V. G. Pravdin
the "NTC BIO", LLC, 309292 Russia, Belgorod region, Shebekino town, Dokuchayev str., 2

Maria Rajkiewicz
Institute for Engineering of Polymer Materials and Dyes Department of Elastomers and Rubber Technology in Piastów

E. A. Raspopova
University of Chemical Technology of Russia (RChTU) him. D.I. Mendeleev, ch. Biotechnologies, Research Institute of Textile Materials, Moscow, Russian Federation

E. V. Rubtcova
Faculty of Physics, M.V. Lomonosov Moscow State University, Moscow, 119991 Russia, E-mail: E. V. Rubtcova, E-mail: ev.rubcova@physics.msu.ru

V. D. Rumyantseva
M.V. Lomonosov Moscow State University of Fine Chemical Technologies, 119571 Moscow, pr. Vernadskogo, 86; E-mail: vdrum@mail.ru

Marcin Ślączka
AIB Ślączka, Szpura, Dytko spółka jawna, Knurów

A. B. Solovey
Faculty of Physics, M.V.Lomonosov Moscow State University, Moscow, 119991 Russia, E-mail: soloveybird@gmail.com

A. V. Solovey
Semenov Institute of Chemical Physics, Russian Academy of Sciences, Moscow, Russia

S. A. Sudorgin
Volgograd State Technical University, Volgograd, Russia

I. F. Tuktarova
Bashkir State University, 450074, Russia, Republic of Bashkortostan, Ufa, Zaki Validi st., 32. Tel./Fax : +7 (347) 229 97 24; E-mail: tuktarova_irina@rambler.ru

N. A. Ushakova
A. N. Severtsov Institute of Ecology and Evolution, Russian Academy of Sciences, 119071 Russia, Moscow, Leninskij prosp., 33, E-mail naushakova@gmail.com

L. I. Weisfeld
N. M. Emanuel InstituteofBiochemicalPhysicsRAS, Moscow, Russia, 119334 Kosygina str., 4, Moscow, Russia

G. E. Zaikov
N. M. Emanuel Institute of Biochemical physics Russian Academy of Sciences 119334 Moscow, street Kosygina

LIST OF ABBREVIATIONS

AA	Acrylic Acid
ACN	Acetonitrile
AIBN	Azo Isobutyronitrile
ALA	Allylacrylate
ANOVA	Analysis of Variance
APS	Ammonium Persulfate
ASSSC	Aqueous Solution of Sodium Sulfocyanide
BA	Butylacrylate
BFS	Breadth-First Search
BP	Benzoyl Peroxide
BSA	Bovine Serum Albumin
CbS	Cobalt
CETI	Communication with ExtraTerrestrial Intelligence
CFC	Chlorofluorocarbons
CHCl3	Chloroform
CN	Coordination Number
CNT	Carbon Nanotubes
CpS	Copper
CTABr	Cetyltrimethylammonium Bromide
DAC	Dialdehydecellulose
DBTL	Dibutyl Tin Dilaurate
DMF	Dimethylformamide
DMFA	Dimethyl Formamide
DMTA	Dynamic Thermal Analysis
DSC	Differential Scanning Calorimetry
EA	Extended Area
EOE	Ethylene-Octene Elastomer
EPDM	Ethylene-Propylene-Diene Elastomer
FPU	Foamed Polyurethane
FTIR	Fourier Transform Infrared
Gl	Glycerol
Hem	Bovine Hemoglobin
HQ	Hydroquinone
HRP	Hemoglobin Peroxidase
HTC	Hydro-Thermic Coefficient

IA	Itaconic Acid
IS	Iron
LDPE	Polyethylene of Low Density
LS	Lead
MAA	Methacrylic Acid
MAS	Methallyl Sulfonate
MMA	Methylmethacrylate
MR	Microwave Radiation
PAA	Polyacrylamide
PB	Phosphate Buffer
PC	Hepatopancreas Crab
PEG	Polyethylene Glycol
PVA	Polyvinyl Alcohol
PVP	Polyvinylpyrrolidone
RD	Ruling Documents
RED	Renewable Energy Directive
REE	Rare-Earth Elements
SDS	Sodium Dodecyl Sulfate
SSH	Simple Sterile Hybrids
TCQM	Tetracyanoquinodimethane
TEM	Transmission Electron Microscopy
TGA	Thermogravimetric Analysis
TPS	Tver'thermal Power Stations
UHV	Ultra High Vacuum
ZS	Zinc

LIST OF SYMBOLS

A_{ms}	coefficients of the Fourier expansion
t_{Δ}	electron hopping integral
$\varepsilon_{l\sigma}$	energy of the electron
$c_{j\sigma}$	Fermi annihilation
$d_{j\sigma}^{+}$	Fermi creation operators of electrons
$f_s(\mathbf{p},\mathbf{r})$	Fermi distribution function
X_0	inflection point
V_{lj}	matrix element of hybridization
X_{max}	maximum coordinate of the point X
U_0	upper asymptote
Q_{0i}	zero vibration amplitudes
$\sum P$	precipitation amount
$\sum T$	sum of temperatures
A	absorbance of a sample of thickness
HTC	hydrothermal coefficient
N	number of carbon atoms in the lattice
Nimp	number of adsorbed hydrogen atoms
NQ	percentage of the normalized coordinates
px	parallel component of the graphene sheet
R	gas constant
R	regeneration coefficient
T	absolute temperature
V	hybridization potential
v	partial specific gravity
x	concentration of the absorbing species
X	predominantely of proline
Y	hydroxyproline
ρ0	density of the solvent

PREFACE

By providing an applied and modern approach, this volume will help readers understand the value and relevance of studying chemical physics and technology to all areas of applied chemical engineering. It gives readers the depth of coverage they need to develop a solid understanding of the key principles in the field. Presenting a wide-ranging view of current developments in applied methodologies in chemical and biochemical physics research, the papers in this collection, all written by highly regarded experts in the field, examine various aspects of chemical and biochemical physics and experimentation.

The book:

- highlights applications of chemical physics to subjects that chemical engineering students will see in graduate courses
- introduces the types of challenges and real problems that are encountered in industry and graduate research
- provides short chapters that introduce students to the subject in more bite-sized pieces.
- presents biochemical examples and applications
- focuses on concepts above formal experimental techniques and theoretical methods

The book is ideal for upper-level research students in chemistry, chemical engineering, and polymers. The book assumes a working knowledge of calculus, physics, and chemistry, but no prior knowledge of polymers.

CHAPTER 1

ELECTROPHYSICAL PROPERTIES OF CARBON NANOSTRUCTURES WITH DEFECTS IN A STRONG EXTERNAL ELECTRIC FIELD

S. A. SUDORGIN and N. G. LEBEDEV

CONTENTS

ABSTRACT

Examines the influence of defects on the electrical properties of carbon nanostructures in an external electric field. Defects are the hydrogen atoms, which adsorbed on the surface of carbon nanostructures. Carbon nanostructures are considered the single-walled zigzag carbon nanotubes Atomic adsorption model of hydrogen on the surface of single-walled zigzag carbon nanotubes based on the single-impurity Anderson periodic model. Theoretical calculation of the electron diffusion coefficient and the conductivity of zigzag carbon nanotubes alloy hydrogen atoms carried out in the relaxation time approximation. Revealed a decrease in the electrical conductivity and the electron diffusion coefficient with increasing concentration of adsorbed hydrogen atoms. The nonlinearity of the electrical conductivity and the diffusion coefficient of the amplitude of a constant strong electric field at the constant concentration of hydrogen adatoms shown at the figures.

Aim and Background. Investigate the influence of a strong electric field on the electrical, transport and diffusion properties of carbon nanostructures with point defects structure.

1.1   INTRODUCTION

Despite the already long history of the discovery of carbon nanotubes (CNT) [1], the interest in the problem of obtaining carbon nanostructures with desired characteristics unabated, constantly improving their synthesis. Unique physical and chemical properties of CNTs can be applied in various fields of modern technology, electronics, materials science, chemistry and medicine [2]. One of the most important from the point of view of practical applications is the transport property of CNTs.

Under normal conditions, any solid surfaces coated with films of atoms or molecules adsorbed from the environment, or left on the surface in the diffusion process [3]. The most of elements adsorption on metals forms a chemical bond. The high reactivity of the surface of carbon nanotubes makes them an exception. Therefore, current interest is the study of the influence of the adsorption of atoms and various chemical elements and molecules on the electrical properties of carbon nanostructures.

In the theory of adsorption, in addition to the methods of quantum chemistry, widely used the method of model Hamiltonians [3]. In the study of the adsorption of atoms and molecules on metals used primarily molecular orbital approach – self-consistent field, as this takes into account the delocalization of electrons in the metal. Under this approach, the most commonly used model Hamiltonian Anderson [4, 5], originally proposed for the description of the electronic states of impurity atoms in the metal alloys. The model has been successfully applied to study the adsorption of atoms on the surface of metals and semiconductors [6], the adsorption of hydrogen on the surface of graphene [7] and carbon nanotubes [8, 9].

In this paper we consider the influence of the adsorption of atomic hydrogen on the conducting and diffusion properties of single-walled "zigzag" CNTs. Interaction of hydrogen atoms adsorbed to the surface of carbon nanotubes is described in terms of the periodic Anderson model. Since the geometry of the CNT determines their conductive properties, then to describe the adsorption on the surface of CNTs using this model is justified. Transport coefficients (conductivity and diffusion coefficient) CNT electron calculated by solving the Boltzmann equation [10] in the relaxation time approximation.

This technique was successfully applied by authors to calculate the ideal transport characteristics of carbon nanotubes [11] graphene bilayer graphene [12] and graphene nanoribbons [13].

1.2 MODEL AND BASIC RELATIONS

However, with the discovery of new forms of carbon model can be successfully applied to study of the statistical properties of CNTs and graphene. Carbon atom in the nanotube forms three chemical connection σ-type. Lodging with nearest neighbor atoms with three sp^2 hybridization of atomic orbitals. The fourth p-orbital involved in chemical bonding π-type which creates π-shell nanotube describing state of itinerant electrons, that define the basic properties of CNTs and graphene. This allows us to consider the state of π-electron system in the framework of the Anderson model. The model takes into account the kinetic energy of electrons and their Coulomb interaction at one site and neglected energy inner-shell electrons of atoms and electrons involved in the formation of chemical bonds σ-type, as well as the vibrational energy of the atoms of the crystal lattice.

In general, the periodic Anderson model [5] considers two groups of electrons: itinerant s-electrons and localized d-electrons. Itinerant particles are considered free and localized – interact by Coulomb repulsion on a single node. With the discovery of new forms of carbon model can be successfully applied to study the statistical properties of carbon structures are the CNT and the graphene. Carbon atom in the graphene layer has 3 forms chemical bonds σ-type with its immediate neighbors. The fourth orbital p-type forms a chemical bond π-type, describing the state of itinerant electrons. States localized electrons created by the valence orbitals (in this case, the p-type) impurity atoms. This allows us to consider the state of π-electrons in the framework of the Anderson model. The model takes into account the kinetic energy of the electrons in the crystal and impurity electrons interacting through a potential hybridization, and neglects the energy of the electrons of the inner shells of atoms and electrons involved in the formation of chemical bonds σ-type, as well as the vibrational energy of the atoms of the crystal lattice [5].

In the periodic Anderson model state of the electrons of the crystal containing impurities in the π-electron approximation and the nearest neighbor approximation is described by the effective Hamiltonian, having the following standard form [5]:

$$H = \sum_{j,\Delta,\sigma} t_{\Delta}\left(c_{j\sigma}^{+}c_{j+\Delta\sigma} + c_{j+\Delta\sigma}^{+}c_{j\sigma}\right) + \sum_{l,\sigma}\varepsilon_{l\sigma}n_{l\sigma}^{d} + \sum_{l}U n_{l\uparrow}^{d}n_{l\downarrow}^{d} + \\ + \sum_{l,j,\sigma}\left(V_{j}c_{j\sigma}^{+}d_{l\sigma} + V_{j}^{*}d_{l\sigma}^{+}c_{j\sigma}\right) \tag{1}$$

where t_{Δ} is the electron hopping integral between the neighboring lattice sites of the crystal; U is the constant of the Coulomb repulsion of the impurity; $c_{j\sigma}$ and $c_{j\sigma}^{+}$ are the Fermi annihilation and creation operators of electrons in the crystal node j with spin σ; $d_{j\sigma}$ and $d_{j\sigma}^{+}$ are the Fermi annihilation and creation operators of electrons on the impurities l with spin σ; $n_{l\sigma}^{d}$ is the operator of the number of electrons on impurities l with spin σ; $\varepsilon_{l\sigma}$ is the energy of the electron by the impurity l with spin σ; V_{ij} is the matrix element of hybridization of impurity electron l and atom j of the crystal.

After the transition to k-space by varying the crystal by Fourier transformation of creation and annihilation of electrons and crystal use the Green function method, the band structure of single-walled CNTs with impurities adsorbed hydrogen atoms takes the form [8, 9]:

$$E(\mathbf{k}) = \frac{1}{2}\left[\varepsilon_{k} + \varepsilon_{l\sigma} \pm \left((\varepsilon_{k} - \varepsilon_{l\sigma})^{2} + 4\frac{N_{imp}}{N}|V|^{2}\right)^{\frac{1}{2}}\right], \tag{2}$$

where N – number of carbon atoms in the lattice, determines the size of the crystal, N_{imp} – the number of adsorbed hydrogen atoms, V – hybridization potential, $\varepsilon_{l\sigma}$ = – 5.72 eV – electron energy impurities – the band structure of an ideal single-walled nanotubes, for tubes, for example, zigzag type dispersion relation is defined as follows [1]:

$$E(\mathbf{p}) = \pm\gamma\sqrt{1 + 4\cos(ap_{x})\cos(\pi s / n) + 4\cos^{2}(\pi s / n)} \tag{3}$$

where $a = 3d / 2\hbar$, $d = 0.142$ nm is the distance between adjacent carbon atoms in graphene, $\mathbf{p} = (p_{x}, s)$ is the quasimomentum of the electrons in graphene, p_{x} is the parallel component of the graphene sheet of the quasimomentum and $s = 1, 2, ..., n$ are the quantization numbers of the momentum components depending on the width of the graphene ribbon. Different signs are related to the conductivity band and to the valence band accordingly.

Used in the calculation of the Hamiltonian parameters: the value of the hopping integral $t_{0} = 2.7$ eV, hybridization potential $V = – 1.43$ eV estimated from quantum chemical calculations of the electronic structure of CNTs within the semiempirical MNDO [14]. Electron energy impurity $\varepsilon_{l\sigma} = -5.72$ eV was assessed using the method described in Ref. [6, 7].

Consider the effect of the adsorption of atomic hydrogen on the response of single-walled "zigzag" CNTs to an external electric field applied along the x axis is directed along the axis of the CNT (Fig. 1.1).

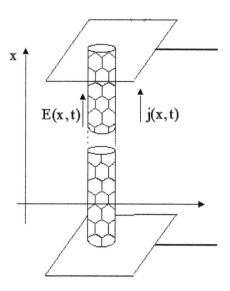

FIGURE 1.1   Geometry configuration. CNT type zigzag is in an external electric field. Field strength vector **E** is directed along the axis of the CNT.

Method of calculating the transport coefficients of electrons in carbon nanotubes described in detail in Refs. [11–13]. Evolution of the electronic system was simulated in the semiclassical approximation of the relaxation time. Electron distribution function in the state with momentum $p = (p_x, s)$ is of the τ – approximation using Boltzmann equation [10]:

$$\frac{\partial f_s(\mathbf{p},\mathbf{r})}{\partial t} + \mathbf{F}\frac{\partial f_s(\mathbf{p},\mathbf{r})}{\partial \mathbf{p}} = \frac{f_s(\mathbf{p},\mathbf{r}) - f_{0s}(\mathbf{p},\mathbf{r})}{\tau}, \tag{4}$$

where $f_s(\mathbf{p},\mathbf{r})$ – the Fermi distribution function $\mathbf{F} = e\mathbf{E}$ – acting on the particle constant electrostatic force.

To determine the dependence of the diffusion and conductive characteristics of CNTs on the external electric field using the procedure outlined in Ref. [15]. The longitudinal component of the current density $j = j_x$ has the following form:

$$j(x) = \sigma(\mathbf{E})\mathbf{E} + D(\mathbf{E})\frac{\nabla_x n}{n} \tag{5}$$

For the case of a homogeneous temperature distribution $T(r)$ = constant in the linear approximation in magnitude [11], expressions for the transport coefficients of single-walled nanotubes: conductivity and diffusivity of electrons. Electrical conductivity of CNT type zigzag given following expression [11]:

$$\sigma(E) = \sum_s \int_{-\pi}^{\pi} \sum_m A_{ms} m f_{0s}(p_x, x) \frac{E}{E^2 m^2 + 1} \left[\sin(mp_x) + Em \cos(mp_x) \right] dp_x \qquad (6)$$

Expression for the diffusion coefficient of electrons in CNT type zigzag has the form [11]:

$$D(E) = \sum_s \int_{-\pi}^{\pi} f_{0s}(p_x, x) \sum_m A_{ms} m \sum_{m'} A_{m's} m' \left\{ \frac{[E^2(m^2 + m'^2) + 1][EmR + M]}{K} + \right.$$

$$\left. + \frac{[E^3(m'^3 - 2m^2 m') + Em']T}{K} \right\} dp_x + \sum_s \int_{-\pi}^{\pi} f_{0s}(p_x, x) \sum_m A_{ms} m \sum_{m'} A_{m's} m' \frac{F}{P} dp_x, \qquad (7)$$

where the following notation:

$$K = [E^4(m^4 + m'^4 - 2m^2 m'^2) + 2E^2(m^2 + m'^2) + 1][E^2 m^2 + 1],$$

$$P = [E^2 m^2 + 1]^2 [E^2 m'^2 + 1],$$

$$R = \cos(mp_x)\sin(m'p_x) + \cos(mp_x)\cos(m'p_x) - \sin(mp_x)\sin(m'p_x),$$

$$M = \sin(mp_x)\sin(m'p_x) + \sin(mp_x)\cos(m'p_x) + \cos(mp_x)\sin(m'p_x),$$

$$T = [\cos(mp_x)\cos(m'p_x) - Em\sin(mp_x)\cos(m'p_x)],$$

$$F = [\sin(m'p_x) + Em\cos(m'p_x)][\sin(mp_x) + 2Em\cos(mp_x) - E^2 m^2 \sin(mp_x)],$$

$A_{ms}, A_{m's}$ are the coefficients of the Fourier expansion of the dispersion relation of electrons in CNT, m and m' order Fourier series. For the convenience of visualization and qualitative analysis performed procedure and select the following dimensionless relative unit of measurement of the electric field $E_0 = 4.7 \times 10^6$ V/m.

1.3   RESULTS AND DISCUSSION

To investigate the influence of an external constant electric field on the transport properties of single-walled CNT type zigzag with adsorbed hydrogen atoms selected the following system parameters: temperature T ≈ 300 K, the relaxation time is $\tau \approx 10^{-12}$ s in accordance with the data [16]. For numerical analysis considered type semiconducting CNT (10,0).

It should be noted that a wide range of external field behavior of the specific conductivity $\sigma(E)$ for nanotubes with hydrogen adatoms has the same qualitative nonlinear dependence as for the ideal case of nanoparticles, which was discussed in detail in Ref. [11]. In general, the dependence of conductivity on the electric field has a characteristic for semiconductors form tends to saturate and decreases monotonically with increasing intensity. This phenomenon is associated with an increase in electrons fill all possible states of the conduction band. Behavior of electrical conductivity under the influence of an external electric field is typical for semiconductor structures with periodic and limited dispersion law [17].

Figure 1.2 shows the dependence of conductivity $\sigma(E)$ on the intensity of the external electric field E for ideal CNT (10,0) and CNT (10,0) with adsorbed hydrogen at relatively low fields. The graphs show that the addition of single adsorbed atom (adatom) hydrogen reduces the conductivity by a small amount (about 2×10^{-3} S/m). Lowering the conductivity of the hydrogen atom in the adsorption takes place due to the fact that one of the localized electron crystallite forms a chemical bond with the impurity atom and no longer participates in the charge transport by CNT.

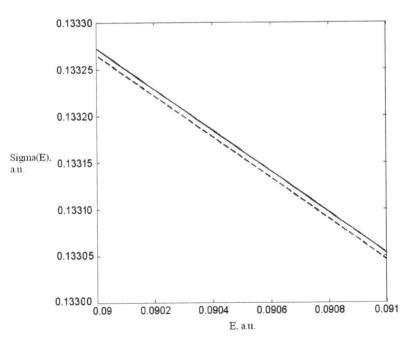

FIGURE 1.2 Dependence of the conductivity $\sigma(E)$ on the magnitude of tension external electric field E: for ideal CNT (10,0) – solid line and the CNT (10,0) with hydrogen adatom – dashed line. x-axis is a dimensionless quantity of the external electric field E (unit corresponds to 4.7×10^{6} V/m), the y-axis is dimensionless conductivity $\sigma(E)$ (unit corresponds to 1.9×10^{3} S/m).

Also analyzed the dependence of the conductivity $\sigma(E)$ on the intensity of the external electric E for CNT (10,0) type, containing different concentrations of hydrogen adatoms (Fig. 1.3). The increasing of the number of adsorbed atoms reduces the conductivity of zigzag CNT proportional to the number of localized adsorption bonds formed. When you add one hydrogen adatom conductivity of CNT type (10,0) is reduced by 0.06%, adding 100 adatoms by 0.55%, adding 300 adatoms by 1.66%, adding 500 adatoms by 2.62%.

Figure 1.4 shows that this behavior is typical for semiconductor conductivity of CNTs with different diameters. With the increasing diameter of the nanotubes have high electrical conductivity, since they contain a larger amount of electrons which may participate in the transfer of electrical charge. The graphs in Fig. 1.4 shows for the (5,0), (10,0) and (20,0) CNT with the addition of 100 hydrogen adatoms.

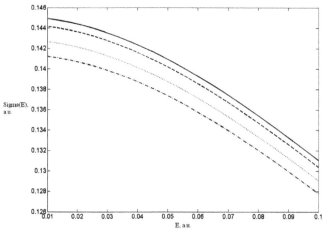

FIGURE 1.3   Dependence of the conductivity $\sigma(E)$ on the magnitude of tension E external electric impurity for CNT (10,0) one hydrogen adatom – solid line; 100 adatoms – dashed line; 300 adatoms – dotted line; 500 adatoms – dash-dot line. x-axis is a dimensionless quantity of the external electric field E (unit corresponds to 4.7×10^6 V/m), the y-axis is dimensionless conductivity $\sigma(E)$ (unit corresponds to 1.9×10^3 S/m).

The electron diffusion coefficient $D(E)$ from the electric field in the single-walled zigzag CNT with adsorbed hydrogen atoms has a pronounced nonlinear character (Fig. 1.5). Increase of the field leads to an increase in first rate, and then to his descending to a stationary value. This phenomenon is observed for all systems with intermittent and limited electron dispersion law [17]. Electron diffusion coefficient can be considered constant in the order field amplitudes $E \approx 5 \times 10^6$ V/m. The maximum value of the diffusion coefficient for semiconductor CNTs observed at field strengths of the order of $E \approx 4.8 \times 10^5$ V/m.

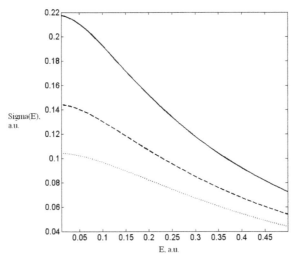

FIGURE 1.4 Dependence of the conductivity σ(E) on the magnitude of tension external electric E for different types of CNTs with the addition of hydrogen adatoms 100 (20,0) – solid line, (10,0) – dashed line; (5,0) – the dotted line. *x*-axis is a dimensionless quantity of the external electric field E (unit corresponds to 4.7×10^6 V/m), the *y*-axis is dimensionless conductivity σ(E) (unit corresponds to 1.9×10^3 S/m).

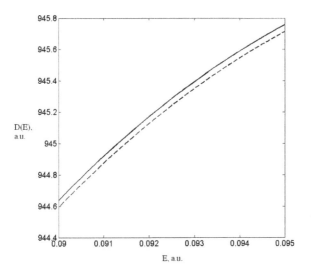

FIGURE 1.5 Dependence of the electron diffusion coefficient $D(E)$ on the intensity of the external electric field E: for CNT (10,0) ideal – solid line and hydrogen adatom – dashed line. *x*-axis is a dimensionless quantity of the external electric field E (unit corresponds to 4.7×10^6 V/m), the *y*-axis is a dimensionless diffusion coefficient $D(E)$ (unit corresponds to 3.5×10^2 A/m).

When adding the adsorbed hydrogen atoms the electron diffusion coefficient, as well as the conductivity is reduced by 0.05% (Fig. 1.5). This behavior of the diffusion coefficient in an external electric field is observed for different concentrations of hydrogen adatoms (Fig. 1.6) and semiconductor CNTs with different diameters by adding 100 adatoms (Fig. 1.7).

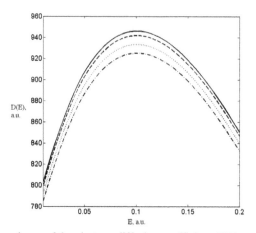

FIGURE 1.6   Dependence of the electron diffusion coefficient $D(E)$ on the intensity of the external electric E for impurity CNT (10,0) one hydrogen adatom – solid line; 100 adatoms – dashed line; 300 adatoms – dotted line; 500 adatoms – dash-dot line. x-axis is a dimensionless quantity of the external electric field E (unit corresponds to 4.7×10^6 V/m), the y-axis is a dimensionless diffusion coefficient $D(E)$ (unit corresponds to 3.5×10^2 A/m).

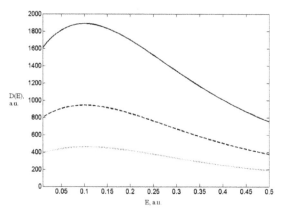

FIGURE 1.7   Dependence of the electron diffusion coefficient $D(E)$ on the intensity of the external electric E for different types of CNTs with the addition of hydrogen adatoms 100 (20,0) – solid line, (10,0) – dashed line; (5,0) – the dotted line. x-axis is a dimensionless quantity of the external electric field E (unit corresponds to 4.7×10^6 V/m), the y-axis is a dimensionless diffusion coefficient $D(E)$ (unit corresponds to 3.5×10^2 A/m).

The presented results can be used for the preparation of carbon nanotubes with desired transport characteristics and to develop the microelectronic devices, which based on carbon nanoparticles.

1.4 CONCLUSION

The main results are formulated in the conclusion.
- The method for theoretical calculation of the semiconducting zigzag CNT transport properties with adsorbed hydrogen atoms developed. Analytical expressions for the conductivity and the electron diffusion coefficient in zigzag CNT with hydrogen adatoms in the presence of an electric field.
- Numerical calculations showed nonlinear dependence of the transport coefficients on the electric field. For strong fields coefficients tend to saturate.
- Atomic hydrogen adsorption of the semiconducting zigzag CNT reduces their conductivity by several percent. The electron diffusion coefficient also decreases with increasing concentration of adsorbed hydrogen atoms, and a decrease of the diffusion coefficient is more pronounced than the decrease of electrical conductivity for each of the above types of semiconducting CNTs at a larger number of adatoms.
- Transport properties of nanotubes with adatoms increase with the diameter. A physical explanation for the observed effect.

1.5 ACKNOWLEDGMENT

This work was supported by the Russian Foundation for Basic Research (grant № 13-03-97108, grant № 14-02-31801), and the Volgograd State University grant (project № 82-2013-a/VolGU)

KEYWORDS

- **Adsorption**
- **Carbon nanostructures**
- **Conductivity**
- **Diffusion coefficient**
- **"zigzag" nanotubes**

REFERENCES

1. Diachkov, P. N. (2010). Electronic properties and applications of nanotubes. M. Bin, *the Laboratory of Knowledge*, 488 p.
2. Roco, M. C., Williams, R. S. & Livisatos, A P. (2002). Nanotechnology in the next decade. Forecast the direction of research. Springer-Verlag, 292 p.
3. Bolshov, L. A., Napartovich, A. P., Naumovets, A. G. (1977). & Fedorus A G. *UFN, T. 122(1)*, 125.
4. Anderson, P. W. (1961). *Phys. Rev, 124*, 41.
5. Izyumov, A., Chashchin, I. I. & Alekseev, D. S. (2006). The theory of strongly correlated systems. Generating functional method. Moscow-Izhevsk: *NITs "Regular and Chaotic Dynamics."* 384.
6. Davydov, S. U. & Troshin, S. V. (2007). *Phys, T. 49(8)*, 1508.
7. Davydov, S. Y. & Sabirov, G. I. (2010). *Letters ZHTF, T. 36(24)*, 77.
8. Pak, A. V. & Lebedev, N. G. (2012). *Chemical Physics, 31(3)*, 82–87.
9. Pak, A. V. & Lebedev, N. G. (2013). *ZhFKh, T. 87*, 994.
10. Landau, L. D. & Lifshitz, E. M. (1979). Physical Kinetics. M. *Sci. Lit*, 528 c.
11. Belonenko, M. B., Lebedev, N. G. & Sudorgin, S. A. (2011). *Phys, T. 53*, 1841.
12. Belonenko, M. B., Lebedev, N. G. & Sudorgin, S. A. (2012). *Technical Physics, 82(7)*, 129–133.
13. Sudorgin, S. A., Belonenko, M. B. & Lebedev, N. G. (2013). *Physica Scripta, 87(1)*, 015–602 (1–4).
14. Stepanov, N. F. (2001). Quantum Mechanics and Quantum Chemistry. Springer-Verlag, 519 sec.
15. Buligin, A. S., Shmeliov, G. M. & Maglevannaya, I. I. (1999). *Phys, T. 41*, 1314.
16. Maksimenko, S. A. & Slepyan, G. Ya. (2004). Nanoelectromagnetics of low-dimensional structure. In Handbook of nanotechnology. Nanometer structure: theory, modeling, and simulation. Bellingham: SPIE Press, 145 p.
17. Dykman, I. M. & Tomchuk, P. M. (1981). Transport phenomena and fluctuations in semiconductors. *Kiev. Sciences. Dumka*, 320 c.

CHAPTER 2

TOPOLOGY AND GEOMETRY OF PROTEIN WATER SHELL BY COMPUTER MODELING

E. V. RUBTCOVA, A. B. SOLOVEY, and V. I. LOBYSHEV

CONTENTS

ABSTRACT

Water shells of fibrillar and globular proteins are studied. The simulations are held in TIP3P water potential using CHARMM force field. The water shell topology and internal parameters have been analyzed in terms of statistical characteristics and specific examples of water structures. Some of the structures obtained correspond to the Bulienkov parametrical model of bonded water.

Aim and Background. The idea of the chapter is to study topological characteristics of protein water shell, to compare fibrillar and globular protein water shell structures and to discover if some topological invariants exist in H-bond water network of protein water shell. We consider H-bond water network from the graph theory point of view [1, 2]. In addition we compare the obtained structures with "ideal" parametric bonded-water of Bulienkov's model [3], to understand if ideal or distorted th-cycles exist in water shells.

2.1   INTRODUCTION

Most of proteins are active only in the presence of a sufficient amount of water [4]. Water molecules and H-bonds network are essential structural components in various biomacromolecular systems such as nucleic acids complexes, membranes, fibrillar and globular proteins. The topology of water shell macromolecules structure is important for understanding many of the processes occurring in a living cell [5]. The experimental studies of the hydration process are complex enough because the signal from bulk water sometimes interrupts a signal from water shell water molecules. Theoretical approach and the corresponding simulation brings us closer to a better understanding of protein hydration. It was shown that surface water molecules play a crucial role in biomolecular interactions and processes that lead to protein folding [6]. The interactions between macromolecules are carried throughout the water surface with the help of "binding bridge" (as in the case of interaction between the protein–DNA) [7].

The surface structure of the protein-water is the subject of conflicting theoretical studies. It requires both direct experimental and theoretical studies. For many proteins their crystal structures are well defined. Surface analysis of these proteins shows that there is first water shell, wherein the average density is 10% greater than in the bulk water under similar conditions [8]. Comparison with the other studies suggests that this fact may be a general property of water surfaces.

Determination of the water shell topology is a challenge both in terms of computing, and in terms of understanding of the physical process components. Computer simulation parameters should have a physical meaning, the results of simulation should correspond to the experimental data, and requirements for computing power should remain within the reasonable limits.

Physical properties of the first-shell water of macromolecular water shell differ from the bulk water properties. Some of them can be measured experimentally. They are: the phase transition temperature, dielectric constant, NMR relaxation parameters, calorimetric characteristics, kinetic characteristics, such as self-diffusion coefficient and some others. The main factors that define protein structure are residue sequence, hydration, ionic and hydrophobic interactions.

In this work we perform a computer modeling of water molecules with fragments of proteins containing the amino acid sequence Gly Pro-Pro. These fragments are the parts of the core structural unit of collagen, structured into supramolecular triple helix.

Collagen is one of the most common and important structural proteins, it is characterized by a specific residue sequence [-Gly X-Y-] where X is predominantly of proline and Y of hydroxyproline or other aminoacids. Structure and function of collagen are well known.

The collagen water shell structure was chosen for study because the protein itself has rather simple residue sequence. One can define topological patterns of H-bond water network more easily on regular proteins such as collagen than on common globular proteins, due to its regular residue structure. The obtained results were compared with the same characteristics of ubiquitin to check, if the results are similar and features that were studied could be common for proteins of different type.

Experimental data on the structure of native collagen give an idea about the basic motives of tertiary spiral, but limited resolution makes it impossible to localize bound water molecules on the protein surface.

There is a detailed analysis of the hydration structure for a short fragment of collagen molecule in Ref. [12]. The water molecules around carbon group and a hydroxyproline hydroxyl group are organized in some specific spatial structure. There are repeating pattern of water bridges that bind oxygen atoms with a single protein chain between different chains. Water molecules form the cluster structures that surround fibrils and connect them to the triple helix structure. It has been established that water plays an important role in the "support" structure of collagen native conformation. Using a variety of experimental techniques, for example NMR and dielectric relaxation, it was shown that bound water molecules in structure of collagen are less mobile than in bulk water. These results may be interpreted as water molecules are closely linked with specific sites of the collagen chains, and/or they fill space between the chains forming water bond net.

Adsorption of water on collagen fibers was studied by NMR [13]. It was confirmed that water molecules form H-bonds chains around the collagen molecules. By the experimental data it was indicated that there is some inner hydrated layer, containing approximately 24% water in ratio to the weight of dry collagen, and this layer is highly ordered.

There are areas inside some globular proteins where water is strongly associated [14, 15]. Water molecules in these areas have a sedentary lifetime from 10^{-8} to 10^{-2}

s. These water molecules on the protein surface have lifetimes that are about or less than 10^{-9} s. Bound water properties both on the protein surface and inside of the globule differ from the properties of bulk water appearing in their kinetic characteristics, molar volume, and several macroscopic thermodynamic parameters.

There is a problem of bound water both in lifeless nature and in biological systems at various levels of the structural hierarchy. In this regard we can talk about the existence of the phenomenon of bound water with different from bulk water structures and properties [17–19]. Let us consider a model of water associated with biopolymer. The basis of this model is the fact that water molecule can be represented as a distorted tetrahedron because an ideal tetrahedron "bond angle" is 109.28, and free water molecule bond angle is 104.5. The model assumes that the water may produce a continuous three-dimensional network of tetrahedral particles.

2.2   EXPERIMENTAL PART

In our study we use the following structures:
- X-ray crystallographic data of a collagen-like peptide with the repeating sequence [Gly Pro-Pro]$_7$, PDB code: 1A3I
- average crystal structure of [Pro-Pro-Gly]$_9$ at 1.0 Å resolution, PDB code: 1ITT
- structure of ubiquitin refined at 1.8 Å resolution, PDB code: 1UBQ

We put water shell of various sizes around proteins and analyze the size-dependent results. SOLVATE software is used to obtain this water shell [20].

2.2.1   MINIMIZATION

First of all we need to obtain a stable system "protein-water." Stable means having a local energy minimum. We held minimization process in vacuum without periodic boundary conditions with time step 1 fs, the whole minimization time was 1 ns. Minimization was performed using the software package NAMD [21], using CHARMM force field, simulations were held in water model TIP3P. Minimization of the energy structure in this package implemented the conjugate gradient method.

Various projections of collagen with water shell are shown in Figs. 2.1 and 2.2, and ubiquitin in Fig. 2.3. Water shell near collagen as a system of H-bond hexacycles is shown in Fig. 2.4.

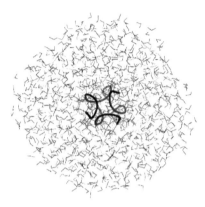

FIGURE 2.1   Gly Pro-Pro triple helix as a model of collagen structure with 8 Å water shell. View 1.

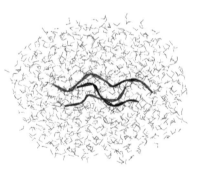

FIGURE 2.2   Gly Pro-Pro triple helix as a model of collagen structure with 8 Å water shell. View 2.

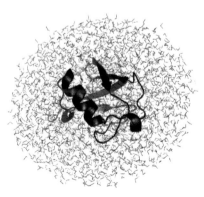

FIGURE 2.3   Ubiquitin with 6 Å water shell after the minimization procedure.

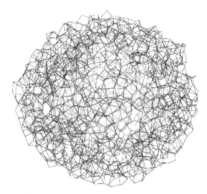

FIGURE 2.4   Water shell of collagen as a system of H-bond hexacycles.

To verify the minimization procedure a plot of total energy is given (Figs. 2.5 and 2.6) for both collagen. The minimization procedure and the structure achieved seem to be good enough from the energetic point of view.

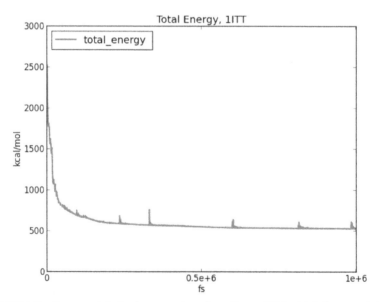

FIGURE 2.5   Energy minimization procedure for collagen 1ITT with 8 Å water shell.

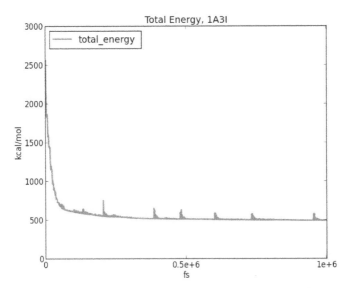

FIGURE 2.6 Energy minimization procedure for collagen 1A3I with 8 Å water shell.

2.3 RESULTS AND DISCUSSION

2.3.1 STUDY OF STRUCTURES TOPOLOGY

Let us represent water shell structure as a graph where oxygen atoms in water molecules are represented by the vertices and H- bonds between them by the edges.

To restore the H-bonds we can apply various criteria for the H-bond existence such as energetic, geometric and dynamic criteria, sometimes combined criterion "energy + geometric" is used [22]. In the case of energy criterion it is considered that the bond is formed, if the energy of the H-bond (O-H.O) is more than some selected value, which is usually (2–5) kcal/mol. In dynamic criterion it is supposed not to take into account the very short time periods when a bond appears or disappears.

We use geometric criterion of H-bond existence [22]. According to it we assume that two water molecules are connected by H-bond if the distance between their oxygen atoms is not greater than $R\_OO^{\times}$ and the distance between oxygen atom of first molecule and hydrogen atom of second molecule is not greater than $R\_OH^{\times}$, that is,

- $R\_OO^{\times} < 3.3$ Å
- $R\_OH^{\times} < 2.6$ Å

Each water molecule can forms up to four H-bonds – two of them form an oxygen atom and two hydrogen atoms. By taking into account over coordinated

H-bonds (bifurcated bonds) the number of water bonds increases up to 5. Typical classes of water clusters with various numbers of H-bonds are shown in Fig. 2.7.

FIGURE 2.7 Topologically different water clusters. 3, 4 and 5 H-bonds with the central molecule.

Thus water molecules in the above represent topologically different particles. The distribution of various particles number is given in Table 2.1.

TABLE 2.1 1ITT, 1A3I, 1UBQ Proteins With Water Shell of Different Thickness T, Number of Water Molecules N, Total Number of H-Bonds H, Number of H-Bonds B in Topologically Different Clusters

	1ITT T = 8 Å N = 764 H = 1447	1ITT T = 10 Å N = 1120 H= 2182	1ITT T = 12 Å N = 1556 H = 3049	1A3I T = 8 Å N = 1069 H = 2093	1A3I T = 10 Å N = 1539 H = 2992	1A3I T = 12 Å N = 2092 H = 4141	1UBQ T = 6 Å N = 2327 H = 4463
B = 2	254	394	531	385	542	758	789
B = 3	202	291	400	360	390	550	545
B = 4	37	60	95	70	95	114	143
B = 5	4	6	8	7	0	13	14

This distribution shows that molecular count ratio with B = 3 to B = 4 decreases with increasing solvent particles count. With increasing the total number of water molecules the ratio of surface molecules to bulk molecules decreases. The more water molecules solvates the protein and take part in simulation—the more of them have 4 bonds—so they are full-coordinated.

We consider the distribution of angles O–O–O (oxygen atoms) for all triples of water molecules (A-B-C). In this triple there are 2 H-bonds: molecules A, B are bonded and molecules B, C are bonded too (see Fig. 2.8, angle ABC). We achieved a geometric picture of the entire water shell and explored the angle O–O–O distribution. We mark these O–O–O angles as "valence angles." Note that it is not a valence H-O-H angle of a single water molecule, but it is an angle between two H-bonds of oxygen neighbors.

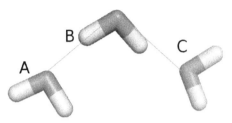

FIGURE 2.8 Three water molecules, angle O–O–O.

Each water molecule with two H-bonds may be in three various configurations according to the donor-acceptor nature of the H-bond: proton-donor for every neighbor, proton-acceptor for every neighbor, proton-donor for one neighbor and proton-acceptor for another. The first case (proton-donor for all neighbors) is shown on Fig. 2.8.

The "valence angle" distribution has two peaks for water shells shown in Fig. 2.9. All molecules have been divided into categories depending on how many neighbors (H-bonds) they have. We have enough data to state the nonstochastic distribution of valence angles.

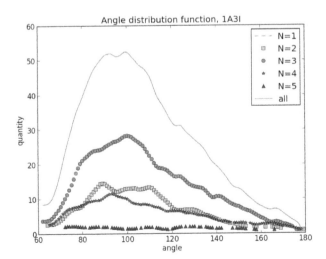

FIGURE 2.9 O–O–O angle distribution for various N in 12Å water shell minimized for 1A3I.

The approximately 90 wide maximum can be realized when the central oxygen is a proton acceptor. The angle O–O–O could be less than tetrahedral without any significant energy consuming because of large flexibility of electronic density of oxygen atom.

We suppose that the polymodality of O–O–O angle distribution is associated with the donor-acceptor nature of the H-bond. The possible reason of the smaller maximum is that the central molecule is a proton acceptor; and the possible reason of the bigger maximum is that the central oxygen is donor. We assume the cause of this polymodality is the fact that it is rigid H–O–H angle equals to 104.5 in TIP3P water potential. So for the acceptor H-bonds O–O–O angle tend to be close to the H-O-H angle (Figs. 2.10 and 2.11).

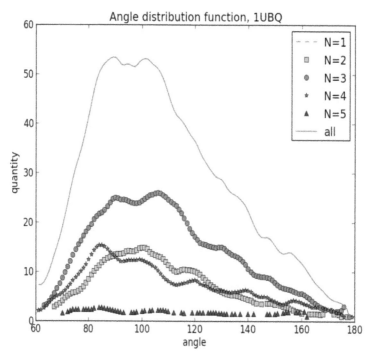

FIGURE 2.10 O–O–O angle distribution for various N in 6Å water shell minimized for 1UBQ.

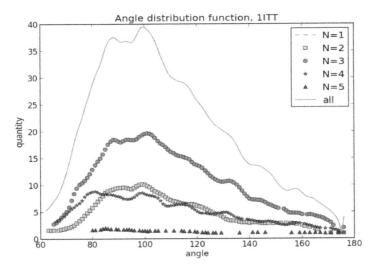

FIGURE 2.11 O–O–O angle distributions for various N in 12Å water shell minimized for 1ITT.

The dihedral O–O–O–O angle distributions were also defined for ubiquitin and collagen water shells (Fig. 2.12 and 2.13).

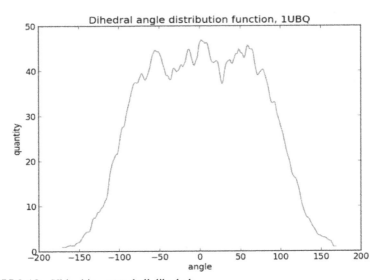

FIGURE 2.12 Ubiquitin water shell dihedrals.

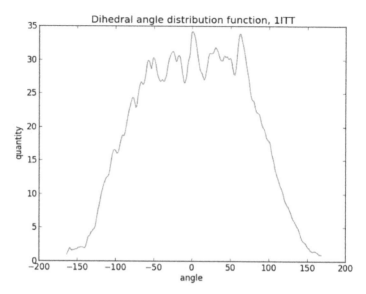

FIGURE 2.13   Collagen water shell dihedrals.

Ubiquitin water shell shows a very interesting distribution. The three maxima at −60, 0 and 60 indicates a boat-like and chair-like hexacycle structures. This is because all 6 dihedrals in a chair hexacycle are gauche(60) but 2 of 6 diredrals in a boat hexacycle are cis(0) and 4 of 6 dihedrals in boat hexacycle are gauche(60). Gauche and cis dihedrals correspond to boat-like and chair-like hexacycles and these hexacycles correspond to the ice-1structure. Minimized water means ice (Fig. 2.14).

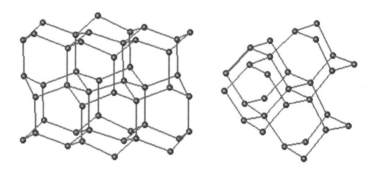

FIGURE 2.14   Hexagonal and cubic ice structures (Ice-1h and Ice-1c).

But for collagen water shell we see another distribution—there are no three maxima—there are five maxima. That's why we cannot consider the ice-1 structure in collagen bound water. The intermediate dihedrals on +/– 30 degrees do not point to boat-like or chair-like hexacycles, on contrary they point to a twist-boat hexacycle. The twist-boat hexacycle does not exist in ice structure. The collagen water shell seems to be more disturbed than ubiquitin water shell and not quite similar to ice-1.

To describe H-bond network parameters some statistical calculations could be provided: angles distributions, particles count statistics (full-coordinated, nonbonded, 1,2,3,4,5-bonded). But if we have only statistical functions and calculations and have no structural model we can hardly restore the entire structure of protein water shell. The structural model of protein water shell is necessary to understand various protein functions and the details of biochemical reactions. That's why we need a structural model of water shell.

There are few models of bonded water. The most common thermodynamics model considers bonded water to be ice-like. To clarify this suggestion we analyze models including thousands water molecules. This amount is necessary to see real structural properties–topology of cycles and common laws of H-bond connections.

We use the Bulienkov parametric bound water model to describe the protein water shell. Water molecules are bonded into H-bond network. This H-bond network can be performed as a system of hexacycles in "twist-boat" conformation. The twist-boat hexacycles provide non-Euclidean geometry parameters as it should be in crystal structure such an ice [18]. In ice structure the internal parameters of all hexacycles are equal–intermolecular distances, valence and torsion angles are constant and can vary only by a thermal motion. So if any hexacycle system is constructed using only twist-boat pattern then geometrical parameters must distort [23].

The basic structural element of building twist-boat is chiral hexacycle that is composed of tetrahedral particles Twist-boat is chiral, that is, exists in the left and right form (Fig. 2.15).

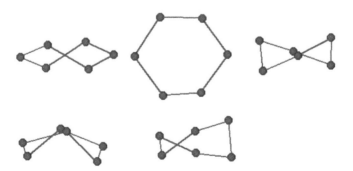

FIGURE 2.15   Twist-boat hexacycle (th-cycle) in various projections.

We will call a hexacycle structure constructed only of twist-boats connected with each other as th-cycle [18]. The th-cycle is a system of cycles that are built of tetrahedral particles (water molecules connected by H-bonds). The tetrahedral particles are arranged in a twist-boat hexacycles of the same chirality. The twist-boats are linked with each other in special way [18] and form a th-cycle. The main characteristics of th-cycle are:

1. three-dimensional Euclidean space couldn't be filled with them;
2. hierarchy structures are built by the same topological rules from tetrahedral particles.

The model of the protein water shell is a system of tetrahedral water molecules bonded into twist-boat hexacycles, so it is a th-cycle.

Th-cycles have basic geometric and topological characteristics that are the same as characteristics of biosystems at the molecular hierarchy level – macromolecules and macromolecular complexes. Biosystems are also discontinual, hierarchical and possess "invariants." It is assumed that water shell of the macromolecule should have similar geometric and topological characteristics.

Th-cycle is just an example of several possible hierarchical tetrahedral systems built on "biological principles." Twist-boat associated with the crystal structures of water (hexagonal and cubic ice) can be obtained by a simple dispiration transformation. Enthalpy of ice–th-cycle transition is negligible and caused by H-bonds distortion. This fact consistent with the enthalpy of transition ice-1h – liquid water [25].

H-bonds of th-cycles form a network that consists of twist-boats with definite chirality. L–cluster is one of the th-cycle examples (Fig. 2.16).

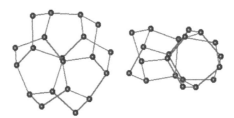

FIGURE 2.16   L–cluster.

Equilibrium protein water shell obtained by minimizing of the system in chosen TIP3P potential can be studied from the point of accordance of its topological model to th-cycles model. We consider its geometrical and topological characteristics statistically and with the help of concrete examples.

At first we found all simple cycles, that is, a path with nonrepeated vertices and edges except the start and end point that are the same.

Traditional BFS algorithm (breadth-first search):

1. Put the starting vertex in the empty queue.
2. Extract from the queue vertex u and mark it as visited.

3. Add to the end of the queue all adjacent vertices of a vertex u, that have not yet been visited and are not in the queue.

4. If the queue is empty, then the algorithm is completed – all the vertices, reachable from the start have already been visited.

To find cycles we run the modified BFS from each vertex. Also we create an array, where for each vertex we will put the parent one. Then we change 2 step of the BFS algorithm. If the adjacent vertex has been already visited, then try to complete cycle, walking up through the parents array to the starting vertex. If the cycle length is greater than 2 and less or equal MAX_CYCLE_LENGTH, (equal to 6 in our case) and no vertex is not repeated twice in a cycle (except the starting one), we found a cycle and add it to the set of cycles. Otherwise – go ahead. If the adjacent vertex is not visited, then add it to the queue, memorizing the parent vertex.

We are interested in cycles with number of vertices up to N = 6. The resulting histograms of cycles with various amount of vertices are shown in Figs. 2.17 and 2.18.

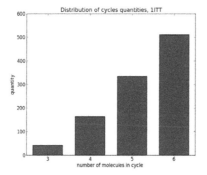

FIGURE 2.17   Distribution of cycles containing 3, 4, 5 and 6 vertices in 1ITT with water.

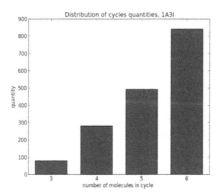

FIGURE 2.18   Distribution of cycles containing 3, 4, 5 and 6 vertices in 1A3I with water.

Some representative examples of simple cycles with various number of vertices are shown in Fig. 2.19.

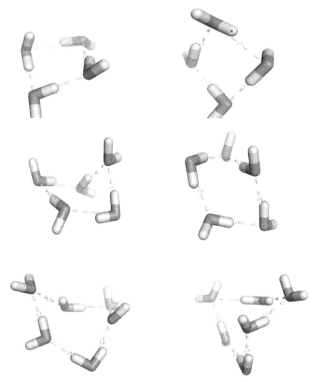

FIGURE 19 Representative examples of simple cycles with various number of vertices.

Cycles of various topology were found. The water shell can be represented as a hexacycle graph. More that 90% molecules construct this graph. So when we speak about protein water shell we speak about the hexacycle graph.

The distribution of dihedral angles in hexacycles is also analyzed. Using dihedral angle distribution it is possible to define some known hexacycles:

- chair (all dihedral angle are equal to 60, Fig. 2.20, left);
- boat (4 angles are equal to 60, 2 angles are equal to 0, the sequence order (60, 60, 0, 60, 60, 0), Fig. 2.20, right).
- twist-boat (4 angles are equal to 36, 2 angles are equal to −79, the sequence order (36, 36, −70, 36, 36, −79) or (−36, −36, 70, −36, −36, 79), Figs. (2.21 and 2.22)).

Pictures of twist-boat in our simulation and ideal theoretical similar structure are shown in Fig. (2.21–2.22).

FIGURE 2.20   Hexacycles in chair (left) and boat (right) conformation.

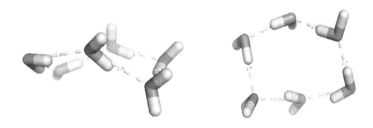

FIGURE 2.21   Twist-boat hexacycle of water molecules from collagen water shell

FIGURE 2.22   Ideal twist-boat hexacycle from Bulienkov model.

Statistics of hexacycle conformations:

- protein 1ITT (Table 2.2) with water shell with various thick T, number of hexacycles in water shell H_0, number of twist-boats H_1.

TABLE 2.2 Proteins 1ITT, 1A3I, 1UBQ with Water Shell with Various Thicknesses T, Number of Hexacycles in Water Shell H_0, Number of Twist-Boats H_1

	1ITT	1ITT	1ITT	1A3I	1A3I	1A3I	1UBQ
T	8	10	12	8	10	12	6
H_0	511	794	1049	842	1019	1550	1539
H_1	7	9	14	10	20	35	22

Let Q—the common number of molecules, Q_0—number of molecules contained in at least one hexacycle. Then the fraction of molecules in hexacycles will be equal to $p = Q\_0/Q$ (Table 2.3).

TABLE 2.3 Fraction of Molecules in Hexacycles for Various Proteins

1ITT	0.97
1A3I	0.96
1UBQ	0.95

Practically all molecules belong to at least one hexacycle. Let us consider the way how hexacycles are linked with each other. Two hexacycles could have various number of common vertices V: 2, 3, 4 (examples are shown in Fig. 2.23).

FIGURE 2.23 Possible examples of linked hexacycles (with 2, 3 and 4 common vertices)

TABLE 2.4 Number of Common Vertices V for Various Proteins

	V= 2	V= 3	V= 4	V= 5
1A3I	2793	969	414	135
1ITT	2315	780	340	126
1UBQ	4438	1508	709	263

The dominant pattern of hexacycles compound is connection with two common vertices. We examined the way of how hexacycles are connected. Common vertices in hexacycle could be adjacent (number of edges between connecting with another cycle vertices is E = 1, or common vertices could follow in 1 or in 2 vertices (E =2 or E = 3), Fig. 2.24. (Table 2.5).

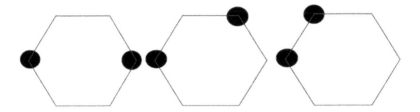

FIGURE 2.24   Possible ways for hexacycles connections (by 2 points).

TABLE 2.5   Number of Edges Between Common Point for Two Connected Hexacycles

	E = 1	E = 2	E = 3
1A3I	10 990	96	86
1ITT	9084	100	76
1UBQ	17 376	236	140

2.4   CONCLUSIONS

The water shell of collagen and ubiquitin is calculated using computer simulation with CHARMM force field and TIP3P water molecules.

H-bond network can be considered as hexacycle grid, because 95% of water molecules belong at least to 1 hexacycle. These hexacycles can be classified into a set of categories based on their dihedral angles pattern. Twist-boats hexacycles were found.

The dihedral angles distribution for collagen and ubiquitin shows three maxima in ubiquitin water shell and a wide and not sharp maxima in collagen shell. These maxima are the same that in boat hexacycle, so it seems to be ice-like. Ubiquitin water shell seems to be more ice-like than collagen water shell.

KEYWORDS

- H-bond network
- Parametric structures of bound water
- Protein water shell
- Topological properties

REFERENCES

1. Dexter Kozen. (1992). The Design and Analysis of Algorithms, Springer.
2. Bondy, A. J. & Murty, S. U. (1976). Graph theory with applications, North Holland.
3. Lobyshev, I. V., Solovey, B. A. & Bulienkov, A. N. (2003). Computer construction of modular structures of water, *J Molecular Liquids, 106(2–3),* 277–297.
4. Levy, Y. & Onuchic, J. N. (2006). Water mediation in protein folding and molecular recognition. *Annu. Rev. Biophys. Biomol. Struct. 35,* 389–415.
5. Ball, P. (2008). Water as an active constituent in cell biology. *Chem. Rev., 108,* 74–108.
6. De Simone, A., Dodson, G. G. & Fraternali, F. (2000). Prion and water: tight and dynamical hydration sites have a key role in structural stability, *Proceedings of the National Academy of Sciences of the United States of America, 102,* 7535–7540.
7. Fuxreiter, M., Mezei, M. & Osman, R. (2005). Interfacial water as a "hydration fingerprint" in the noncognate complex of BamHI. *Biophys. J., 89,* 903–911.
8. Svergun, D. I., Richard, S., Koch, M. H. J., Sayers, Z., Kuprin, S. & Zaccai, G. (1997). Protein hydration in solution: Experimental observation by x-ray and neutron scattering, *Proceedings of the National Academy of Sciences of the United States of America, 95(5),* 2267–2272.
9. Franci Merzel and Jeremy C. Smith. (2001). Is the first water shell of lysozyme of higher density than bulk water? *Proceedings of the National Academy of Sciences of the United States of America, 99(8),* 5378–5383.
10. Gottfried Otting. & Kurt Wuethrich. (1989). Studies of protein hydration in aqueous solution by direct NMR observation of individual protein-bound water molecules, *J. Am. Chem. Soc., 111(5),* 1871–1875.
11. Soper, K. A. (2000). Structures of high-density and low-density water. *Physical Review Letters. 84(13),* 2881–2884.
12. Jordi Bella, Barbara Brodsky. & Helen M Berman. (1995). Hydration structure of a collagen peptide, *Structure, 3(9),* 893–906.
13. Chapman, E. G., Danyluk, S. S. & Mc Lauchlan, A. K. (1971). A Model for Collagen Hydration, Proceedings of the Royal Society of London, *Series B, Biological Sciences, 178(1053),* 465–476.
14. Edsall, T. J. & McKenzie, A. H. (1983). Water and proteins. II. The location and dynamics of water in protein systems and its relation to their stability and properties, Adv. *Biophys., 1(6),* 53–183.
15. Kuntz, D. I. & Kauzmann, W. (1974). Hydration of proteins and polypeptides, Adv. *Prot. Chem., 28,* 239–345.

16. Soper, A. K. (2000). The radial distribution function of water and ice from 220 to 673 K and at pressures up to 400 MPa. *Chemical Physics, 258,* 121–137.
17. Pethig, R. (1992). Protein-water interactions determined by dielectric methods. Annu. *Rev. Phys. Chem., 43,* 177–205.
18. Bulienkov, A. N. (2005). The role of modular design in the study of the self-organization of biological systems [in Russian]. *Biophysics, 50(5),* 934–958.
19. Jacob, N. (1991). Israelashvili. Intermolecular and Surface Forces, *Academic press,* 456.
20. http://www.mpibpc.mpg.de/grubmueller/solvate
21. http://www.ks.uiuc.edu/Research/namd/
22. Voloshin, P. V., Zheligovskaya, A. E., Malenkov, G. G., Naberukhin, I. Ju. & Tytik, D.I. (2001). Structures and dynamics of the H-bonds of water molecules in the condensed water systems [in Russian], Russian Chemical *Journal, XLV, 3(31–37).*
23. Marchenko, O. A., Solovey, B. A. & Lobyshev, V. I. (2013). Computer modeling of parametric structures of water, *Biophysics, 58(1),* 27–35.
24. Luigi Vitagliano., Rita Berisio. & Alfonso De Simone. (2011). Role of Hydration in Collagen Recognition by Bacterial Adhesins, *Biophys J, 100(9),* 2253–2261.
25. Solovey, B. A. (2006). Computer modeling of bound water structures [in Russian]. *Ph.D. Thesis*, Moscow.
26. Bulienkov, N. A. & Zheligovskaya, A. E. (2006). Functional modular dynamic model of water surface [in Russian], "*J Physical Chemistry,*" *80(10),* 1784–1805.

CHAPTER 3

OPTIMIZATION OF MOLECULAR MODELS FOR CALCULATING QUANTUM YIELDS OF PHOTOCHEMICAL REACTIONS

V. I. BARANOV, I. V. MIKHAILOV, and N. I. POTESHNAYA

CONTENTS

ABSTRACT

The recently proposed semi empirical method for simulation of photochemical processes and calculation of quantum yields of reactions has been modified. The specific form of double minimum potential energy surfaces of the molecules involved in the conversion has been considered in a more correct manner. It has been shown that the number of molecular models parameters can be reduced by two-thirds compared with the approach used previously. The quantum yields of photochemical transformations of six dienes into their cyclic isomers have been calculated. Substantially better quantitative agreement of the calculated values with the experimental data has been achieved for all the reactions. It has been shown that the model parameters have good transferability in a series of related molecules.

Aim and Background. To account for the asymmetry (anharmonicity) of potential wells of the molecules involved in the photochemical conversion in a more correct manner a new version of the method for calculating quantum yields has been developed. The practicability and advantage of the modified approach has been exemplified with the photochemical transformations of some diene compounds into their cyclic isomers.

3.1   INTRODUCTION

The optically induced processes of chemical transformations of polyatomic molecules are wide spread in the nature and have a broad application area. The diversity and importance of photochemical processes call for development of methods suitable for both qualitative and quantitative theoretical prediction of the proceeding and final results (quantum yield of the products) of reactions for actual molecules with different characteristics of optical excitation. The design of a photochemical experiment and the choice of the most optimal conditions of its setting from a broad variety of options are quite complicated without such the oretical estimates.

Qualitative theoretical photochemistry was created long ago and its basic concepts have been outlined in a large number of fundamental monographs and review articles [1–16]. The possibility of the a priori evaluation of the course and results of photochemical reactions has been studied to a much lesser extent. Theoretical estimation of the most important characteristics, such as, quantum yields of reactions, was impossible until recently even on the level of order of magnitude.

Certain progress in this line of research has been made as a result of development of the general quantum theory of molecular transformations [17], which formed the basis for formulation of the parametric method for the evaluation of photochemical processes and a prognostic computer simulation procedure [18–22].

Our consideration is restricted here to the case of structural isomerization reactions.

During a photochemical transformation, strong changes in the relative arrangement of the nuclei take place (e.g., a molecule converts from one isomeric form into another). For each of these isomeric structures, Schrödinger equation can be solved in a limited area of the nuclear coordinate space, which is close to the geometry of an isometric structure of interest. To describe the structural transformation, however, it is necessary to combine the characteristics of the reacting systems and to introduce a relation between them. The matrix formalism (the construction of a generalized matrix $\mathbf{H}$ with the basis set of the functions of the reacting substructures) turns out to be more rational. We will obtain a linear combination of functions for the isomers. These functions themselves are linear combinations of the basis functions of individual isomeric forms.

The problem of relative arrangement of the isomer basis functions in the nuclear coordinate space – normal coordinates for vibrations of the isomeric forms – was solved earlier [17]. The following expression for the transformation of the normal coordinates of the isomers Q' and Q was obtained:

$$Q' = \mathbf{A}Q + \mathbf{b}$$

and it was shown how to calculate the coordinate-system rotation matrix $\mathbf{A}$ and the well-minimum displacement vector $\mathbf{b}$.

During a photochemical process, both optical transitions (absorption and emission of electromagnetic waves) and reaction transitions (structural transformations of molecules) take place. For the model of an isolated molecular system of a given atomic composition, the reaction transitions can be only non radiative; that is, they occur without a change of the energy state. The time development of the process with the possibility of spontaneous nonradiative transitions from the initial isomeric from to the final one can be obtained by solving the following equation,

$$i\hbar\mathbf{S}\dot{\mathbf{c}} = \mathbf{H}\mathbf{c}.$$

Here $\mathbf{S}$ is the matrix of the overlap integrals of the basis functions and $\mathbf{c}$ is the column matrix of coefficients in their linear combinations. Nondiagonal elements of the energy matrix $\mathbf{H}$ characterize relations between the substructures. The multi dimensional problem can be reduced to a set of one-dimensional problems via the simultaneous bringing of the matrices $\mathbf{S}$ and $\mathbf{H}$ to the diagonal form. In the monograph [23] Gribov and Orville–Thomas have shown that such an operation is feasible in the case of over loaded basis as well. Since the vibronic functions of the same isomer can be considered ortho normalized and the overlap of the functions is small, it is admissible to assume from the very beginning that the integral overlap matrix $\mathbf{S}$ is the identity matrix. Solving equation

$$i\hbar\dot{\mathbf{c}} = \mathbf{H}\mathbf{c}$$

with the corresponding initial conditions, we can find the coefficients $c(t)$ and can describe the progress of the photochemical structural transformation.

The reaction is feasible if the total energy matrix **H** contains equal or very close diagonal elements and there are corresponding nonzero off-diagonal elements that characterize the mixing of the wave functions of the isomers that affect the rate of the chemical reaction. To cover transitions of different types within the frame of the same concept, it is most reasonable to introduce the idea of migration of a wave packet localized initially in the potential well of the reactant molecular entity to that of the product entity. To fulfill the condition of energy invariability in the process of nonradiative transitions we can introduce an n-fold degenerated level (with the average value of the energy) common for both isomers. Then the migration of the wave packet in the resonance region of energy levels of isomeric forms is not accompanied by energy change.

In the simplest case of two resonating states with energies E_1 and E_2 and stationary Eigen functions Ψ_1 and Ψ_2, we obtain for the square of the common wave function the following expression,

$$\Psi^2(t) = \Psi_1^2 \cos^2 \frac{\Omega t}{2} + \Psi_2^2 \sin^2 \frac{\Omega t}{2}.$$

Here Ω is the quantum beat frequency. In a good approximation Ω can be taken as $2E_{12}S_{12}/\hbar$, where $E_{12} = (E_1 + E_2)/2$ and S_{12} is the overlap integral of the corresponding functions Ψ_1 and Ψ_2. It is easy to see that the population of such a level via upward transitions upon absorption of light energy by the reactant molecule will result, with a certain delay, in transitions on to lower levels of the forming molecular entity. This will be recorded in the experiment as the results of the photochemical reaction. Thus the value of Ω associated with the rate of migration of the wave packet determines probability of chemical transformation. In a more complex case, when more than two levels resonate, the form of the wave packet becomes substantially different. The theory for the general multilevel case is detailed in Ref. [24].

Following the principle of a rational choice of appropriate simplified models, we can retain in the matrix H only the block corresponding to the groups of resonating levels of isomeric forms. In the simplest case of a pair of isomers, this can be a second-order block. For this block it is possible to find mixed functions for the resonating levels. Note that the initial values of the coefficients in the column c ($t = 0$) are always known exactly. They are determined by the character of optical excitation of the stable reactant isomerine the photochemical process. Then, the parabolic potential well of its own can be ascribed to excited (as well as ground) states of each of the reacting isomers. The position of the well bottom both in the space of relative nuclear coordinates and on the energy scale can be found according to semiempirical methods developed in the theory of vibrational and vibronic spectra of polyatomic molecules [17, 25–27].

The structural-dynamic models of molecules in the ground state are defined by experimental data on the geometry and force constants obtained by simulation of their IR spectra [28]. The correctness of these models was confirmed by qualitative and quantitative reproduction of not only the frequencies but also the intensities of the experimental spectra. Complex organic molecules are characterized by small differences in the potential functions between the electronically excited and the ground states, at least for the lower excited states, so that they need small corrections in nature [27].

In Fig. 3.1, illustrating the general idea of the molecular photo transformation A→B the vibronic energy levels of molecules A and B are sketched and the arrows show optical excitation and some of the possible radiative and nonradiative transitions between the levels.

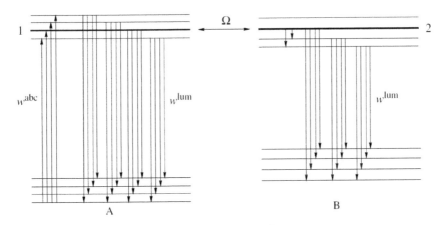

FIGURE 3.1 Radiative (optical absorption w^{abs} and luminescence w^{lum}) and nonradiative (quantum beat Ω) transitions between vibronic energy levels of isomers A and B. The bold lines refer to the resonating levels 1 and 2 of the isomers.

Experimental data on the excitation energy at which the abrupt rise in the reaction rate begins and which is associated with the height of the potential barrier between the potential wells of the isomers makes it possible to determine the crossing region of the parabolic potential surfaces for the combining isomeric forms. Anharmonic energy levels for the corresponding vibration frequency range can be found with the use of the "broadened" harmonic functions or functions for the Pöschl–Teller "skewed parabola" [17, 29, 30]. Then the quadratic matrix (which includes time-dependent elements) of the probabilities **P** is formed according to the known (calculated) probabilities of intramolecular optical transitions in the isomers and mixed functions for the resonating levels. The entire kinetics of the process is described by the set of differential equations (in matrix form),

$$\frac{d\mathbf{n}}{dt} = \mathbf{Pn}.$$

Here **n** is the column matrix of the energy level population of the molecular ensemble composed of both reactant (A) and product (B) molecules. If the system of rate equations is solved, it is easy to find the quantum yield of the photochemical reaction. By varying the parameters to be introduced, it is always possible to fit the obtained results with the experimental data. Note that such fitting does not make sense unless there is transferability, at least within reasonable limits, of parameters in a series of related molecules.

The selecting the resonating levels for vibration modes corresponding to the structural rearrangement of the molecules during isomerization is based on the geometry of the combining of isomers, the condition of the greatest overlap of vibrational functions, and the experimental data. In Fig. 3.2, the area of the largest overlap of the vibrational functions for two isomers is shown schematically. The vicinity of the minimal energy point (point X) in the normal coordinate space on the crossing curve of the paraboloids of the potential functions of the isomers is shaded.

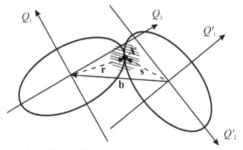

FIGURE 3.2   Cross-section of two-dimensional potential energy surfaces corresponding to the minimum on the intersection curve of the paraboloids of the potential functions of the isomers (point X). The area of the maximal overlap of the vibrational functions of the isomers is shaded. r and s are the vector radii of point X in the (Q_1, Q_2) and (Q'_1, Q'_2) coordinate systems, respectively. B is the displacement vector of the minima of the potential wells of the isomers in the normal coordinate space.

For the known relation between the normal coordinates of isomers the coordinates $X_i (X'_i)$ of point X can be calculated. The quantum numbers $v_i (v'_i)$ of the vibrational functions of the isomers corresponding to the greatest overlap is related to these coordinates by the simple expressions,

$$v_i = \left[\frac{1}{2} \left(\frac{X_i}{Q_{0i}} \right)^2 \right] \text{ and } v'_i = \left[\frac{1}{2} \left(\frac{X'_i}{Q'_{0i}} \right)^2 \right],$$

where the subscript i numbers the normal coordinates, Q_{0i} and Q'_{0i} are the zero vibration amplitudes, and the brackets (Iverson notation for the floor function) indicate that the integer part of the value is taken. In the general, multidimensional case, these vibrations are composite and have many quantum numbers $v_i \neq 0$, $v'_i \neq 0$. Then, as a rule, on the basis of the experimental data it is necessary to pass to energetically lower vibrational states with total vibrational quantum numbers of $v_i, v'_i \leq 2(3)$ by reducing proportionally the values of v_i and v'_i. The resulting states are the vibration modes characterized by nonzero quantum numbers corresponding to the structural rearrangement relevant to an isomeric transition.

The next step in the choice of the model is to takes into account anharmonicity. The simplest version of the approach employing the harmonic functions obtained via potential-well broadening is quite natural, since cross sections of the combining potential functions (initial structure and final product) must exhibit distinct asymmetry: they have a more gently steeping form as compared to parabolic functions in the region of the barrier between the wells (Fig. 3.3).

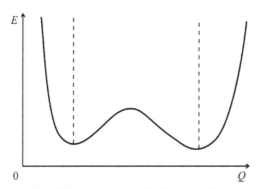

FIGURE 3.3 Section of combining potential well of two reacting structures along a normal coordinate Q.

For each cross section (along normal coordinates Q) of each well, its feature can be described by two conjugated half-parabolas with different focal parameters (the distance between the focus and the directrix of a parabola) coupled at the point of their common minimum. It is reasonable to take the focal parameter p_0 of the half-parabola on the edge of the well, that is, the one opposite to the half-parabola that approximates the barrier wall, to be the same as the value of focal parameter corresponding to the IR spectrum of the initial structure (let us call it the base half-parabola). Then the focal parameter p_u of the half-parabola at the barrier wall can be obtained by multiplying the focal parameter of the base half-parabola by a factor u more than unity: $p_u = p_0 \cdot u$. This factor, the broadening (asymmetry) parameter $u = p_u / p_0$, was introduced as an empirical quantity common for all the cross sections of potential energy surfaces of both isomers. Its value was determined from the condition of the best fit of calculation results to experimental data.

Although passing to broadened half-parabolas gives incorrect description of the form of the potential function and, hence, the vibrational function in the area opposite the barrier along the corresponding coordinate, it does not play a significant role, since the quantity sought is the overlap integral of the functions in the barrier region and this integral is independent of the form of the functions far from this region [17].

Varying the magnitude of broadening, we can obtain model vibrational functions that most adequately describe their properties in the region of interest in the internal-coordinate space. A set of calculations for actually observed photochemical reactions [18–22] showed that the oretically predicted quantum yields of all the reactions considered are in satisfactory agreement with the experimental values (see Table 3.1). The deviation of the predicted from the experimental values for the reactions with quantum yields in the range of 0.01÷1 was 10÷60%. In cases when reactions do not proceed according to the experimental data, the calculated quantum yields turned out to be less than 10^{-6}. The ratio of the quantum yields of the considered isomer transformations was also quantitatively predicted at an acceptable level, including the case of similar photochemical processes with quantum yields differing qualitatively by a few orders of magnitude.

TABLE 3.1 Calculated φ_{calc} and Experimental φ_{exp} [1, 2] Quantum Yields of Photochemical Reactions

	Reaction	φ_{calc}	φ_{exp}
1	2,4-dimethylpentadiene-1,3→ trimethylcyclobutene	10^{-13}	0
2	2,3-dimethylbutadiene-1,3→ dimethylcyclobutene	0.54	0.12
3	pentadiene-1,3 → 3-methylcyclobutene	10^{-4}	0.03
4	*cis*-butadiene-1,3 → cyclobutene	0.6	0.03, 0.3
5	2-methylbutadiene-1,3 → 1-methylcyclobutene	0.03	0.09
6	1-methoxybutadiene-1,3 → methoxycyclobutene	10^{-7}	0
7	*o*-xylene → *m*-xylene	0.011	0.013
8	*m*-xylene → *o*-xylene	0.009	0.006
9	*m*-xylene → *p*-xylene	0.021	0.024
10	*o*-diethylbenzene → *m*-diethylbenzene	0.001	0.03
11	cyclopropanecarboxaldehyde → 2-butenal	0.17	0.35
12	cyclopropylethanone®3-pentenone-2	0.27	0.30

The calculation results showed that the scatter in optimal values of the asymmetry (broadening) parameter u at which the best agreement with the experiment is attained was no more than $15 \div 20\%$. The stable value of u demonstrated that this parameter possessed the inherent property of transferability. For four transformations of dienes into their cyclic isomers (Eqs. (2)–(5)) the optimal values of the broadening parameter u lay in an arrow range with a scatter about the mean ($u_{av} = 6.25$) less than 15% to be $u_{opt} = 5.5, 7.25, 5.25$, and 6.5, respectively. The calculated values of the quantum yields for Eqs. (1)–(6) at the same average value of the parameter u_{av} are presented in Table 3.2. For Eqs. (1) and (6) in a wide range of $u = 5 \div 7.5$ the calculated values of the quantum yield are less than 10^{-10} and 10^{-5} respectively and it was accepted that $u_{opt} = u_{av}$. The most illustrative fit was observed for reactions of the same type Eqs. (1) and (5). The occurrence of Eq. (5) is experimentally detectable and it has a relatively high quantum yield, whereas similar Eq. (1) does not occur. This situation was quantitatively reproduced in the calculation with a high accuracy.

TABLE 3.2 Experimental Values φ_{exp} [2] of the Quantum Yields for Eqs. (1)–(6) and the Yields Calculated at $u = u_{av}$

Reaction	φ_{exp}	$\varphi_{calc}(u_{av})$
1	0	10^{-13}
2	0.12	0.6
3	0.03	10^{-4}
4	0.03	0.6
5	0.09	0.03
6	0	10^{-7}

3.2 THEORETICAL PART

To improve molecular models used in the method for calculating quantum yields, variants of a more correct allowance (differentiated with respect to the normal coordinates) for the asymmetry (anharmonicity) of potential wells have been examined.

The potential energy surface of a molecular system with two stable states between which the isomer–isomer transition is possible is shown schematically in Fig. 3.4. For greater clarity, the simplest case of the two-dimensional coordinate system (the normal coordinates $Q_1, Q_2 (Q_1', Q_2')$) is sketched. Potential well minima are displaced along only one coordinate $Q_1 (Q_1')$. In this case the displacement vector $\mathbf{b}$ in the expression for transformation of the normal coordinates $Q' = AQ + \mathbf{b}$ have a nonzero value only for the first component and the square matrix $\mathbf{A}$ of coordinate rotation and scale variation is diagonal (no entanglement of the normal coordinates of the isomers).

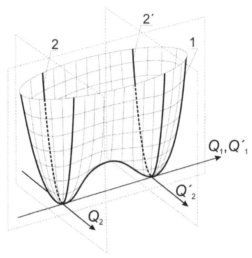

FIGURE 3.4   Potential energy surface of a molecular system with two structural isomeric states, as well as sections of the surface along the coordinates (1) Q_1 (Q_1'), (2) Q_2 and (2') Q_2'.

Curve 1 in Fig. 3.4 clearly shows that the potential energy surface of the system is significantly asymmetric along the normal coordinates Q_1 because of the presence of the barrier. To account for this particular property, in the approach discussed above broadened wave functions of the harmonic oscillator were used. However, it is clearly seen that the potential functions along the coordinates Q_2 and Q_2' are very close to parabolic (curves 2 and 2') and the presence of the potential barrier has a little effect on their forms.

To take into account this feature in the molecular models we can make an introducing correction (broadening) of the potential function dependent on the degree of anharmonic transformation of each vibration mode. The value of the corresponding elements of the displacement vector **b** would seem to be a natural criterion for the selection of the correction magnitude. However, the position of the area of the maximal overlap of the vibrational wave functions in the case of multidimensional displacement turns out to be a more adequate characteristic. This area is characterized by the point X (the top of the potential barrier, see Fig. 3.2) with coordinates X_i (X_i') in space of the normalized normal coordinates.

We can use a correction (or broadening) function $u(X_i)$ instead of the common asymmetry (broadening) parameter u. This function associates the value of a normalized coordinate X_i of the point X with the value of the ratio of the focal parameters for the corresponding cross section of the potential energy surface. Suppose that $u(X_i)$ satisfy the following conditions: $u(X_{max}) = U_0$ and $u(0) = 1$, where X_{max} is the maximum coordinate of the point X.

Among the types of functions (linear function, power functions, exponential functions, step (staircase) function, sigmoid functions) that have been tested the last two variants are of particular interest. Step function has only two parameters: the upper limit U_0 and the point X_0 of jump discontinuity (a certain "critical" value of the normalized coordinate X_i):

$$u(Y_i) = \begin{cases} U_0, & X_i > X_0 \\ 1, & X_i \le X_0 \end{cases}.$$

Smoothed step functions (rational sigmoid, arctangent, hyperbolic tangent, error function, etc.) have three parameters: the upper asymptote U_0, the inflection point X_0 (more precisely, the point where growth rate is maximum) and third parameter s defining the degree of smoothing (see Fig. 3.5).

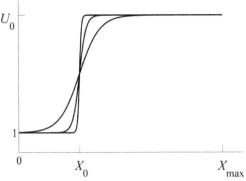

FIGURE 3.5 Graph of the function $u(X_i) = \dfrac{U_0 - 1}{2} \tanh[s(X_i - X_0)] + \dfrac{U_0 + 1}{2}$. The less parameter s the more smoothed the stair.

The most important factors that define the choice of broadening functions are following. The function is supposed to reduce the actual number of model parameters. This adequately reflects the specific features of the structural transformations of polyatomic molecules in which a relatively small number of coordinates out of their total number describing the structure have been shown [17] to transform during the reactive transition. Moreover, broadening must not lead to a significant shift in space of the normalized coordinates of the barrier position, the point X (top of the barrier). This greatly simplifies the procedure for determining the resonant states through which the isomer–isomer transition is accomplished and makes this procedure unambiguous. Only the staircase functions turn out to satisfy both of these conditions.

The results of the model calculations with the use of the step broadening functions have shown that the displacement of the position of the maximum overlap of the wave functions is related to the critical point $X_0 (X_0')$ in a peculiar manner (Fig. 3.6). The displacement value has been defined by

$$\xi = \arccos\left(\frac{(\mathbf{r}, \mathbf{r}_u)}{|\mathbf{r}| \cdot |\mathbf{r}_u|}\right) \text{ or } \xi = \arccos\left(\frac{(\mathbf{s}, \mathbf{s}_u)}{|\mathbf{s}| \cdot |\mathbf{s}_u|}\right),$$

That is, the angle of rotation of the vector $\mathbf{r}$ ($\mathbf{s}$) by introducing the broadening ($\mathbf{r}_u$, $\mathbf{s}_u$), in the systems of normalized coordinates.

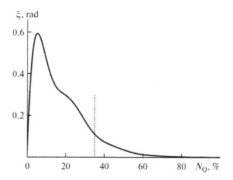

FIGURE 3.6   Typical pattern of the dependence for the displacement of the top of the potential barrier ξ upon the fraction N_Q of the normalized normal coordinates for which the potential function is broadened. The dotted line shows the notional boundary between large and small values of ξ; its position corresponds to $X_0 = 1/\sqrt{2}$.

For clarity, Fig. 3.6 presents the dependence of the rotation angle ξ upon N_Q, rather than X_0, where N_Q is the percentage of the normalized coordinates for which the potential function is broadened to take account of its asymmetry. Note that an increase in N_Q corresponds to a decrease in X_0. From the data in Fig. 3.6, it is seen that there are two distinct regions of values of ξ: (1) small rotation angles ($\xi < 0.1$ rad) and (2) their rapid growth from $\xi \approx 0.1$ rad to a maximum value of $\xi_{max} \approx 0.6$ rad. The boundary between these regions is shown by the dotted line and its position corresponds to the value of $X_0 = 1/\sqrt{2}$. It is clear that the sought critical value of X_0 should lie within the region of small values of ξ.

To select the critical value of X0 from this region, we will make use of the experience of simulation of vibronic spectra of complex molecules. Such "intraisomeric" vibronic transitions are characterized by a small displacement of the potential functions of combining states and by a large overlap of the vibrational functions, unlike interisomeric transitions in which the functions overlap by their "tails" (Fig. 3.7).

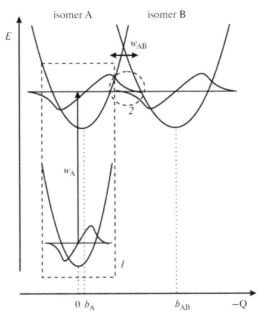

FIGURE 3.7 Schematic representation of potential and vibrational functions of isomers A and B, showing the distinctive features of intraisomeric optical transitions (isomer A, w_A is the probability of absorption) and interisomeric radiation less transformations (w_{AB} is the probability of the isomer-isomer transition A $\leftrightarrow$ B). The dashed lines specify the area of overlap of the vibrational functions that define the transition probabilities for (1) optical absorption and (2) radiationless structural transformation. b_A and b_{AB} are the displacements of the combining potential functions for the intra and the interisomeric transformation, respectively; $b_A < b_{AB}$.

Thus, we make the number of normal modes that are little affected by the presence of the interisomeric potential barrier and have no substantial effect on the structural transformation of the molecules to include the modes for which $X_i < X_0$ ($X_i' < X_0$). The point on the potential energy function equal to half the energy of the corresponding zero-point vibration ($X_0 = 1/\sqrt{2}$) definitely corresponds to the case typical of intraisomeric transitions and to small displacements of the position of the potential barrier. When $X_0 > 1/\sqrt{2}$ significant displacements are possible, indicating indirectly the inclusion of the normal coordinates involved in the isomeric transition in the number of the coordinates that do not significantly affect the structural transformation of molecules.

3.3   RESULTS AND DISCUSSION

To verify the efficiency of the proposed model, the quantum yields of photochemical transformation of some diene molecules into their cyclic isomers (shown in the schemes below) have been calculated.

(1) 2,4-dimethylpentadiene-1,3 →trimethylcyclobutene

(2) 2,3-dimethylbutadiene-1,3 → dimethylcyclobutene

(3) pentadiene-1,3 → 3-methylcyclobutene

(4) cis-butadiene-1,3 → cyclobutene

(5) 2-methylbutadiene-1,3 → 1-methylcyclobutene

(6) 1-methoxybutadiene-1,3 → methoxycyclobutene

These photochemical reactions represent the same type of structural transformations; therefore, their simulation makes it possible to evaluate the degree of transferability of the parameters used in the method (U_0 and X_0) in the series of the molecules and, hence, the predictive potential of the theory.

Published experimental data [2, 3] show that the photo isomerization reactions of interest occur at a relatively low temperature (35 °C) with excitation into the first absorption band; that is, upon optical population of vibrational sublevels with total quantum numbers of $v \leq 2$ of lower electronically excited states.

The deviations of the calculated energies for the ground and excited states of the isomers are on the order of a vibrational quantum energy, which has a slight effect on the photoisomerization kinetics; therefore, these differences can be ignored in the system of levels of isomer models. The selected resonating vibrational states conform in the type of the vibration process to the structural rearrangement of molecules upon isomerization. The quantum numbers, frequencies, and forms of vibrations of the resonating states are given in. Table 3.3.

TABLE 3.3 Quantum Numbers (v_i), Frequencies (ω_i, cm^{-1}), and Forms of Vibrations of the Resonating Levels of Isomers

v_i	ω_i	Form	v_i	ω_i	Form
2,4-dimethylpentadiene-1,3 →trimethylcyclobutene					
1	129.54	α, β, γ	1	302.36	η
2,3-dimethylbutadiene-1,3 → dimethylcyclobutene					
1	187.44	γ, δ	1	126.05	Q_d
1	298.0	α, β, ε, ζ	1	963.38	Q
pentadiene-1,3 → 3-methylcyclobutene					
1	189.62	α, β	1	198.44	Q_d
cis-butadiene-1,3 → cyclobutene					
1	284.4	α, β	1	262.2	Q_d
1-methoxybutadiene-1,3 → methoxycyclobutene					
1	244.29	α, β, γ	1	172.02	Q_d

For the Notation of CCC Angles, see the Schemes above. Q_d is the Counter Rotation of CH$_2$ Groups Relative to the Ring Plane.

The numerical values of the parameters of broadening functions $u(X_i)$ were determined from the best fit of the theoretical to the experimental results. For this purpose, the quantum yields φ were plotted against the parameters U_0 and X_0. Examples of dependence $\varphi(U_0)$ at $X_0 = 1/\sqrt{2}$ for Eqs. (2), (3), and (5) are given in Fig. 3.8.

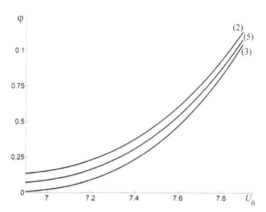

FIGURE 3.8   Dependence of the quantum yields of Eqs. (2), (3), and (5) upon the parameter U_0 of the step broadening functions ($X_0 = 1/\sqrt{2}$).

Optimal values of U_0 (at $X_0 = 1/\sqrt{2}$) corresponding to the best fit of the calculated quantum yields to the experimental data for each isomerization reaction separately lie within quite a narrow range with a spread of values about the mean ($U_{0, av} = 7.7$) less than 10%. For Eqs. (1) and (6) in a wide range of $U_0 = 6.5-8.0$, the calculated values of the quantum yield φ_{calc} are $<10^{-7}$, in agreement with the experiment. On these grounds, it was taken that optimal values of $U_0 = U_{0, av}$ for the reactions in question.

Table 3.4 presents the calculated values of the quantum yields for the same average value of $U_{0, av}$ ($X_0 = 1/\sqrt{2}$). It can be seen that the deviation from the experimental values is no more than 50% (on average, about 35%). The most striking result is the exact reproduction by calculation of the qualitative differences between reactions of the same type: the calculated quantum yields close to zero for Eqs. (1) and (6) versus quite large quantum yields of the other reactions, as well as the rise in φ_{calc} consistent with the experimental observation through the series of Eqs. (3), (5), (2), and (4).

TABLE 3.4   Results of Calculation of the Quantum Yields of Eqs. (1)–(6) for the Optimized Molecular Model ($U_{0, av} = 7.7$; $X_0 = 1/\sqrt{2}$)

Reaction	N_0, %	φ_{exp} [2, 3]	φ_{calc}
1	32	0	10^{-15}
2	35	0.12	0.09
3	34	0.03	0.06
4	34	0.03, 0.3	0.6
5	43	0.09	0.07
6	44	0	10^{-8}

This is evidence for both the stability of the parameter U_0 in a series of reactions and its inherent property of transferability in the given optimized molecular model. Moreover, the degree of transferability of U_0 is even higher than for the broadening (asymmetry) parameter u in previous model with the full broadening of the potential surface along the all coordinates – the spread of optimum values for reactions of the same type is less than by a factor of 1.5.

The values of the quantum yields predicted with the use of the optimized model better agree with the experimental values for all the reactions (cf. Tables 3.2 and 3.4), with the improvement being very significant – the calculated quantities were close in magnitude to the experimental values in all cases. For example, the new and old models give the following values of the quantum yields for Eq. (3): 0.06 and 10^{-4} ($\varphi_{exp} = 0.03$); and for Eq. (2): 0.09 and 0.6 ($\varphi_{exp} = 0.12$). Earlier, we speculated only about qualitative agreement with the experiment, but there is every reason to claim a quantitatively robust prediction by the advanced model.

It is noteworthy that the substantially better predicting potential of the advanced model is achieved with a smaller number of the normal coordinates for which the transformation of the potential energy surface is performed and, hence, a smaller number of model parameters (U_0), with the decreases in number being significant, by 55–70% (Table 3.4). This indicates that the molecular characteristics are more correctly described with the new model.

The result obtained for the quantum yields of there actions suggests that the calculated time development of photoisomerization correctly renders the specific features of the processes that occur in reality. The rate curves for level populations and the spectral curves of the time dependence of emission from resonating levels via transition to the ground states of the combining isomers are exemplified in Figs. 3.9 (for Eq. (2)) and 3.10 (for Eq. (3)), respectively.

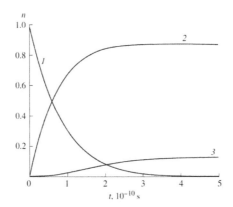

FIGURE 3.9 Rate curves for the reaction of photoisomerization of 2,3-dimethylbutadiene-1,3 into dimethylcyclobutene. Time change in the population (n) of (1) the resonance level and the ground electronic states of (2) 2,3-dimethylbutadiene-1,3 and (3) dimethylcyclobutene.

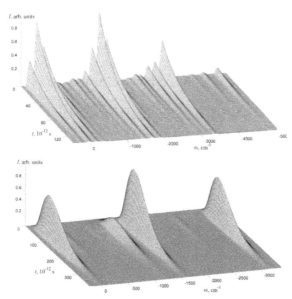

FIGURE 3.10   Time-dependent fluorescence spectra upon the reaction of photoisomerization of pentadiene-1,3into 3-methylcyclobutene corresponding to transitions from the resonant level to the vibronic levels of the ground electronic states of pentadiene-1,3 (upper spectrum) and 3-methylcyclobutene (lower spectrum).Each spectrum is normalized to the maximum intensity value. The frequency ω is counted from the frequency of the 0–0 transition.

3.4   CONCLUSION

An improved model taking into account the asymmetry (anharmonicity) of the potential wells of molecules involved in photochemical transformation has been proposed, which suggests a lesser curvature of the wells in the potential barrier region than at the edges. In contrast to the approach developed previously, in which the harmonic vibrational functions obtained with the same potential well broadening along all the normal coordinates were used, the model more correctly considers the form of the double-minimum potential surface. The fitting (broadening) of the potential function and, hence, the vibrational wave function is performed only for the normal modes, for which such an anharmonic transformation is significant because of the potential barrier. A criterion for selection of such vibration (normal coordinates) has been proposed. It has been shown that their number (and, consequently, the number of model parameters) is reduced by two-thirds compared with the previously used model.

The quantum yields of photochemical transformations of six molecules from the diene series into their cyclic isomers have been calculated. Substantially better quantitative agreement of the calculated values with the experimental data has been achieved for all the reactions than in the case of the previous model: the mean relative deviation from the experiment for reactions with $\varphi_{exp} \neq 0$ is smaller by more than a factor of 2 and the values of φ_{calc} for reactions that do not go ($\varphi_{exp} = 0$) are reduced by a few orders of magnitude. The example of these reactions shows that the parameters of the method have a high degree of transferability within a series of related molecules, there by ensuring the predictive nature of calculation. The results obtained in this study give reason to believe that our approach will make it possible to simulate photochemical processes and predict the quantum yields of reactions at the quantitative level.

KEYWORDS

- **Anharmonicity**
- **Photochemical reactions**
- **Potential energy surface**
- **Quantum yield**

REFERENCES

1. Turro, N. J. (1965). Molecular Photochemistry, New York: Columbia Univ.
2. Calvert, J. G. & Pitts, J. N. (1965). Photochemistry, New York: Wiley.
3. Terenin, A. N. (1967). Fotonikamolekulkrasitelei (Photonics of Dye Molecules), Leningrad: Nauka.
4. Bagdasar'yan, Kh. S. (1976). Dvukhkvantovayafotokhimiya (Two Photon Photochemistry), Moscow: Nauka.
5. Molin, Yu. N., Panfilov, Yu. N. & Petrov, A. K. (1985). Infrakrasnayafotokhimiya (Infrared Photochemistry), Novosibirsk: Nauka.
6. Wayne, R. P. (1988). *Principles and Application of Photochemistry*, Oxford: Oxford Univ. Press,
7. Michl, J. & Bonacic Koutecky, V. (1991). Electronic Aspects of Organic Photochemistry, New York: Wiley.
8. Klessinger, M. & Michl, J. (1995). Excited States and Photochemistry of Organic Molecules, New York: Wiley-VCH.
9. Surface Photochemistry, Anpo, M. (1996). Ed., New York: Wiley.
10. Maier, G. V., Artyukhov, V. Ya., Bazyl,' O. K., Kopylova, T. N., Kuznetsova, R. T., Rib, N. R. & Sokolova, I. V. (1997). Elektronno-vozbuzhdenny-esostoyaniyaifoto-khimiya-organic-heskikhsoedinenii (Electronically Exited States and Photochemistry of Organic Compounds), Novosibirsk: Nauka.
11. Coyle, J. D. (1998). Introduction to Organic Photochemistry, New York: Wiley,
12. Neckers, D. C., Jenks, W. S. & Wolff, T., (2007). Eds. Advances in Photochemistry, Hoboken, NJ: Wiley, 29.
13. Turro, N. J., Ramamurthy, V. & Scaiano, J. C. (2009). Principles of Molecular Photochemistry: An Introduction, Mill Valley, CA: University Science Books.

14. Wardle, B. (2009). Principles and Applications of Photochemistry, New York: Wiley.
15. Klan, P. & Wirz, J. (2009). Photochemistry of Organic Compounds: From Concepts to Practice, Chichester: Wiley.
16. Ramamurthy, V. & Inoue, Y. (2011). Supramolecular Photochemistry: Controlling Photochemical Processes, Eds., Hoboken, NJ: Wiley.
17. Gribov, L. A. & Baranov, V. I. (2006). Teoriyaimetodyraschetamolekulyarnykhprotsessov: spektry, khimicheskieprevrashcheniyaimolekulyarnayalogika (Theory and Methods of Calculation of Molecular Processes: Spectra, Chemical Transformations, and Molecular Logics), Moscow: KomKniga.
18. Baranov, V. I., Gribov, L. A., Dridger, V. E., Iskhakov, M. Kh. & Mikhailov, I. V. (2009). A Method for Simulation of Photochemical Processes and Calculation of Quantum Yields of Reactions, *High Energy Chemvol. 43(5)*, 362.
19. Baranov, V. I., Gribov, L. A., Dridger, V. E., Iskhakov, M. Kh. & Mikhailov, I. V. (2009). Simulation of Photochemical Processes and Calculation of the Quantum Yields of Isomerization Reactions of Substituted Dienes, *High Energy Chem., 43(6)*, 489.
20. Baranov, V. I., Gribov, L. A., Dridger, V. E. & Mikhailov, I. V. (2010). Calculation of the Quantum Yield of the PhotochemicalIsomerization Reaction Methoxybutadiene → Methoxycyclobutene, *High Energy Chem., 44(3)*, 181.
21. Baranov, V. I., Gribov, L. A., Iskhakov, M. Kh. & Mikhailov, I. V. (2010). Simulation of the Cyclopropanecarbaldehyde and Cyclopropylethanone Photoisomerization Processes and Calculation of the Quantum Yields of the Reactions, High. *Energy Chem., 44(4)*, 277.
22. Baranov, V. I., Gribov, L. A., Iskhakov, M. Kh. & Mikhailov, I. V. (2011). Simulation of Photochemical Reactions of Butadiene and Calculation of the Quantum Yields, *High Energy Chem, 45(6)*, 486.
23. Gribov, L. A. & Orville-Thomas, W. J. (1988). Theory and Methods of Calculation of Molecular Spectra, Chichester: Wiley.
24. Gribov, L. A. & Baranov, V. I. (2009). General Method to Simulate Molecular Processes Involving Complex Interactions between Combining Subsystems, *Journal of Structural Chemistry, 50(1)*, 10.
25. Gribov, L. A. (2008). Kolebaniyamolekul (Molecular Vibrations), Moscow: *Knizhnyi Dom "Librokom."*
26. Baranov, V. I., Gribov, L. A. & Djenjer, V. O. (1996). Semiempirical Parametric Method in the Theory of Vibronic Spectra of Polyatomic Molecules, *J Structural Chemistry, 37(3)*, 367.
27. Gribov, L. A., Baranov, V. I. & Zelentsov, D. Yu. (1997). Elektronno-kolebatel'nyespektrymnogoatomnykhmolekul: Teoriyaimetodyrascheta (UV Spectra of Polyatomic Molecules: Theory and Methods of Calculation), Moscow: Nauka.
28. Gribov, L. A., Dement'ev, V. A. & Todorovskii, A. T. (1986). Interpretirovannyekolebatel'nyespektryalkanov, alkenoviproizvodnykhbenzola (Interpreted VibrationalSpectra of Alkanes, Alkenes, and Benzene Derivatives), Moscow: Nauka.
29. Gribov, L. A. & Pavlyuchko, A. I. (1998). Variatsionnyemetodyresheniyaangarmonicheskikhzadach v teoriikolebatel'nykhspektrovmolekul (Variational Methods of Solution of Unharmonic Sums in the Theory of Vibrational Spectra of Molecules), Moscow: Nauka.
30. Gribov, L. A. & Dement'ev, V. A. (2004). *Russ. J. Phys. Chem, 78(1)*, 99.

CHAPTER 4

FURACILINUM RELEASE FROM POLYHYDROXYBUTYRATE-SCHUNGITE FILMS

A. A. OLKHOV, A. L. IORDANSKII, and G. E. ZAIKOV

CONTENTS

ABSTRACT

For the purpose of creation of biode composed polymeric materials for delivery of medicinal substances three fold system PHB-schungite furacilinum has appeared rather perspective. Schungite 0–5% occurs to maintenance growth reduction of values of speed of release M.F. from composites in 10 times. In ternary films maintenance increase furacilinum in absence schungite raises hardness in 1.5 times; the maintenance increase schungite in absence furacilinum raises hardness in 2.5 times. Formation of associates PHB-schungite-furacilinum assumed by us result into substantial growth of hardness of composite films (in 2–3 times) in comparison with initial and two componential films.

4.1   INTRODUCTION

In the last decade, significant number of the works devoted to the quantitative description of processes of release of low molecular weight medicinal substances from polymeric matrixes. One of the major problems of modern medicine is creation of new methods of the treatment based on purposeful local introduction of medical products in a certain place with set speed. The basic requirement shown to the carrier of active material, its destruction and a gradual conclusion from an organism [1, 2].

Medicinal forms (m.f.) with controllable release (medicinal forms with operated liberation, medicinal forms with programmed liberation) group of medicinal forms with the modified liberation, characterized by elongation of time of receipt m.f. in a bio phase and its liberation, corresponding to real requirement of an organism.

The decision of a problem of controllable release m.f. will allow designing a polymeric matrix of various degree of complexity, setting there by programmed speed of allocation of a medicine in surrounding biological environment. Regulation of transport processes in polymers taking into account their morphological features is one of actual problems in polymers. Research of interaction of polymeric materials with water important for many reasons, but the main things there are two: this interaction which plays the important role in the processes providing ability to live of the person. It influences operational properties of polymeric materials.

The mechanism of address delivery m.f. includes diffusive process of its release of a polymeric system [3]. It is known, that essential impact makes on diffusive properties of polymeric films morphology and crystallinity of the components forming a composite. Besides, at composite formation the nature of the filler changing not only structure of the most polymeric matrix is very important.

The purpose of this chapter was studying of influence finely divided schungite on structure, mechanical characteristics and kinetics of release of a medicinal drug furacilinum from films poly hydroxyl butyrate.

4.2   EXPERIMENTAL PART

As a polymeric matrix in work used polyhydroxybutyrate (PHB) bio decomposed polymer. Thanking these properties PHB it is applied as packing, to the biomedical appointment, self-resolving fibers and films, etc.

Basic properties PHB: melting point of 173–180wasps, temperature of the beginning of thermal degradation of 150wasps, degree of crystallinity 65–80%, molecular weight 10^4–10^6 g/MOLE, ultimate tensile strength 40 MPa, an elastic modulus 3.5 GPa, tensile elongation 6–8%.

The basic lack of products from PHB is low specific elongation at the expense of high crystallinity and formation of large sphemlitic aggregates.

Introduction in PHB, for example, finely divided mineral particles of filler can reduce fragility at the expense of formation of fine-crystalline polymer structure.

As a medical product in work used antibacterial means furacilinum (m.f.). Filler fine grained schungite technical specifications 2169–001–5773937–natural formation, on 30% consisting of carbon and 70% of silicates, the Zagozhinsky deposit [4].

It is necessary to note high probability of presence in schungite carbon of appreciable quantities, their chemical derivative and molecular complexes can play an important role as structural plastifiers.

In work investigated a film drug of matrix type which can be used at direct superposition on a wound or internally, for example, at intravitreal in a fabric [5]. Film made as follows

Powder PHB filled in with chloroform (500–600 mg PHB on 15–20 mL $CHCl_3$) and agitated on a magnetic stirrer before formation of homogeneous weight (~10 min). Then a solution lead up to boiling and at a working agitator brought furacilinum, and behind it schungite, agitated even 10 min). A hot solution filtrated through two beds of kapron and poured out in Petri dish D=9 sm which seated in a furnace at temperature 25°C, densely covered with the second Petri dish and dried up to constant weight.

For all samples defined the maintenance furacilinum, release furacilinum in water, density of samples, physical-mechanical characteristics on corresponding state standard specification, technical specifications and to laboratory techniques [6].

4.3   RESULTS AND DISCUSSION

In Fig.4.1, typical dependence of size of release (M_∞/M_t) furacilinum from time is shown at the various maintenance schungite. From it follows, that all values of desorption monotonously increase in due course releases.

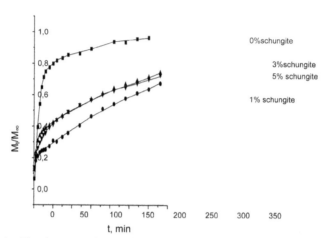

FIGURE 4.1   Kinetic curve releases of a medical product from films PHB at concentration furacilinum 3%.

Apparently from drawing, schungite and, hence, appreciable impact on diffusion rate f.m. makes on its speed of liberation. To maintenance growth schungite there is a significant falling of diffusivities, that is reduction of speed of allocation of medicinal substance. This effect can be connected with immobilization (interaction) of molecule m.f. with a surface schungite (with oxygen groups Si–O$_2$). It is follow-up possible to expect, that impenetrable particles schungite represent a barrier to diffusion M.F., and, hence, speed of transport in a composite drops.

On Fig. 4.2a, dependence of density of films PHB on the maintenance schungite in absence furacilinum for single-stage and two-phasic films is represented.

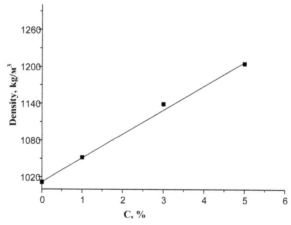

FIGURE 4.2A   Dependence of density of films PHB on the maintenance schungite without furacilinum.

From drawing it is visible, that to maintenance growth schungite there is directly proportional increase in density of films. Increase of density of composite films is caused by increase in their concentration of more dense filler.

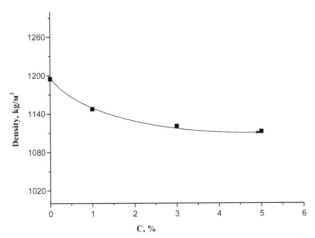

FIGURE 4.2B   Dependence of density of films PHB on the maintenance schungite at 5% furacilinum.

On Fig. 4.2b dependence of density of films PHB on the maintenance schungite is presented at 5% furacilinum. From drawing it is visible, that with maintenance increase schungite, the density decreases. It is possible to explain density decrease by a synergism resulting interference schungite and furacilinum. Yielded to explain formation of large associate furatsilin-shungite. Associates can arise at the expense of adsorption of molecules furacilinum on particles schungite, having an active surface, in solvent at formation of composite films. Associates can interfere with crystallization PHB and by that to increase quantity of the loosened amorphous phase of polymer.

Under condition of a share constancy schungite in composites with maintenance increase furacilinum the density varies like the curve on Fig. 4.2b.

So at introduction 0, 5% furacilinum are more narrows there is a reduction of density of a composition by 1%, and then at the further increase the maintenance furacilinum to 5% the density of films drops on 3–4%. Density reduction is caused, possibly, due to the structures PGB and formation of associates furatsilin-shungite which, increasing quantity of an amorphous phase, and create additional free volume in polymer a matrix.

The establishment of influence of the maintenance schungite and furacilinum on mechanical characteristics of composite films was one of the purposes supplied in work. Dependence of ultimate tensile strength of films PHB on maintenance M.F. is presented in Fig. 4.3a.

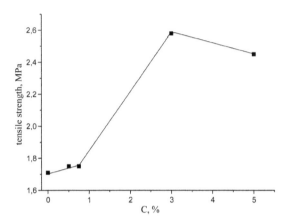

FIGURE 4.3A   Dependence of tensile strength of films PHB on the maintenance furacilinum without schungite.

As follows from Fig. 4.3a, hardness increases from the minimum value 1, 71 MPa in absence schungite to 2, 5 MPa, i.e. increases in 1.5 times. The hardness increase, possibly, is caused by formation of hydrogen bridges between furan groups LV and it groups PHB. In more details the mechanism of formation of associates will be stated further.

On Fig. 4.3b, dependence of hardness of films PHB on the maintenance schungite is presented at 5% furacilinum in films.

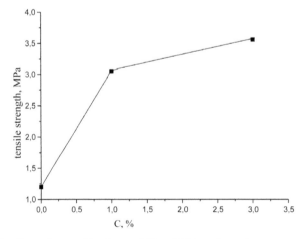

FIGURE 4.3B   Dependence of hardness of films PHB on the maintenance schungite at 5% furacilinum.

From Fig. 4.3b, the increase in hardness with maintenance increase schungite more, than in 2 times is visible, that also speaks formation of fine-crystalline structure of films a preparation stage. As it was already marked above, additional hardening also follows the account of formation of associates PHB- furacilinum-schungite.

Presence schungite in a composition result into hardening of films PHB data (Fig. 4.4).

On Fig. 4.4 dependence of hardness of films PHB on the maintenance schungite is presented.

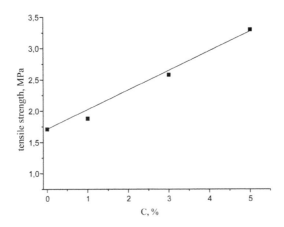

FIGURE 4.4   Dependence of hardness of films PHB on the maintenance schungite without furacilinum.

The increase in hardness of films schungite occurs to maintenance increase in 2.5 times. Apparently, at a stage of preparation of films schungite forms more perfect heterogeneous fine-crystalline structure. Schungite, being the nucleation center, creates set of nuclei, thereby, doing structure of films more ordered, that promotes hardness increase [7].

Considering collaterally (Figs. 4.3 and 4.4) it is possible to assume following structural changes of a composite at influence on PHB schungite and M. F.

Apparently, in composites PHB schungite furacilinum difficult associates, that is, formation of hydrogen bridges between oxygen containing groups furacilinum and similar groups PHB are formed. As a result formed communications act in a role of the cross linking localized in amorphous regions of a polymeric matrix. Presence of such cross linking should result in to increase in hardness of composite films. Under our assumptions, at first molecules furacilinum form hydrogen bridges with trailer. It groups, and then absorbed by other end on a surface schungite (Fig. 4.5).

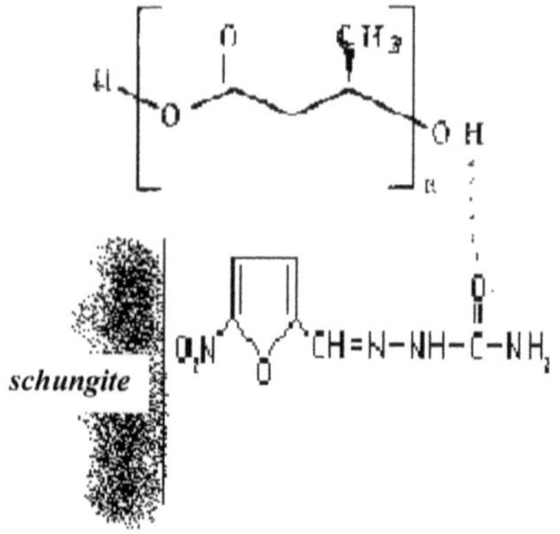

FIGURE 4.5 Schema of formation of associate PHB furacilinum schungite.

Growth of number of a cross linking does not occur in direct ratio since there is a restriction by quantity of end groups PHB. The yielded assumption is confirmed by means of IR-spectroscopy methods, DSC and EPR.

Thus, as a result of the made work the following has been positioned:

1. For the purpose of creation of bio decomposed polymeric materials for delivery of medicinal substances threefold system PHB schungite furacilinum has appeared rather perspective.
2. Schungite 0–5% occurs to maintenance growth reduction of values of speed of release M. F. from composites in 10times. That is schungite the prolonged an effect has on release kinetics furacilinum.
3. In ternary films maintenance increase furacilinum in absence schungite raises hardness in 1.5 times; the maintenance increase schungite in absence furacilinum raises hardness in 2.5 times.
4. Formation of associates PHB-schungite furacilinum assumed by us result ins to substantial growth of hardness of composite films (in 2–3 times) in comparison with initial and two-componental films.

KEYWORDS

- Furacilinum
- Polyhydroxybutyrate
- Schungite
- Structure
- Two componental films

REFERENCES

1. Plate, N. A., & Vasiliev, A. E. (1986). Physiologically Active Polymers Chemistry, Moscow.
2. Feldshtein, M. M., & Plate, N. N. (1999). In: *"Nuclear, Biological and Chemical Risks"* (Eds. Sohn, T., & Voicu, V. A.), Kluver Academic Publ. Dordrecht Boston London; *25,* 441–458. "A structure property relationship and quantitative approach to the development of universal transdermal drug delivery system."
3. Zaikov, G. E., Iordanskii, A. L., & Markin, V. S. (1988). Diffusion of Electrolytes in Polymers Ser, New Concepts in Polymer Science VSP Science Press Utrecht Tokyo Japan, 321p.
4. [http://www.xumuk.ru/farmacevt/1420.html] Furacilinum.
5. Koros, W. J. (Ed) (1990). Barrier Polymers and Structures, In: *ACS Symposium,* series 423. Washington, DC. *Am. Chem. Soc.*
6. Ju, S. (2008). Simonova Influence Schungite on structure, mechanical characteristics and Furacilinum release from films poly hydroxy butyrate, *Magister Dissertation,* Moscow State University of Fine Chemical Technology, *60,* 82 p.
7. Kuleznev, V. N., & Shershenev, V. A. (1988). Chemistry and Physics of Polymers: Studies for chemicotechnol Vuzov Higher School, Moscow, 312 p.

CHAPTER 5

STRUCTURE OF UV-PROTECTIVE NANOCOMPOSITES FILMS BASED ON POLYETHYLENE

A. A. OLKHOV, M. A. GOLDSHTRAKH, G. NYSZKO, K. MAJEWSKI, J. PIELICHOWSKI, and G. E. ZAIKOV

CONTENTS

ABSTRACT

High-strength polyethylene films containing 0.5–1.0 wt. % of nanocrystalline silicon (nc-Si) were synthesized. Samples of nc-Si with an average core diameter of 7–10 nm were produced by plasmochemical method and by laser-induced decomposition of monosilane. Spectral studies revealed almost complete (up to ~95%) absorption of UV radiation in 200–400 nm spectral region by 85 micron thick film if the nc-Si content approaches to 1.0 wt. %. The density function of particle size in the starting powders and polymer films containing immobilized silicon nanocrystallites were obtained using the modeling a complete profile of X-ray diffraction patterns, assuming spherical grains and the lognormal distribution. The results of X-ray analysis shown that the crystallite size distribution function remains almost unchanged and the crystallinity of the original polymer increases to about 10% with the implantation of the initial nc-Si samples in the polymer matrix.

5.1   INTRODUCTION

In recent years, considerable efforts have been devoted for search new functional nanocomposite materials with unique properties that are lacking in their traditional analogs. Control of these properties is an important fundamental problem. The use of nanocrystals as one of the elements of a polymer composite opens up new possibilities for targeted modification of its optical properties because of a strong dependence of the electronic structure of nanocrystals on their sizes and geometric shapes. An increase in the number of nanocrystals in the bulk of composites is expected to enhance long-range correlation effects on their properties. Among the known nanocrystals, nanocrystalline silicon (nc-Si) attracts high attention due to its extraordinary optoelectronic properties and manifestation of quantum size effects. Therefore, it is widely used for designing new generation functional materials for nanoelectronics and information technologies. The use of nc-Si in polymer composites calls for a knowledge of the processes of its interaction with polymeric media. Solid nanoparticles can be combined into aggregates (clusters), and, when the percolation threshold is achieved, a continuous cluster is formed.

An orderly arrangement of interacting nanocrystals in a long-range potential minimum leads to formation of periodic structures. Because of the well-developed interface, an important role in such systems belongs to adsorption processes, which are determined by the structure of the nanocrystal surface. In a polymer medium, nanocrystals are surrounded by an adsorption layer consisting of polymer, which may change the electronic properties of the nanocrystals. The structure of the adsorption layer has an effect on the processes of self-organization of solid-phase particles, as well as on the size, shape, and optical properties of resulting aggregates. According to data obtained for metallic [1] and semiconducting [2] clusters, aggregation and adsorption in three-phase systems with nanocrystals have an effect on the optical properties of the whole system. In this context, it is important to reveal the structural features of systems containing nanocrystals, characterizing aggregation and adsorption processes in these systems, which will make it possible to establish a correlation between the structural and the optical properties of functional nanocomposite systems.

Silicon nanoclusters embedded in various transparent media are a new, interesting object for physicochemical investigation. For example, for particles smaller than 4 nm in size, quantum size effects become significant. It makes possible to control the luminescence and absorption characteristics of materials based on such particles using of these effects [3, 4]. For nanoparticles about 10 nm in size or larger (containing $\sim 10^4$ Si atoms), the absorption characteristics in the UV and visible ranges are determined in many respects by properties typical of massive crystalline or amorphous silicon samples. These characteristics depend on a number of factors: the presence of structural defects and impurities, the phase state, etc. [5, 6]. For effective practical application and creation on a basis nc-Si the new polymeric materials possessing useful properties: sun-protection films [7] and the coverings [8] photoluminescent and electroluminescent composites [9, 10], stable to light dyes [11], embedding of these nanosized particles in polymeric matrixes becomes an important synthetic problem.

The method of manufacture of silicon nanoparticles in the form of a powder by plasma chemical deposition, which was used in this study, makes possible to vary the chemical composition of their surface layers. As a result, another possibility of controlling their spectral characteristics arises, which is absent in conventional methods of manufacture of nanocrystalline silicon in solid matrices (e.g., in $\tilde{\alpha}SiO_2$) by implantation of charged silicon particles [5] or radio frequency deposition of silicon [2]. Polymer composites based on silicon nanopowder are a new object for comprehensive spectral investigation. At the same time, detailed spectral analysis has been performed for silicon nanopowder prepared by laser-induced decomposition of gaseous SiH_4 (see, for example, [6, 12]). It is of interest to consider the possibility of designing new effective UV protectors based on polymer containing silicon nanoparticles [13]. An advantage of this nanocomposite in comparison with other known UV protectors is its environmental safety, that is, ability to hinder the formation of biologically harmful compounds during UV-induced degradation of components of commercial materials. In addition, changing the size distribution of nanoparticles and their concentration in a polymer and correspondingly modifying the state of their surface, one can deliberately change the spectral characteristics of nanocomposite as a whole. In this case, it is necessary to minimize the transmission in the wavelength range below 400 nm (which determines the properties of UV-protectors [13]) by changing the characteristics of the silicon powder.

5.2   OBJECTS OF RESEARCH

In this study, the possibilities of using polymers containing silicon nanoparticles as effective UV protectors are considered. First, the structure of nc-Si obtained under different conditions and its aggregates, their adsorption and optical properties was studied in order to find ways of control the UV spectral characteristics of multiphase polymer composites containing nanocrystalline silicon. Also, the purpose of this work was to investigate the effect of the concentration of silicon nanoparticles embedded in polymer matrix and the methods of preparation of these nanoparticles on the spectral characteristics of such nanocomposites. On the basis of the data obtained, recommendations for designing UV protectors based on these nanocomposites were formulated.

nc-Si consists of core–shell nanoparticles in which the core is crystalline silicon coated with a shell formed in the course of passivation of nc-Si with oxygen and/or nitrogen. nc-Si samples were synthesized by an original procedure in an argon plasma in a closed gas loop. To do this, we used a plasma vaporizer/condenser operating in a low-frequency arc discharge. A special consideration was given to the formation of a nanocrystalline core of specified size. The initial reagent was a silicon powder, which was fed into a reactor with a gas flow from a dosing pump. In the reactor, the powder vaporized at 7000–10,000 °C. At the outlet of the high-temperature plasma zone, the resulting gas–vapor mixture was sharply cooled by gas jets, which resulted in condensation of silicon vapor to form an aerosol. The synthesis of nc-Si in a low-frequency arc discharge was described in detail in Ref. [3].

The microstructure of nc-Si was studied by transmission electron microscopy (TEM) on a Philips NED microscope. X-ray powder diffraction analysis was carried out on a Shimadzu Lab XRD-6000 diffractometer. The degree of crystallinity of nc-Si was calculated from the integrated intensity of the most characteristic peak at $2\theta = 28°$. Low-temperature adsorption isotherms at 77.3 K were measured with a Gravimat-4303 automated vacuum adsorption apparatus. FTIR spectra were recorded on in the region of 400–5000 cm^{-1} with resolution of about 1 cm^{-1}.

Three samples of nc-Si powders with specific surfaces of 55, 60, and 110 m^2/g were studied. The D values for these samples calculated by Eq. (2) are 1.71, 1.85, and 1.95, respectively; that is, they are lower than the limiting values for rough objects. The corresponding D values calculated by Eq. (3) are 2.57, 2.62, and 2.65, respectively. Hence, the adsorption of nitrogen on nc-Si at 77.3 K is determined by capillary forces acting at the liquid–gas interface. Thus, in argon plasma with addition of oxygen or nitrogen, ultra disperse silicon particles are formed, which consist of a crystalline core coated with a silicon oxide or oxynitride shell. This shell prevents the degradation or uncontrollable transformation of the electronic properties of nc-Si upon its integration into polymer media. Solid structural elements (threads or nanowires) are structurally similar, which stimulates self-organization leading to fractal clusters. The surface fractal dimension of the clusters determined from the nitrogen adsorption isotherm at 77.3 K is a structurally sensitive parameter, which characterizes both the structure of clusters and the morphology of particles and aggregates of nanocrystalline silicon.

As the origin materials for preparation film nanocomposites served polyethylene of low-density (LDPE) marks 10803-020 and ultradisperse crystal silicon. Silicon powders have been received by a method plazmochemical recondensation of coarse-crystalline silicon in nanocrystalline powder. Synthesis nc-Si was carried out in argon plasma in the closed gas cycle in the plasma evaporator the condenser working in the arc low-frequency category. After particle synthesis nc-Si was exposed microcapsulating at which on their surfaces the protective cover from SiO_2, protecting a powder from atmospheric influence and doing it steady was created at storage. In the given work powders of silicon from two parties were used: nc-Si-36 with a specific surface of particles ~36 m^2/g and nc-Si-97 with a specific surface ~97 m^2/g.

Preliminary mixture of polyethylene with a powder nc-Si firms "Brabender" (Germany) carried out by means of closed hummer chambers at temperature 135±5°C, within 10 min and speed of rotation of a rotor of 100 min^{-1}. Two compositions LDPE + nc-Si have been prepared: (i) composition PE + 0.5% ncSi-97 on a basis ncSi-97, containing 0.5 weights silicon %; (ii) composition PE + 1% ncSi-36 on a basis ncSi-36, containing 1.0 weights silicon %.

Formation of films by thickness 85±5 micron was spent on semi-industrial extrusion unit ARP-20-150 (Russia) for producing the sleeve film. The temperature was 120–190°C on zones extruder and extrusion die. The speed of auger was 120 min^{-1}. Technological parameters of the nanocomposites choose, proceeding from conditions of thermostability and the characteristic viscosity recommended for processing polymer melting.

5.3 EXPERIMENTAL METHODS

Mechanical properties and an optical transparency of polymer films, their phase structure and crystallinity, and also communication of mechanical and optical properties with a microstructure of polyethylene and granulometric structure of modifying powders nc-Si were observed.

Physicomechanical properties of films at a stretching (extrusion) measured in a direction by means of universal tensile machine EZ-40 (Germany) in accordance with Russian State Standard GOST-14236-71. Tests are spent on rectangular samples in width of 10 mm, and a working site of 50 mm. The speed of movement of a clip was 240 mm/min. The 5 parallel samples were tested.

Optical transparency of films was estimated on absorption spectra. Spectra of absorption of the obtained films were measured on spectrophotometer SF-104 (Russia) in a range of wavelengths 200–800 nanometers. Samples of films of polyethylene and composite films PE + 0.5% ncSi-36 and PE + 1% ncSi-36 in the size 3 3 cm were investigated. The special holder was used for maintenance uniform a film tension.

X-ray diffraction analysis by wide-angle scattering of monochromatic X-rays data was applied for research phase structure of materials, degree of crystallinity of a polymeric matrix, the size of single-crystal blocks in powders nc-Si and in a polymeric matrix, and also functions of density of distribution of the size crystalline particles in initial powders nc-Si.

X-ray diffraction measurements were observed on Guinier diffractometer: chamber G670 Huber [14] with bent Ge (111) monochromator of a primary beam which are cutting out line $K\alpha_1$ (length of wave $\lambda = 1.5405981$ Å) characteristic radiation of x-ray tube with the copper anode. The diffraction picture in a range of corners 2θ from $3°$ to $100°$ was registered by the plate with optical memory (IP-detector) of the camera bent on a circle. Measurements were spent on original powders nc-Si-36 and nc-Si-97, on the pure film LDPE further marked as PE, and on composite films PE + 0.5% ncSi-97 and PE + 1.0% ncSi-36. For elimination of tool distortions effect diffractogram standard SRM660a NIST from the crystal powder LaB_6 certificated for these purposes by Institute of standards of the USA was measured. Further it was used as diffractometer tool function.

Samples of initial powders ncSi-36 and ncSi-97 for X-ray diffraction measurements were prepared by drawing of a thin layer of a powder on a substrate from a special film in the thickness 6 microns (MYLAR, Chemplex Industries Inc., Cat. No: 250, Lot No: 011671). Film samples LDPE and its composites were established in the diffractometer holder without any substrate, but for minimization of structure effect two layers of a film focused by directions extrusion perpendicular each other were used.

Phase analysis and granulometric analysis was spent by interpretation of the X-ray diffraction data. For these purposes the two different full-crest analysis methods [15, 16] were applied: (i) method of approximation of a profile diffractogram using analytical functions, polynoms and splines with diffractogram decomposition on making parts; (ii) method of diffractogram modeling on the basis of physical principles of scattering of X-rays. The package of computer programs WinXPOW was applied to approximation and profile decomposition diffractogram ver. 2.02 (Stoe, Germany) [17], and diffractogram modeling at the analysis of distribution of particles in the sizes was spent by means of program PM2K (version 2009) [18].

5.4   RESULTS AND DISCUSSION

Results of mechanical tests of the prepared materials are presented to table 5.1 from which it is visible that additives of particles nc-Si have improved mechanical characteristics of polyethylene.

TABLE 5.1   Mechanical Characteristics of Nanocomposite Films Based of LDPE and Nc-Si

Sample	Tensile strength, kg/cm²	Relative elongation-at-break, %
PE	100 ± 12	200–450
PE + 1% ncSi-36	122 ± 12	250–390
PE + 0.5% ncSi-97	118 ± 12	380–500

The results presented in the table show that additives of powders of silicon raise mechanical characteristics of films, and the effect of improvement of mechanical properties is more expressed in case of composite PE + 0.5% ncSi-97 at which in comparison with pure polyethylene relative elongation-at-break has essentially grown.

Transmittance spectra of the investigated films are shown on Fig. 5.1.

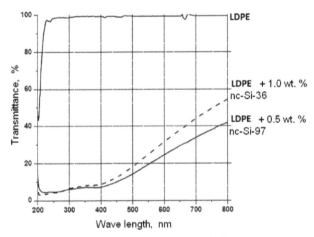

FIGURE 5.1   Transmittance spectra of the investigated films LDPE and nanocomposite films PE + 0.5% ncSi-97 and PE + 1.0% ncSi-36.

It is visible that additives of powders nc-Si reduce a transparency of films in all investigated range of wavelengths, but especially strong decrease transmittance (almost in 20 times) is observed in a range of lengths of waves of 220–400 nanometers, that is, in UV areas.

The wide-angle scattering of X-rays data was used for the observing phase structure of materials and their component. Measured X-ray diffractograms of initial powders ncSi-36 and ncSi-97 on intensity and Bragg peaks position completely corresponded to a phase of pure crystal silicon (a cubic elementary cell of type of diamond–spatial group $Fd\bar{3}m$, cell parameter $a_{Si} = 0.5435$ nanometers).

For the present research granulometric structure of initial powders nc-Si is of interest. Density function of particle size in a powder was restored on X-ray diffractogram a powder by means of computer program PM2K [18] in which the method [19] modeling's of a full profile diffractogram based on the theory of physical processes of diffraction of X-rays is realized. Modeling was spent in the assumption of the spherical form of crystalline particles and logarithmically normal distributions of their sizes. Deformation effects from flat and linear defects of a crystal lattice were considered. Received function of density of distribution of the size crystalline particles for initial powders nc-Si are represented graphically on Fig. 5.2, in the signature to which statistical parameters of the found distributions are resulted. These distributions are characterized by such important parameters, as *Mo(d)* – position of maximum (a distribution mode); $<d>_V$ – average size of crystalline particles based on volume of the sample (the average arithmetic size) and *Me(d)* – the median of distribution defining the size *d*, specifying that particles with diameters less than this size make half of volume of a powder.

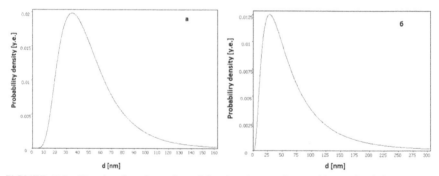

FIGURE 5.2   Density function of particle size in powders ncSi, received from X-ray diffractogram by means of program PM2K.

The results represented on Fig. 5.2, show that initial powders nc-Si in the structure have particles with the sizes less than 10 nanometers which especially effectively absorb UV radiation. The both powders modes of density function of particle size are very close, but median of density function of particle size of a powder ncSi-36 it is essential more than at a powder ncSi-97. It suggests that the number of crystalline particles with diameters is less 10 nanometers in unit of volume of a powder ncSi-36 much less, than in unit of volume of a powder ncSi-97. As a part of a powder ncSi-36 it is a lot of particles with a diameter more than 100 nanometers and even there are particles more largely 300 nanometers whereas the sizes of particles in a powder ncSi-97 don't exceed 150 nanometers and the basic part of crystalline particles has diameter less than 100 nanometers.

(a)–ncSi-97   Mo(d) = 35 nm   Me(d) = 45 nm $<d>_V$ = 51 nm;

(6)–ncSi-36   Mo(d) = 30 nm   Me(d) = 54 nm $<d>_V$ = 76 nm.

The phase structure of the obtained films was estimated on wide-angle scattering diffractogram only qualitatively. Complexity of diffraction pictures of scattering and structure don't poses the quantitative phase analysis of polymeric films [20]. At the phase analysis of polymers often it is necessary to be content with the comparative qualitative analysis, which allows watching evolution of structure depending on certain parameters of technology of production. Measured wide-angle X-rays scattering diffractograms of investigated films are shown on Fig. 5.3. Diffractograms have a typical form for polymers. As a rule, polymers are the two-phase systems consisting of an amorphous phase and areas with distant order, conditionally named crystals. Their diffractograms represent [20] superposition of intensity of scattering by the amorphous phase, which is looking like wide halo on the small-angle area (in this case in area 2 θ between 10° and 30°), and intensity Bragg peaks scattering by a crystal phase.

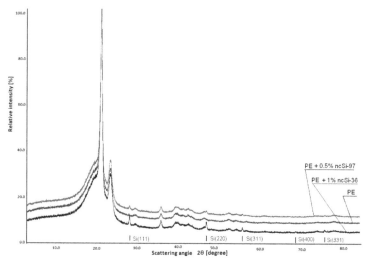

FIGURE 5.3 Diffractograms of the investigated composite films in comparison with diffractogram of pure polyethylene. Below vertical strokes specify reference positions of diffraction lines of silicon with their interference indexes (hkl).

Data on Fig. 5.3 is presented in a scale of relative intensities (intensity of the highest peak is accepted equal 100%). For convenience of consideration curves are represented with displacement on an axis of ordinates. The scattering plots without displacement represented completely overlapping of diffractogram profiles of composite films with diffractogram of a pure LDPE film, except peaks of crystal silicon, which weren't present on PE diffractogram. It testifies that additives of powders nc-Si practically haven't changed crystal structure of polymer.

The peaks of crystal silicon are well distinguishable on diffractograms of films with silicon (the reference positions with Miller's corresponding indexes are pointed below). Heights of the peaks of silicon with the same name (i.e., peaks with identical indexes) on diffractograms of the composite films PE + 0.5% ncSi-97 and PE + 1.0% ncSi-36 differ approximately twice that corresponds to a parity of mass concentration Si set at their manufacturing.

Degree of crystallinity of polymer films (a volume fraction of the crystal ordered areas in a material) in this research was defined by diffractograms Fig. 5.3 for a series of samples only semiquantitative (more/less). The essence of the method of crystallinity definition consists in analytical division of a diffractogram profile on the Bragg peaks from crystal areas and diffusion peak of an amorphous phase [20], as is shown in Fig. 5.4.

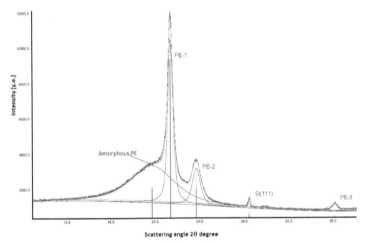

FIGURE 5.4   Diffractogram decomposition on separate peaks and a background by means of approximation of a full profile by analytical functions on an example of the data from sample PE+1%ncSi-36 (Fig. 5.3). PE-n designate Bragg peaks of crystal polyethylene with serial numbers n from left to right. Si (111)–Bragg silicon peak ncSi-36. Vertical strokes specify positions of maxima of peaks.

Peaks profiles of including peak of an amorphous phase, were approximated by function pseudoFoigt, a background 4 order polynoms of Chebysheva. The nonlinear method of the least squares minimized a difference between intensity of points experimental and approximating curves. The width and height of approximating functions, positions of their maxima and the integrated areas, and also background parameters were thus specified. The relation of integrated intensity of a scattering profile by an amorphous phase to full integrated intensity of scattering by all phases except for particles of crystal silicon gives a share of amorphy of the sample, and crystallinity degree turns out as a difference between unit and an amorphy fraction.

It was supposed that one technology of film obtaining allowed an identical structure. It proved to be true by coincidence relative intensities of all peaks on diffractograms Fig. 5.3, and samples consist only crystal and amorphous phases of the same chemical compound. Therefore, received values of degree of crystallinity should reflect correctly a tendency of its change at modification polyethylene by powders nc-Si though because of a structure of films they can quantitatively differ considerably from the valid concentration of crystal areas in the given material. The found values of degree of crystallinity are represented in Table 5.2.

TABLE 5.2 Characteristics of the Ordered (crystal) Areas in Polyethylene and Its Composites with Nc-Si

PE			PE + 1% ncSi-36			PE + 0.5% ncSi-97		
Crystallinity	**46%**		**47.5%**			**48%**		
2θ [°]	d [E]	ε	2θ [°]	d [E]	ε	2θ [°]	d [E]	ε
21.274	276	8.9	21.285	229	7.7	21.282	220	7.9
23.566	151	12.8	23.582	128	11.2	23.567	123	11.6
36.038	191	6.8	36.035	165	5.8	36.038	162	5.8
Average values	206	9.5×10^{-3}		174	8.2×10^{-3}		168	8.4×10^{-3}

One more important characteristic of crystallinity of polymer is the size d of the ordered areas in it. For definition of the size of crystalline particles and their maximum deformation ε *in* X-ray diffraction analysis [21] Bragg peaks width on half of maximum intensity (Bragg lines half-width) is often used. In the given research the sizes of crystalline particles in a polyethylene matrix calculated on three well-expressed diffractogram peaks Fig. 5.3. The peaks of polyethylene located at corners 2 θ approximately equal 21.28°, 23.57° and 36.03° (peaks PE-1, PE-2 and PE-3 on Fig. 5.4 see) were used. The ordered areas size d and the maximum relative deformation ε of their lattice were calculated by the joint decision of the equations of Sherrera and Wilson [21] with use of half-width of the peaks defined as a result of approximation by analytical functions, and taking into account experimentally measured diffractometer tool function. Calculations were spent by means of program *WinX*[POW] *size/strain*. Received d and ε, and also their average values for investigated films are presented in Table 5.2. The updated positions of maxima of diffraction peaks used at calculations are specified in Table 5.2.

The offered technology allowed the obtaining of films LDPE and composite films LDPE + 1% ncSi-36 and LDPE + 0.5% ncSi-97 an identical thickness (85 microns). Thus concentration of modifying additives ncSi in composite films corresponded to the set structure that is confirmed by the X-ray phase analysis.

By direct measurements it is established that additives of powders ncSi have reduced a polyethylene transparency in all investigated range of lengths of waves, but especially strong transmittance decrease (almost in 20 times) is observed in a range of lengths of waves of 220–400 nanometers, that is, in UV areas. Especially strongly effect of suppression UV of radiation is expressed in LDPE film + 0.5% ncSi-97 though concentration of an additive of silicon in this material is less. It is possible to explain this fact to that according to experimentally received function of density of distribution of the size the quantity of particles with the sizes is less 10 nanometers on volume/weight unit in a powder ncSi-97 more than in a powder ncSi-36.

Direct measurements define mechanical characteristics of the received films–durability at a stretching and relative lengthening at disrupture (Table 5.1). The received results show that additives of powders of silicon raise durability of films approximately on 20% in comparison with pure polyethylene. Composite films in comparison with pure polyethylene also have higher lengthening at disrupture, especially this improvement is expressed in case of composite PE + 0.5% ncSi-97. Observable improvement of mechanical properties correlates with degree of crystallinity of films and the average sizes of crystal blocks in them (Table 5.2). By results of the X-ray analysis the highest crystallinity at LDPE film + 0.5% ncSi-97, and at it the smallest size the crystal ordered areas that should promote durability and plasticity increase.

5.5 ACKNOWLEDGMENT

This work is supported by grants RFBR № 10-02-92000 and RFBR № 11-02-00868 also by grants FCP "Scientific and scientific and pedagogical shots of innovative Russia," contract № 2353 from 17.11.09 and contract № 2352 from 13.11.09.

KEYWORDS

- **Nanocrystalline silicon**
- **Polyethylene**
- **Polymer nanocomposites**
- **Spectroscopy**
- **UV-protective film**
- **X-ray diffraction analysis**

REFERENCES

1. Karpov, S. V. & Slabko, V. V. (2003). Optical and Photophysical Properties of Fractally Structured Metal Sols, Novosibirsk: Sib. *Otd. Ross. Akad. Nauk.*

2. Varfolomeev, A. E., Volkov, A. V., Godovskii, D. Yu., et al. (1995). *Pis'ma Zh. Eksp. Teor. Fiz., 62,* 344.
3. Delerue, C., Allan, G. & Lannoo, M. (1999*). Lumin. J. 80,* 65.
4. Soni, R. K., Fonseca, L. F., Resto, O., et al. (1999). *Lumin, J. 83–84,* 187.
5. Altman, I. S., Lee, D., Chung, J. D., et al. (2001). Phys. Rev. B: Condens. *Matter Mater. Phys. 63,* 161402.
6. Knief, S. & Niessen, W. von. (1999). *Phys. Rev. B: Condens. Matter Mater. Phys. 59,* 12940.
7. Olkhov, A. A., Goldschtrakh, M. A. (2009). Ischenko, A. A. RU Patent № 2009145013.
8. Bagratashvili, V. N., Tutorskii, I. A. & Belogorokhov, A. I. (2005). et al. Reports of Academy of Sciences. *Physical Chemistry, 405,* 360.
9. Kumar, V. (Ed.) (2008). Nanosilicon. Elsevier Ltd.–xiii + 368 p.
10. (2009). Nanostructured Materials. Processing, Properties, and Applications. Edited by Carl, C., Koch. NY: William Andrew Publishing, 752.
11. Ischenko, A. A., Dorofeev, S. G., Kononov, N. N., et al. (2009). RU Patent №2009146715.
12. Kuzmin, G. P., Karasev, M. E., Khokhlov, E. M., et al. (2000). *Laser Phys. 10,* 939.
13. Beckman, J. & Ischenko, A. A. (2003). RU Patent No. 2-227-015.
14. Stehl, K. (2000). The Huber G670 imaging-plate Guinier camera tested on beamline I711 at the MAX II synchrotron *J. Appl. Cryst, 33,* 394–396.
15. Fetisov, G. V. (2010). The X-ray phase analysis. Chapter 11, 153–184. Analytical chemistry and physical and chemical methods of the analysis. T. 2. Red. A.A. Ischenko. M.: ITC Academy, 416 p.
16. Scardi, P. & Leoni, M. (2006). Line profile analysis: pattern modeling versus profile fitting. *J. Appl. Cryst, 39,* 24–31.
17. WINX^{PoW} Version 1.06. STOE & CIE GmbH Darmstadt/Germany–1999.
18. Leoni, M., Confente, T. & Scardi, P. (2006). PM2K: a flexible program implementing Whole Powder Pattern Modelling Z. *Kristallogr. Suppl., 23,* 249–254.
19. Scardi, P. (2008). Recent advancements in whole powder pattern modeling Z. *Kristallogr. Suppl, 27,* 101–111
20. Strbeck, N. (2007). X-ray scattering of soft matter. Springer-Verlag Berlin Heidelberg–xx + 238 p.
21. Iveronova, V. I. & Revkevich, U. P. (1978). The theory of scattering of X-rays. M.: MGU, 278 p.

CHAPTER 6

IDENTIFICATION OF THE ACIDIC AIR POLLUTION BY IR SPECTROSCOPIC STUDY OF EPIPHYTIC LICHENS

A. F. MEYSUROVA, S. D. KHIZHNYAK, and P. M. PAKHOMOV

CONTENTS

ABSTRACT

Changes in chemical composition of various species of the epiphytic lichens treated with the acidic pollutants under the laboratory conditions are investigated by means of Fourier transform infrared (FTIR) spectroscopy. It is established that alkyl nitrates, sulfates or both types of these substances are formed in the lichen samples under the influence of HNO_3, H_2SO_4 or mixture of these acids. Hydrochloric acid and its mixtures with HNO_3 and H_2SO_4 appear to be highly toxic and cause mainly destruction processes in the lichens. It is found out that the mid-resistant species Hypogymnia physodes (L.) Nyl and Parmelia sulcata Tayl exhibit the best indicator properties. The application of FTIR spectroscopy in the monitoring of urban lichens allowed us to identify sulfur- and nitrogen-containing pollutants in the atmosphere around the Thermal Power Stations and along the motoways of industrial town of Tver.

    Aim and Background. It is well known that the amount of pollutants penetrating into atmosphere increases over the years due to the industrial revolution and human activities [1], therefore, the development of various approaches allowing one to perform the efficient monitoring of air contamination is one of the main task of the environmental chemistry. The goal of this chapter is to develop the spectroscopic technique to control air quality and to study the effect of the acidic pollutants (sulfuric, nitric and hydrochloric acids) and their mixtures on four species of epiphytic lichens such as *Hypogymnia physodes, Parmelia sulcata* Tayl., *Evernia mesomorpha* (Flot.) Nyl. and *Xanthoria parietina* (L.) Belt.) under laboratory conditions by means of FTIR spectroscopy. The results of the laboratory experiments will allow to identify changes in the chemical composition of the lichens caused by the acidic pollutants and to compare indicator capability of the widely spread lichen species. Spectroscopic analysis of lichen samples collected in contaminated sites of industrial town of Tver will give information about airborne acidic pollutants and, therefore, allow to judge about atmospheric pollutant levels.

    Epiphytic lichens have a special position among indicator species as long-living, sessile organisms with a high sensitivity to environments, which can be monitored during the entire year. They synthesize a variety of chemical compounds in response to their environmental conditions. Lichen indication combined with various techniques such as Fourier-transform Raman spectroscopy [2, 3], Fourier-transform infrared spectroscopy and SEM (scanning electron microscopy) [4], inductively coupled plasma mass spectroscopy [5] creates a powerful tool for obtaining information about chemical responses of lichens to stressed environments.

6.1   INTRODUCTION

It has been established [6, 7, 8] that FTIR spectroscopy can be applied successfully in the monitoring of atmosphere to identify various airborne pollutants (SO_2, NO_2)

in the widely spread lichen species *Hypogymnia physodes* (L.) Nyl, because the effect of sulfuric and nitric acids on lichen samples depends on the acid' concentration and exposition time. It is very interesting to evaluate the indicator capability of different lichen species and to clarify the mechanisms of their interactions with various pollutants. Based on the results of the spectroscopic studies, one could select the most sensitive lichen species for using them to control the air quality, taking into account specific effects of different pollutants on the lichens to enhance the efficiency of air monitoring. For further development of the biomonitoring with the help of FTIR spectroscopy, investigation of the lichen samples collected at urban areas will be performed.

6.2   EXPERIMENTAL PART

Four species of the epiphytic lichens such as *Hypogymnia physodes* (L.) Nyl., *Parmelia sulcata* Tayl., *Evernia mesomorpha* (Flot.) Nyl. and *Xanthoria parietina* (L.) Belt. are chosen for the study. A distinguish feature of the lichens is a various sensitivity to the pollutants. According to Trass' classification [9] there are sensitive (*Evernia mesomorpha),* mid-resistant (*Hypogymnia physodes, Parmelia sulcata)* and resistant (*Xanthoria parietina)* species. For the modeling experiment the lichen samples were collected in a relatively uncontaminated area in 60 km from the city of Tver—an industrial center, which is situated on a motorway between Moscow and St. Petersburg. The lichens were taken on the trees of one location at the same height (1.3–1.5 m). The FTIR spectra of the collected lichens of each species before treatment are use as a control (reference).

The treatment of the freshly collected lichens with the pollutants was performed by the following way: a wet lichen sample with size of about 1.5 1.5 cm² was kept in 1 l glass over 30 mL of 0.5% acid or mixture of the acids for a week at a room temperature (22–24°C) (Table 6.1). A sample was fixed at the inner surface of a tightly closed glass cover. During the experiments all glasses were kept at the sun-exposed site. It is known [10] that gaseous SO_2, NO_2 and HCl form sulfuric, nitric and hydrochloric acids, respectively, dissolving in water and this process is reversible, thus, by this way an attempt to reproduce the effect of the acidic rains was performed. An averaged concentration of SO_2 in the glasses 5–8 after a week exposition of the samples measured by means of portable gas analyzer "Miran SapphIRe XL" (USA) was equal 6 mg/m³.

The chosen concentration of the acidic pollutant (0.5%) was determined by conditions of the modeling experiment, namely, by the necessity of obtaining a rapid response of the lichen to an ecotoxicant. It should be remarked that under natural conditions, the pollutant concentration in the air is much lower, and its effect on lichens has a cumulative character.

TABLE 6.1   Experiment Schedule

Lichen species	Number of a sample					
	Acidic pollutant (0.5%)			Mixed acidic pollutant (0.5%)		
	HNO_3	H_2SO_4	HCl	$HNO_3+H_2SO_4$	HNO_3+HCl	H_2SO_4+HCl
Hypogymnia physodes	1	5	9	13	17	21
Parmelia sulcata	2	6	10	14	18	22
Evernia mesomorpha	3	7	11	15	19	23
Xanthoria parietina	4	8	12	16	20	24

To test the results of the modeling experiment and to monitor the atmospheric pollutions, FTIR was applied for the analysis of widely spread epiphytic lichens *Hypogymnia physodes* collected around Tver' thermal power stations (TPS-1, TPS-3, TPS-4) and along the motorways with dense traffic. Table 6.2 gives the location and characteristics of the sampling site.

TABLE 6.2   Characteristics of the Sampling Sites in Tver

Sample Number	Sampling site	Source of pollution	Main pollutants [11]
25	*District Tsentralny:* Tverskoy av.	Dense traffic: gasoline engine vehicles, aging of basic vehicle fleet, using of the exhaust system below "Euro-2"	CO_x, NO_x, C_xH_y, SO_2
26	*District Proletarsky:* Kalinin av.		
27	*District Moskovsky:* Gagarin sq.		
28	*District Proletarsky:* TPS-1	Industrial zone: fuel incineration processes, fuel oil, gas	SO_2, NO_x, solid particles
29	200 m to TPS-1		
30	*District Zavolzhsky,* in the vicinity of TPS-3	Industrial zone: coal, fuel oil, peat, gas	
31	*District Moskovsky* in the vicinity of TPS-4	Fuel oil, peat, gas	

The spectra of the lichens in the form of pellets with KBr were recorded on a FTIR spectrometer "Equinox 55" ("Bruker") in the range of 400–4000 cm^{-1} at a spectral resolution of 4 cm^{-1}, 32 scans. For the spectroscopic analysis a lichen sample was oven-dried at 32 °C for 48 h and ground; 22 mg of the sample was mixed with KBr (0.7 g) and pelletized in a hydraulic press.

Quantitative estimations of the spectroscopic data were based on the equation known as the Lambert-Beer law

$$A = kcx, \qquad (1)$$

where A is the absorbance of a sample of thickness x, c is the concentration of the absorbing species and the absorption coefficient k is independent of x or c. To avoid the influence of the sample thickness, the absorbance ratio of the analyzed and the internal standard bands has been taken into consideration; the base line correction procedure of all spectra was performed previously. In our case the absorption band at 2925 cm^{-1} corresponding to the asymmetric stretching mode of a CH$_2$ groups was used as the standard band in all calculations. Thus, the A_{1385}/A_{2925} and A_{1313}/A_{2925} absorbance ratios were used to evaluate the alkyl nitrates and sulfates contents, respectively.

6.3   RESULTS AND DISCUSSION

6.3.1   RESULTS OF MODELING

The spectroscopic analysis of lichens undergone either separated or combined effects of the pollutants gave the following results. The IR spectra of samples 1–24 differ essentially in dependence on lichen species and the type of ecotoxicant. The IR spectroscopic study of changes in the chemical composition of lichens allowed us to distinguish conventionally two main modes of the pollutant effect on lichens. The first one is related with the absorption of pollutant by the thallus with substantial changes in the chemical composition of the lichen. The second mode is not related with the accumulation of pollutant; there are no changes in the chemical composition of the lichen (or changes are insignificant), but a high-rate decomposition of the thallus takes place.

The combined effect of HNO$_3$ and H$_2$SO$_4$ is referred to the first version. The spectral analysis showed changes in the chemical composition due to the absorption of pollutant by the thalli of indicator species.

6.3.1.1   EFFECT OF HNO$_3$

Newly appeared absorption bands at 1384 v_s(-O-NO$_2$), 875 and 779 cm^{-1} δ(O–N–O) assigned to alkyl nitrates (R–O–NO$_2$) (Fig. 6.1a) [6, 12] were detected in the spectrum of H. physodes (1). Similar changes were revealed in the spectra of other indicator species (Fig. 6.1b–d). Alkyl nitrates are formed in the lichen's thallus through the interaction of free hydroxyl groups of lichenin (L_1), the main component of cell walls, with HNO_3 according to the following reaction:

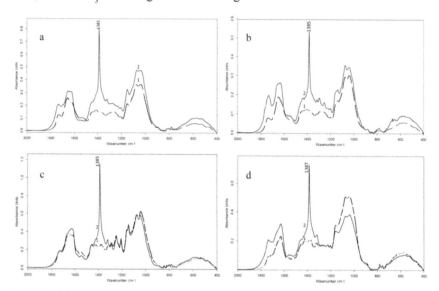

FIGURE 6.1 IR spectra of *Hypogymnia physodes* (a), *Parmelia sulcata* (b), *Evernia mesomorpha* (c) and *Xanthotia parietina* (d) samples: 1–control; 2–treated with 0.5% HNO_3.

$$L_1\text{-OH} + \text{OH-NO}_2 = \text{HOH} + L_1\text{-O-NO}_2 \tag{2}$$

The formation of alkyl nitrates in lichens is accompanied by the oxidation of free OH-groups of lichenin to carbonyl ($>C=O$) and carboxyl ($>COOH$) groups [13]. Thus, an increase of absorbance of C=O absorption bands in the infrared spectra of lichens (1725–1735 cm^{-1}) testify to decomposition processes in thallus.

The quantitative analysis of the infrared spectra of samples 1–4 has shown that among the indicator species the relative content of alkyl nitrates formed under the given conditions of modeling experiment is the most high in the sensitive species *Evernia mesomorpha* (A_{1385}/A_{2925} = 4.0) and the least amount of alkyl nitrates is found in the sample of the resistant species *Xanthoria parietina* (A_{1387}/A_{2925} = 2.13).

The exposition of lichens over 0,5% HNO_3 acid caused changes in the coloration of thallus and density of the core layer. The thalli of *Hypogymnia physodes*, *Parmelia sulcata* and *Evernia mesomorpha* became yellow-and-fawn colored in patches. Small-area necrotic spots were revealed in *Parmelia sulcata*. The alteration of the core layer density seems to be an adaptive reaction, which reduces toxicity of the stressor action and decreases the level of pollutant penetration. Thus, it was found that all studied species of the lichens are suitable for monitoring of the atmospheric pollution by airborne HNO_3. The lichens are able to accumulate the pollutant extensively. However, despite a high sensitivity of *Evernia mesomorpha*, there are good reasons for using the species *Parmelia sulcata* and *Hypogymnia physodes*. The thallus decompositions and rapid necrosis make these species good indicators of the air pollution by HNO_3.

6.3.1.2   EFFECT OF H_2SO_4

The changes caused by the formation of sulfates ($R–O–SO_2–OR_1$) in lichens exposed over 0.5% H_2SO_4 were revealed in infrared spectra of mid-resistant species *Hypogymnia physodes* (Fig. 6.2). The presence of sulfates in the sample (5) was evidenced by newly appeared absorption bands at 1313 $v_a(SO_2)$, 781, 663 and 518 cm^{-1} $v(S–O–C)$ [7, 12]. Moreover, the formation of sulfates in the lichens is accompanied by the thallus decomposition and alterations in its appearance.

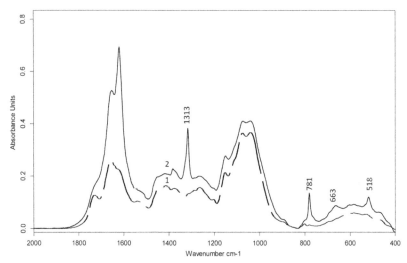

FIGURE 6.2   IR spectra of *Hypogymnia physodes* samples: 1–control; 2–treated with 0,5% H_2SO_4.

Among the studied species of the lichens, the samples of *Hypogymnia physodes* and *Parmelia sulcata* occurred to be the most damaged. The rate of the appearing of external signs allows one to suggest that the pollutant tolerance threshold of *Parmelia sulcata* being higher than that of *Hypogymnia physodes*.

Not too many changes in IR spectra of the *Evernia mesomorpha* and *Xanthoria parietina* samples are observed in comparison to the spectra of the mid-resistant species. A low absorption at 1356 cm^{-1} $v_a(SO_2)$ was detected in the spectrum of *Evernia mesomorpha* (Sample 7). *Xanthoria parietina* appeared to be the most resistant to the action of H_2SO_4, as in the IR spectrum of the sample 8 there were no bands related to the pollutant.

The data on the changes in the chemical compositions as well as in the external manifestations allow one to suggest the following sequence of lichen sensitivity to H_2SO_4: *Parmelia sulcata–Hypogymnia physodes–Evernia mesomorpha–Xanthoria parietina*. The resistance to the action of the given pollutant grows from the left to the right. Thus, it is reasonable to use *Parmelia sulcata* and *Hypogymnia physodes* species in monitoring of air by H_2SO_4 pollution. A high tolerance to H_2SO_4 makes *Xanthoria parietina* inappropriate for the evaluation of air contamination by this pollutant.

6.3.1.3   EFFECT OF HNO_3 AND H_2SO_4

The spectroscopic analysis of lichen sample (13) exposed over a mixture of the two acids showed changes in the chemical composition caused by the formation of two kinds of compounds, namely, sulfates ($R-O-SO_2-OR_1$) and alkyl nitrates ($R-O-NO_2$) (Fig. 6.3). The absorption bands at 1318, 779, 666 and 520 cm^{-1} indicate on the presence of sulfates in the lichens *Hypogymnia physodes,* while the band at 1384 cm^{-1} is characteristic for alkyl nitrates. The exposition of other species of lichens (15–16) over two acids did not cause any changes in the chemical composition, which could be related to the sulfates formation. In the IR spectra of samples *Evernia mesomorpha* (15) and *Xanthoria parietina* (16), weakly pronounced absorption at 1384 cm^{-1} indicated on the presence of alkyl nitrates. The values of A_{1384}/A_{2925} in the spectra of *Evernia mesomorpha* and *Xanthoria parietina* were nearly equal–0.84 and 0.83, respectively.

A combined action of ecotoxicants resulted in alterations of external appearance of lichens. The most damaged were the samples of mid-resistant species. Pearl-gray surface of processes of *Parmelia sulcata* becomes dingy-gray with spots of brownish and reddish-brawn color. Glaucous upper surface of processes *Hypogymnia physodes* acquires pearl-gray color; process' edges are fawn-colored in places. There are microcracks on external parts of the upper core. External appearance alterations in *Evernia mesomorpha* are pronounced in a less degree. *Xanthoria parietina* sample demonstrates a high tolerance to the action of acidic deterioration. According to sensitivity to acidic pollutants, one might construct the following sequence: *Parmelia sulcata, Hypogymnia physodes–Evernia mesomorpha–Xanthoria parietina.* The most appropriate species for biomonitoring of air pollution by the given type of pollutants are *Parmelia sulcata* and *Hypogymnia physodes* ones.

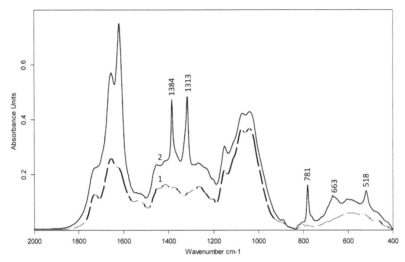

FIGURE 6.3 IR spectra of *Hypogymnia physodes* samples: 1–control; 2–treated with mixture of 0,5% H₂SO₄ and HNO₃.

6.3.1.4   EFFECT OF HCL, HNO₃ AND HCL, H₂SO₄ AND HCL

An example of the pollutant effect of the second type is the action of HCl as well as its combination with either HNO_3 or H_2SO_4 on lichens. In the case of such an effect, no changes due to pollutant absorption were found in the IR spectra (Fig. 6.4). There are no bands at 1461 and 720 cm⁻¹ v(C–Cl) [12], which are responsible for the accumulation of chlorine-containing polluting agents.

In the IR spectra of samples of mid-resistant lichens (9–10, 17–18, 21–22), insignificant changes in the intensities of previously existing absorption bands were observed. In the case of *Hypogymnia physodes* and *Parmelia sulcata*, these are the bands at ~1730 v(C=O) and 1619 cm⁻¹ δ (OH) [14]. The increase of the absorbance value of these bands might be connected with the decomposition caused by the oxidation of OH–groups of lichenin to carbonyl groups (>C=O) and carboxyl ones (>COOH). There are no changes in the spectra of *Evernia mesomorpha* and *Xanthoria parietina* lichens.

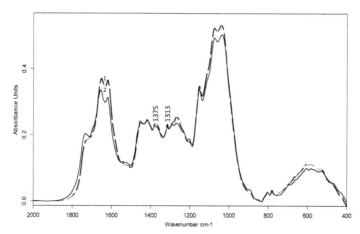

FIGURE 6.4   IR spectra of *Hypogymnia physodes* samples: 1–control; 2–treated with 0.5% HCl.

The lack of the changes in the chemical composition of lichens is combined with substantial alterations in the thallus appearance. Under the action of HCl, the thalli discolored and mellowed to the end of the experiment; the slimming of the upper and lower core occurred thus reducing its barrier function relatively detrimental compounds. The effect of HCl in combination with other acids turned out to be toxic for lichens: all thalli became dingy-gray or brownish. There appear multiple small cracks and tears on the surface of upper core. It seems that the use of these indicator species in the case of air pollution by low-concentrated chlorine-containing pollutants is inappropriate.

Thus, the IR spectra of lichens differ in dependence on the kind of ecotoxicant effect. One can distinguish two main cases. The first one is connected with the absorption of pollutant by thallus with its subsequent interaction with chemical components of the lichen. The effect of airborne H_2SO_4, HNO_3 and their mixture is referred to this kind. In the case of the second kind effect, the pollutant has no enough time to be absorbed by thallus and to react with its chemical components. The main cause is a high toxicity of pollutants for the lichen's thallus. The latter one is undergone to the strong decomposition before the pollutant absorption begins. An example of the effect of this kind is given by action of chlorine-containing compounds as well as those in combination with other ecotoxicants, that is HCl, HNO_3 and HCl, H_2SO_4, and HCl.

The performed analysis of the effect of pollutants on lichens, which consisted of the analysis of the chemical properties of nascent compounds in thallus and the efficiency of their action, morphological alterations, and necrosis appearing, allows us to construct the following sequence of the pollutants according to their toxicity:

$$HNO_3 > H_2SO_4 > H_2SO_4 + HNO_3 > HCl + HNO_3 > HCl + H_2SO_4 > HCl.$$

Among the indicator species, the most pronounced changes in the chemical composition occurred in two mid-resistant species, *Hypogymnia physodes* and *Parmelia sulcata*. These species are the most suitable for using as subjects of research in the practice of acidic pollution monitoring.

6.3.2 FIELD STUDIES

The IR spectroscopic analysis of lichen samples collected in vicinity of all TPSs supplying the city of Tver (TPS-1, TPS-3 and TPS-4) and along density-traffic motor-roads allowed us to detect in air the nitrogen- and sulfur-containing pollutants, such as SO_2 or airborne H_2SO_4 and NO_2, or airborne HNO_3. We failed to identify the chlorine-containing compounds in air with the help of the IR spectroscopic analysis of epiphytic lichens.

Alkyl nitrates were found in the samples of *Hypogymnia physodes* collected along Tverskoy avenue. There was an intensive absorption at 1384 cm^{-1} in the spectrum of lichen (25) (Fig. 6.5, spectrum 4), which indicated on the air polluting by airborne HNO_3. The spectroscopic analysis of lichens collected along Kalinin Avenue and Gagarin Square showed the presence of two compounds, namely alkyl nitrates (1384 cm^{-1}) and sulfates (1313, 781, 666, and 518 cm^{-1}) (Fig. 6.5). The appearance of sulfates in lichens is caused by the activity of the TPS-1 and TPS-4, which use as fuel peat, fuel oil, and in a less extent, natural gas. The quantitative analysis of samples (26–27) showed different content of sulfates. There were more sulfates than alkyl nitrates in sample 26 (A_{1313}/A_{2925} is equal to 1.16), while, on the contrary, there were more alkyl nitrates than sulfates in sample 27 (A_{1313}/A_{2925} is equal 1.43). A difference in distances from TPS-1 to Kalinin avenue and from TPS-4 to Gagarin square caused the difference in the content of sulfates.

In the area of TPS activity, the air pollution by SO_2 and/or H_2SO_4 airosol was identified. Only sulfates were detected in the samples 28–31 (Fig. 6.6). The quantitative estimation of the IR spectra of the samples 28–31 showed a different content of sulfates in the studies lichens.

A high content of sulfates is specific for the samples that were collected in the vicinity of the TPS-1, while a low content was near the TPS-3. The main causes of the difference in the sulfates content are the lifespan of power plants and bad structural ratio of used fuels (peat, coal, fuel oil, gas). For example, the fuel oil is predominantly used at the oldest TPS-1 (production activity from 1912); gas is used in a less degree. The relatively new plants TPS-3 и TPS-4 are natural gas-fired the most of the year.

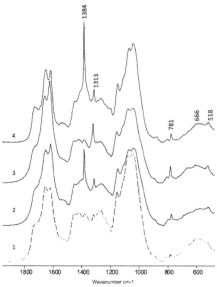

FIGURE 6.5   IR spectra of *Hypogymnia* physodes samples collected in control area (1) and along the streets: 2–Gagarin square; 3–Kalinin avenue; 4–Tver avenue.

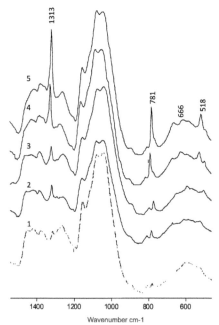

FIGURE 6.6   IR spectra of *Hypogymnia physodes* samples collected in the control area (1) and around various Thermal Power Stations (2–5).

Thus, the spectroscopic analysis of the chemical composition of epiphytic lichens allowed us to identify the air pollution, specify sources of pollution, and determine the content of pollutants in lichens.

6.4 CONCLUSION

The types of interactions of pollutants with lichens were established with the help of IR spectroscopic analysis. The chlorine-containing compounds as well as their combinations with HNO_3 and H_2SO_4 are highly toxic for the lichens. These pollutants are not accumulated by the thallus. The exposure to these pollutants leads to fast decomposition of thallus. The effect of both HNO_3 and H_2SO_4 as well as the effect of combinations of HNO_3 and H_2SO_4 manifest themselves in the formation of alkyl nitrates or sulfates (or both) in thallus.

It was found that the mid-resistant species *Hypogymnia physodes* и *Parmelia sulcata* exhibit the best indicator capability. There were essential changes in the spectra of these species. The changes in the chemical composition are accompanied by exterior alterations in the coloration as well as in the thallus structure. There are good reasons for using these species in the monitoring of urban ecosystems.

Using the IR spectroscopic analysis of *Hypogymnia physodes*, which were collected at town sites, nitrogen- and sulfur-containing compounds were identified in all samples. The pollution of the air by H_2SO_4 in areas of industrial activity is confirmed by the presence of sulfates in IR spectra of the lichens. Alkyl nitrates in the lichens collected along motorways are an evidence of the air pollution by HNO_3. The presence of airborne HNO_3 and H_2SO_4 in air is specific for roads in vicinity of TPS. We found alkyl nitrates and sulfates in the collected samples. A degree of remoteness of motor-roads from TPS determines different contents of the given types of compounds. The obtained results of both laboratory and field studies allow one to apply the FTIR spectroscopic analysis of the lichens for identification of the acidic pollutants in the atmosphere.

KEYWORDS

- **Alkyl nitrates**
- **Evernia mesomorpha**
- **FTIR spectroscopy**
- **Hypogymnia physodes**
- **Lichen**
- **Nitrogen and sulfur dioxides**
- **Parmelia sulcata**
- **Pollutant**
- **Sulfates**
- **Xanthotia parietina**

REFERENCES

1. The state report "On the condition and protection of the environment of the Russian Federation in 2009," Moscow, 2010, 495.
2. Villar, S. E. J., Edwards, H. G. M. & Seaward, M. R. D. (2005). Raman spectroscopy of hot desert, high altitude epilithic lichens *Analyst, Issue, 5(130)*, 730–737.
3. Edwards, H. G. M., Russell, N. C. & Seaward, M. R. D. (1997). Calcium oxalate in lichen biodeterioration studied using FT-Raman spectroscopy. *Spectrochim Acta 53A*, 99–105.
4. Garty, J., Kunin, P., Delarea, J. & Weiner, S. (2002). Calcium oxalate and sulfate-containing structures on the thallial surface of the lichen Ramalina lacera: response to polluted air and simulated acid rain. *Plant, Cell and Environment. 25*, 1591–1604.
5. Freitas, M. C. & Pachero, A. M. G. (2004). Bioaccumulation of Cobalt in Parmelia sulcata *J Atmospheric Chemistry, 49*, 67–82.
6. Meysurova, A. F., Khizhnyak, S. D. & Pakhomov, P. M. (2007). Spectroscopic study of the effect of nitrogen oxides on thallus of lichen Hypogymnia physodes (L.) *Nyl. Ecological Chemistry (Rus), 16(4)*, 27–35.
7. Meysurova, A. F., Khizhnyak, S. D., Dementieva, S. M. & Pakhomov, P. M. (2008). IR spectroscopic studies of the effect of sulfur dioxide on thallus of lichen *Hypogymnia physodes* (L.) Nyl. and their practical application *Ecological Chem. (Rus.), 17(3)*, 181–192.
8. Meysurova, A. F., Khizhnyak, S. D. & Pakhomov, P. M. (2009). **IR spectral analysis of the chemical composition of the lichen** *Hypogymnia physodes* **to assess atmospheric pollution** J. *Applied Spectroscopy, 76(3)*, **420–426.**
9. Trass, H. H. (1985). Classes of paleotolerance of lichens and environment monitoring Problems of environment monitoring and modeling of ecosystems. *7*, 122–137.
10. Goldovskaya, L. F. (2005). *Chemistry of Environment*. Moscow.
11. Kurbatov, A. C., Bashkin, V. N. & Kasimov, N. C. (Eds.). (2004). Urban ecology. Moscow, *Scientific World*, 624.
12. Socrates, G. (Ed.). (1994). Infrared characteristic group frequencies. *Tables and Charts.* London, John Wiley & Sons, 256.
13. Galbrah, L. S. (1996). Cellulose and its derivatives. *J. Soros Educational, 11*, 47–50.
14. Bazarnova, N. G. (2002). Methods of investigations of wood and its derivatives. (Ed.) Altai State University, Barnaul, 160.

CHAPTER 7

THE REACTIVITY OF FLUORO-AND CHLORO-SUBSTITUENTS OF ETHYLENE WITH OZONE: QUANTUM-CHEMICAL CALCULATION

E. A. MAMIN ELDAR ALIEVICH, B. E. KRISYUK, A. V. MAIOROV, and A. A. POPOV

CONTENTS

ABSTRACT

The next calculation methods: DFT B2PLYP, ab initio CASSCF, MRMP2, coupled-cluster CCSD calculations were applied to the study of the reactivity of the C=C bond of 1-monofluorethylene, 1-monochloroethylene, 1,1-difluorethylene, 1,1-dichloroethylene in reaction with ozone, aug-cc-pVDZ basis sets were used. Concerted and nonconcerted additions were investigated. It was shown that method CCSD is the best for modeling of reaction characteristics, MRMP2 results do not correspond to experiment in each case due to partial optimization. Once polar substituent is present, role of nonconcerted mechanism is greater – about 50% in 1,1-difluorethylene, 98% in 1,1-dichloroethylene.

Aims and Background. The aims of this chapter is to study the difference between two reaction ozonolysis mechanisms with respect to the number and nature of the double bond C=C substitutors.

7.1   INTRODUCTION

Addition of the ozone to the double bond is well known reaction and is often used in the analytical chemistry, for example, in determination of caoutchouc composition. Despite of it, there are no detailed comprehensions of that reaction. The investigation of the mechanism of model addition gives a key to following studies of solvent, substituent, structural deformation of the single and double bonds.

In study of this reaction one should take into account its complicated mechanism. The reaction may proceed through as a concerted 1,3-cycloaddition (Criegee mechanism) [1], or through biradical transition state (nonconcerted addition, DeMore) [2]. Question of their competition in case of ethylene is discussed in Ref. [3].

The present study was intended to investigate the influence of substituents on the ozone addition to double bond mechanism.

7.2   EXPERIMENTAL PART

The present study was performed in supercomputer center of IPCP RAS by means of quantum-chemical programs: Gaussian-03 [4], US GAMESS ver.7 [5]; supercomputer center of University of Helsinki VUORI with Gaussian-09 [6]; on personal computer with processor Core i7 in IBCP RAS with Firefly QC [7], with is partially based on [5], for open shell and closed shell (restricted and unrestricted Hartree-Fok and Kohn-Sham methods). Basis sets aug-cc-pVDZ with pure spherical harmonics was used.

Multiconfigurational calculations were done using scanning of intrinsic reaction coordinate (IRC), which makes possible to avoid size-nonconsistentency errors or intruder states arising. Active space choosing was described earlier in Ref. [8]. Calculations were performed in the following way: at first step UHF/basis set was used to find geometry of unfolded transition state TS2.

Obtained geometry and wave function was used as start ones in calculation of energy of saddle point at MCSCF(14,11)/aug-cc-pVDZ level. In existence and correctness of last object was approved by calculation of single imaginary frequency. For symmetrical transition state on the first step RHF instead of UHF was used, the following steps were analogous to calculation with TS2.

For each state scanning of intrinsic reaction coordinate was made on MCSCF(14,11)/aug-cc-pVDZ level from saddle point to reagents and products. Structure corresponding to reagents valley is "remote complex" named at the following discussion. As a result, geometries and wave functions of points at TS-reagents path were found.

Then, by using geometry obtained without its following optimization for each point of IRC curve were found corrections to energy, using MRMP2 method. For estimation of rate constant calculation of frequencies of normal mode vibrations in terms of MCSCF(14,11) method. To obtain thermodynamical functions frequencies of vibrations of ozone with MCSCF(12, 9) and olefins (MCSCF(2,2)). Value of k were found in terms of transition state theory:

$$ k = \frac{k_A RT^2}{hP_A} \exp\left(-\frac{\Delta G^{\neq}}{RT}\right) \qquad (1) $$

where P_A – standard pressure 101325 P_A; $\Delta G^{\neq} = \Delta H^{\neq} - T\Delta S^{\neq} + E_a$, $\Delta H^{\neq}$ и $\Delta S^{\neq}$ – activation enthalpy and entropy, $k_Б$ – Boltzmann constant, h – Plank constant, and R – ideal gas constant.

Values of SO_{298}, H^0_{298} were found with MOLTRAN program [11] by using of normal vibration frequencies. In CCSD calculation we used for this calculation B3LYP-frequencies instead of CCSD-ones, due to very high time consuming CCSD run of frequency calculation. The corresponding error is rather low, because of CCSD and B3LYP is quite close, and SO_{298}, H^0_{298} are enough close for these two methods. CCSD is more reliable because of better describing dynamical correlation and full optimization is performed in that case.

7.3   RESULTS AND DISCUSSION

The data obtained [8] for monochloroethylene and 1,1-dichloroethylene revealed, that it is another possibility for reaction to occur: not only mechanism Criegee but also DeMore mechanism is important and must to be taken into account. In present work we investigated transitions states, with give remarkable contribution in rate constant. Analogous data were obtained for butane-2 in Ref. [9, 10].

These structures are shown in Fig. 7.1(a–d) and will be designated as TS1 and TS2. Transition state cis-TS2 (NTS1) [12] is not obtained in MCSCF level, only in single determinant approximation, which makes its existence is uncertain. As has been found later, this TS is an artifact of some methods, evidence of this fact is presented in section "Calculation Details."

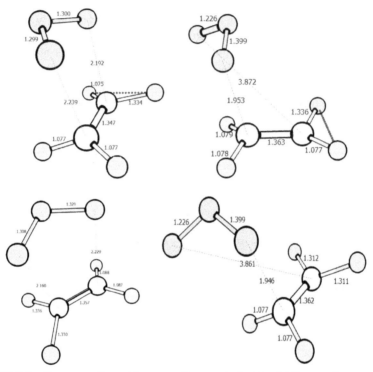

FIGURE 7.1   Geometry of transition state, distances in Å. (a) – TS1b monofluoroethylene; (b) – *trans*-TS2a monofluoroethylene; (c) – TS1 of 1,1-difluoroethylene; (d) – *trans*-TS2– 1,1-difluoroethylene.

For monochloroethylene, monofluoroethylene, 1,1-dichloroethylene and 1,1-fluoroethylene saddle points were found in each case only one imaginary vibration was found. After that, intrinsic coordinate scanning was performed.

These curves are shown on Fig. 7.2. Performing of MRMP2 corrections to each point of MCSCF IRC curve resulted in that despite of different configuration of complexes (far from transition states) energies of these complexes are differs not more than 1 kJ/mol.

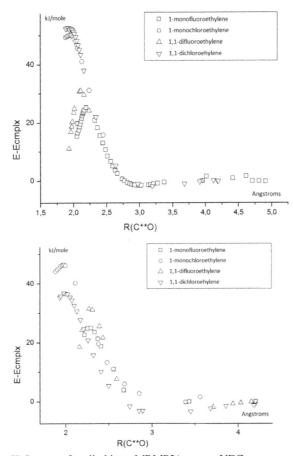

FIGURE 7.2   IRC curves for all objects MRMP2/aug-cc-pVDZ.

One should take into account in comparison of MRMP2 data that for chloroethylenes in Ref. [8] additional scanning on C=C distance was made (for investigation of dependence of rate constant on deformation).

Such scanning revealed that data obtained are not the exact saddle points. Scanning of length of C=C resulted in some deviation of TS on PES and in case of chloroethylenes decrease of activation energy to about 2–7 kJ/mol [8].

This is a result of restricted geometry optimization by using of MRMP2. In case of fluorosubstituents this probably will result in similar corrections, and consequently to increase of rate constant. In other words for fluorosubstituents rate constant is underestimated and correct value is higher. Main results of our work are presented in Tables 7.1 and 7.2. MRMP2 rate constants obtained in terms of corrections to

CASSCF IRC-curve, and complex being correspondent to 4–4.5 Å length between C and O atom. CCSD calculations were performed by finding saddle point and constrained optimization with distances C.O being frozen (4.5 Å, complex).

TABLE 7.1 Kinetic Characteristics of Reaction $O_3+C=C$ For Two Transition States

Method	TS	k_i	E_a	A	E_a^{exp}	k^{exp}
Ethylene + ozone						
MRMP2	TS1	20.1	16.7	$1.17×10^5$		
MRMP2	trans-TS2	0.03	38.3	$9.37×10^5$		
CCSD	TS1	[1]154	18.4	[1]$1.23×10^6$	17.6	930
CCSD	trans-TS2	[1]2	40.1	[1]$1.47×10^7$		[14–17]
B2PLYP	TS1	537	16.4	$1.47×10^6$		
B2PLYP	trans-TS2	0.02	48.4	$1.49×10^7$		
***1*-monofluoroethylene + ozone**						
MRMP2	TS1	3.2	25.3	$7.02×10^4$		
MRMP2	trans-TS2	0.01	51.7	$1.04×10^6$		
CCSD	TS1	55.2	18.9	$3.42×10^5$	30.1	< 420[2]
CCSD	trans-TS2	1.66	36.1	$2.98×10^6$		[20]
B2PLYP	TS1	12	21.6	$2.11×10^5$		
B2PLYP	trans-TS2	0.007	47.8	$4.34×10^6$		
***1*-monochloroethylene + ozone**						
MRMP2	TS1	0.52	25.02	$6.96×10^4$		
MRMP2	trans-TS2	0.0037	44.4	$5.89×10^5$		
CCSD	TS1	14.1	21.9	$2.98×10^5$	-	10–148
CCSD	trans-TS2	6.5	32.9	$3.05×10^6$		[18, 19, 21]
B2PLYP	TS1	1.38	26.2	$1.71×10^5$		
B2PLYP	trans-TS2	0.02	46.7	$4.57×10^6$		

E_a – Electronic Energy of Activation (kJ/mol), k –Rate Constants (L/mol·s) for Monohalogenated Substituents.

---

[1]Frequencies calculated in terms of CCSD, not in B3LYP as in other cases.

[2]Only one work, where according to authors, data should be taken as a top estimation for rate constant.

TABLE 7.2 Kinetic Characteristics of Reaction O_3+C=C For Two Transition States

Method	TS	k_i	E_a	A	E_a^{exp}	k^{exp}
1,1-difluoroethylene + ozone						
MRMP2	TS1[3]	0.032	30.7	$3.05×10^4$		
MRMP2	trans-TS2	$1.6×10^{-4}$	53.9	$1.11×10^6$		
CCSD	TS1	0.476	32.1	$4.91×10^5$		78–118
CCSD	trans-TS2	0.459	40.8	$6.43×10^6$	-	[14, 20]
B2PLYP	TS1	0.02	34.1	$1.77×10^5$		
B2PLYP	trans-TS2	0.001	51.8	$3.60×10^6$		
1,1-dichloroethylene + ozone						
MRMP2	TS1	0.049	31.5	$7.08×10^4$	38.5	1.6–5
MRMP2	trans-TS2	0.090	38.1	$9.65×10^5$		[24, 22]
CCSD	TS1	0.389	32.3	$4.58×10^5$		
CCSD	trans-TS2	18.2	31.6	$6.48×10^6$		
B2PLYP	TS1	$1.3×10^{-2}$	40.0	$3.59×10^5$		
B2PLYP	trans-TS2	$3,69×10^{-2}$	43.8	$5.59×10^6$		

E_a – Electronic Energy of Activation (kJ/mol), k –Rate Constants (L/mol·s) for Dihalogenated Substituents.

Energy of latter is corresponding to energy of reagents valley. Normal mode vibrations were found by using B3LYP for saddle point of corresponding type. CCSD represents experimental values more correctly all four objects concerned. In case of 1,1-dichloroethylene it gave considerable degree of competition between Criegee and DeMore mechanisms, as well as MRMP2, which is not for other objects.

The more precise results are CCSD ones, where not only reliable activation energies are but a good correspondence to experimental rate constants exists. So there is a base to hope to relation of constants (and so ratio of rates) is reliable in terms of CCSD. The latter fact means that role of second channel (nonconcerted mechanism) of reaction arises with the substitution. This is true as for fluoro- so as for chloro-substituents. While in case of ethylene and monosubstituents Criegee mechanism is prevailing one (more than 100 times). Role of DeMore mechanism in case of disub-

---

[3]Calculated through RMP2//MRMP2, due to corresponding saddle point was not found in CASSCF.

stituents is remarkable: k_1/k_2 is unity in 1,1-difluorethylene, 0.02 in case of 1,1-dichloroethylene, that is, in last case nonconcerted channel of addition is prevailing.

Rate contribution of TS2 in the net rate constant and corresponding rate constant is more magnificent. Substitution of one halogen atom causes decreasing net rate constant by approx. 10 for chlorine and for fluorine, mainly owing for rate through TS1 is decreasing.

In addition we used DFT method based on a mixing of standard generalized gradient approximations (GGAs) for exchange by Becke (B) and for correlation by Lee, Yang, and Parr (LYP) with Hartree-Fock (HF) exchange and a pertrubative second-order correlation part (PT2) –B3LYP [13].

Saddle points TS1 and TS2 were localized, IRC curves to regents valley were found. These last calculations resulted in geometries of complex with lengths C.O = 3.8 Å in case of DeMore mechanism and C.O = 2.8Å in case of Criegee mechanism.

In order to verify that they are correspond to reagents, calculations with frozen C.O distances were performed. Obtained data were using to calculation of electronic activation energies and thermodynamical functions in terms of normal mode vibrations in B2PLYP. Kinetic characteristics calculations are presented in Tables 7.1 and 7.2. Results of calculation of rate constants and comparison with literature data are presented in Figs. 7.3 and 7.4.

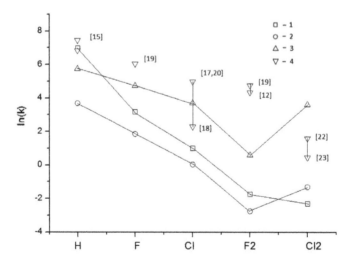

FIGURE 7.3 Values of logarithm of rate constant corresponding to certain object – haloidolefine. 1 –B2PLYP, 2 – MRMP2, 3 – CCSD, 4 – experimental value.

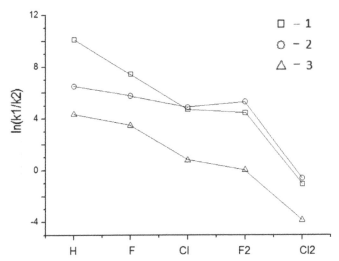

FIGURE 7.4 Values of logarithm of relation of rate constants through TS1 and TS2 corresponding to certain object – haloidolefine. 1 –B2PLYP, 2 – MRMP2, 3 – CCSD.

Comparison of calculation results with literature data [14–24] suggest one to conclude that methods, which we chose are adequate. Also, one can remark, that mechanism of reaction of halogenated olefins is more complicated than that of ethylene reaction [19, 21].

The stabilization of products of decomposition of primary ozonide, and as a consequence secondary reactions with ozone and olefines existence. One can underline that corpuscular chlorine plays role in such reactions with olefines [21]. This fact has been taken into account in Ref. [21] by inducing of radical scavengers. Experimental rate constant is should be considered as an upper edge of rate constant of elementary act, which is an object of present work.

Our data permit us to make a conclusion that by substitution of fluorine the decreasing of rate of both mechanisms occurs. Fluorine atom, as well known, is most electronegative atom, moreover its mass and length of bond CF = 1.34 Å is less declining from that of hydrogen in roe of halogens. So its main influence is in diminishing of electronic density at the double bond. As consequence – decreasing of rate constant of addition of electronegative agent – ozone molecule is approved by calculations.

Chlorine atom is less electronegative, has another number of electrons, different mass, CCl length is enough high CCl=1.73 Å. So there is some other effects are playing role, not only electronegativity.

Values of the preexponential multipliers A in Arrhenius equation differ slightly in different objects if one type of TS is considered, so the influence lies in electronic effects: conjugation p-orbitals of carbon atoms and chlorine atoms, which causes additional stabilization of double bond–reaction center. This effect arises for Criegee mechanism, bond becomes more strong, rate constant decreases (Fig. 7.5a).

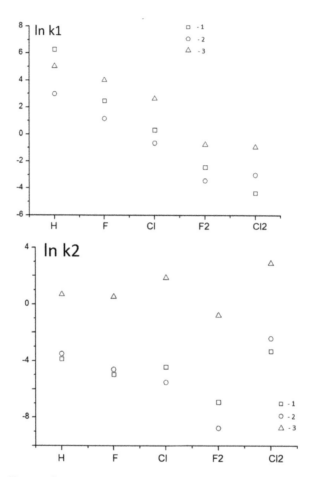

FIGURE 7.5 Change of rate constant of different mechanism (a) Criegee; (b) DeMore.

As for DeMore mechanism (Fig. 7.5b), there exists another, than in case of fluorine dependence on substitutor – rate constant k_2 increases with the substitution as all methods revealed. Such dependence is probably explained by a fact that conjugation stabiles the structure of TS2, and this effect is more pronounced for 1,1-dichloroethylene – conjugation is higher presented in 1,1-dichloroethylene than for monochloroethylene.

7.4 CALCULATION DETAILS

Calculations of ozone reactions with halogenated olefins in some cases resulted in saddle point *cis*-TS2 (NTS1, [12]). Its existence is not approved in all methods and not for all objects. In terms of UHF it has been found for ethylene and mono-substituents. In case of dihalogensubstituents for more reliable methods – CCSD (1,1-difluoethyle), CASSCF (1,1-dichloroethylene). This suggests that it is probably a calculation artifact, which should be additionally verified. For that purpose Gaussian 09, MOLPRO have been used. The verification was done in following way:

- using different active spaces: CAS(10, 9) и CAS(14, 11).
- different correlation-consistent basis sets: cc-pVDZ, aug-cc-pVDZ и cc-pVTZ.
- different methods of determination of saddle point – QSD, GDIIS in MOLPRO; geDIIS in Gaussian09.
- different criteria of verification o saddle point in MOLPRO – Baker, Gaussian.
- in MOLPRO recalculation of hessian every 5 cycles and without recalculation.

According to our data, *cis*-TS2 exists in a few cases for 1,1-difluorethylene – in terms of (10, 9)/aug-cc-pVDZ, (10, 9)/aug-cc-pVDZ, in these cases the biggest CO distance not consistent with other calculations ~2.7, not ~3.0Å in other cases.

Furthermore, *trans*-TS2 and TS1 is approved in all methods: B2PLYP, CCSD, MRMP2, CASSCF with much lesser deviations in geometry. In terms of data of this verification, we conclude that *cis*-TS2 (NTS1) is an calculation artifact and it not need to be taken into account in calculation of net rate constant along with other TSs (TS1, *trans*-TS2).

This conclusion is analogous to one in Ref. [3], where such verification is done for ethylene-ozone reaction.

7.5 CONCLUSIONS

- Comparison of results of implication of three methods taking into account their features allows one to conclude that in case of addition of ozone to 1,1-dichloethylene the competition of Criegee and DeMore mechanisms is presented, and also prevailing of latter.
- The methods implied have some features: MRMP2 underestimates rate constant because of variant we implied does not performs full optimization. (GAMESS), but this method gives god result far away from TSs, in reagents valley, on which indicate close values of energy of remote complex obtained by scanning from TS1 and TS2. CCSD is more exactly to compute correlation energy and so better more reliable relation of channels. However, apart from

TSs there are deficiencies due to restricted basis set aug-cc-pVDZ and its errors at long distances. Method B2PLYP is third and independent on others, but in general provide quite corresponding to CCSD and MRMP2 results. In comparison one should take into account that in distinction on ethylene, the reaction mechanism is more complicated than in case of ethylene: secondary products forming (e.g., corpuscular chlorine) may result in secondary reactions, which increases net rate constant of reagent decomposing while calculation gives an approximation to the rate constant of elementary act.

7.6　ACKNOWLEDGMENT

This work was performed by using of MSU supercomputing complex resources [25].

This research has been supported by Center for International Mobility (CIMO) and the Magnus Ehrnrooth Foundation. CSC – the Finnish IT Center for Science is thanked for computer time. HP CP4000 BL ProLiant supercluster was used. Additional acknowledgment is signified to D. Sundholm and V. Ovchinnikov employees of Laboratory of Structures in Swedish.

KEYWORDS

- **Activation energy**
- **Ozone**
- **Quantum chemical calculation**
- **Rate constant**
- **Substitutors of ethylene**

REFERENCES

1. Criegee, R. (1975). Mechanism of Ozonolysis. *Angew. Chem, 87(21)*, 765.
2. DeMore, W. B. (1969). Arrhenius Constants for the reaction of Ozone with Ethylene and Acetylene. *Intern. J. Chem. Kinetics, 1(1)*, 209.
3. Gadzhiev, O. B., Ignatov, S. K., Krisyuk, B. E., Maiorov, A. V., Gangopadhyay, S. & Masunov, A. E. (2012). Quantum Chemical Study of the Initial Step of Ozone Addition to the Double Bond of Ethylene. *J. Phys. Chem. A*. Oct 25; *116(42)*, 10420–10434. doi: 10.1021/jp307738p. Epub 2012 Oct 10.
4. Frisch, M. J., Trucks, G. W., Schlegel, H. B., Scuseria, G. E., Robb, M. A., Cheeseman, J. R., et al. (2004). Gaussian 03, Revision C.02, Gaussian, Inc., *Wallingford CT*.
5. Boatz, J. A., Elbert, S. T., Gordon, M. S., Jensen, J. H., Koseki, S., Matsunaga, N., Nguyen, K. A., Su, S., Windus, T. L., Dupuis, M. & Montgomery, J. A. (1993). General

Atomic and Molecular Electronic Structure System. Schmidt, M. W., Baldridge. *J. Comput. Chem.* Vol.14. P.1347–1363.

6. Gaussian, K. K., Frisch, M. J., Trucks, G. W., Schlegel, H. B., Scuseria, G. E., Robb, M. A., et al. (2009). *Gaussian, Inc., Wallingford CT.*

7. Granovsky, A. A., Firefly version 8.0.0, http://classic.chem.msu.su/gran/firefly/index.html

8. Krisyuk, B. E., Maiorov, A. V., Mamin, E. A. & Popov, A. A. (2013). Kinetika I Kataliz, *54(2)*, 1–9.

9. Ovchinnikiov, V. A. Krisyuk, B. E., Maiorov, A. V., Mamin, E. A. & Popov, A. A. (2011). *Butlerovskie soobshenia, 25(5)*, 45–51.

10. Krisyuk, B. E., Maiorov, A. V., Mamin, E. A. & Popov, A. A. (2011). *Butlerovskie soobshenia, 28(17)*, 11–16.

11. Ignatov, S. K. (2004). *Moltran v.2.5, Nizhny Novgorod.* http://ichem.unn.ru/tcg/Moltran.htm

12. Saito, T., Nishihara, S., Kataoka, Y., Nakanishi, Y., Kitagawa, Y., Kawakami, T., Yamanaka, S., Okumura, M. & Yamaguchi, K. (2010). Multireference Character of 1,3-Dipolar Cycloaddition of Ozone with Ethylene and Acrylonitrile. *J. Phys. Chem. A.* Nov18; *114(45)*:12116–23. doi: 10.1021/jp108302y. Epub 2010 Oct 27.

13. Grimme, S. (2006). Semiempirical hybrid density functional with pertrubative second-order correlation. *J. Chem. Phys. 124.* 034108.

14. Becker, K. H., Schurath, U. & Seitz, H. (1974). Ozone-Olefin Reactions in the Gas Phase.1. Rate Constants and Activation Energies. *Int. J. Chem. Kinet., (6)*, 725–739.

15. Wei, Y. K. & Cvetanovich, J. (1963). A Study of the vapor phase reaction of ozone with olefins in the presence and absence of molecular oxygen. *Canadian J. Chem. (41)4*, 913–925.

16. Stedman, D. H., Wu, C. H. & Niki, H. (1973). Kinetics of Gas-Phase Reactions of Ozone with Some Olefins. *J. Phys. Chem., 77(21)*, 2511–2514.

17. Herron, J. T. & Huie, R. E. (1974). Rate Constants for the Reactions of Ozone with Ethene and Propene, from 235.0 to 362.0 K. *J. Phys. Chem., 78(21)*, 2085.

18. Gay, B. W., Philip, Jr., Hanst, L., Bufalini, J. J. & Noonan, R. C. **(1976).** Atmospheric Oxidation of Chlorinated Ethylenes. *Environ. Sci. Technol, (10)1*, 58–67.

19. Sanhuesa, E., Hisatsune, I. C. & Heicklen, J. (1976). Oxidation of Haloethylenes. *Chem. Rev., 76(6)*, 801.

20. Adeniji, S. A., Kerr, J. A. & Williams, M. R. (1981). Rate Constants for Ozone-Alkene Reactions under Atmospheric Conditions. *Int. J. Chem. Kinet., 13*, 209–217.

21. Zhang, J., et al. (1983). Rate Constants of the Reaction of Ozone with trans1,2-dichloroethene and Vinyl Chloride in Air. *Int. J. Chem. Kinet., 15Iss.7*, 655–668.

22. Avizanova, et al. (2002). Temperature-Dependent Kinetic Study for Ozonolysis of Selected Tropospheric Alkenes. *Int. J. Chem. Kinet. 34*, 678–684.

23. Leather, K. E., McGillen, M. R., Ghalaieny, M., Shallcross, D. E. & Percival, C. J. (2010). Temperature-Dependent Kinetics for the Ozonolysis Temperature-Dependent Kinetics for the Ozonolysis. *Phys Chem.* Mar 28; *12(12)*, 2935–2943. doi: 10.1039/b919731a. Epub 2010 Feb 2.

24. Hull, L. A., Hisatsune, I. C. & Heicklen, J. (1972). The reaction of O_3 with $C_2H_2Cl_2$. *J. Phys. Chem, 76*, 2659.

25. Voevodin, V. V., Gumatiy, S. A., Sobelev, S. I., Antonov, A. S., Brizgalov, P. A., Nikitenko, D. A., et al. (2012). Praktika Superkompyutera "Lomonosov." Otkritije Sistemi. *Moscow: Izdatelskiy dom "Otkritije Sistemi,"* 7.

CHAPTER 8

USAGE OF SPEED SEDIMENTATION METHOD FOR ENZYME DEGRADATION OF CHITOSAN

V. V. CHERNOVA, E. I. KULISH, I. F. TUKTAROVA, and G. E. ZAIKOV

CONTENTS

ABSTRACT

A possibility of using the viscosimetry and sedimentation and diffusion methods in studying the process of enzyme chitosan hydrolysis is discussed in the chapter. It is shown that a change in the intrinsic viscosity of chitosan may be determined by both the hydrolysis process in the glycoside bonds and transformation of the supramolecule structure of the polymer. Thus, for setting the hydrolysis process it is necessary to use an absolute method for molecular weight determination.

8.1 INTRODUCTION

Chitosan, a polysaccharide of natural origin, as any other natural polymer is capable of biodegradation under the influence of enzyme agents, that is, degradation of the main chain by glycosidic linkages (accompanied by decreasing of its molecular weight). The degradation process of chitosan may be carried out not only under specific enzymes (chitinases and chitosanases), but under the influence of some nonspecific enzymes like lysozyme [1] and celloviridin [2]. If the Mark-Kuhn-Houwink equation is quite known as applied to the studied polymer, then the most convenient method for degradation determination is a viscosimetry method [3–5]. The reduction of polymer viscosity in the solution implicitly indicates to decreasing the chitosan molecular weight during the process of degradation. However, the values of molecular weight received by the viscosimetry method do not match the real ones as this method is relative and requires the usage of gauge dependences [6]. Moreover, the solution viscosity is determined by the macromolecular coil size whereas the coil size and its molecular weight are not one and the same. Thus, reducing a polymer solution viscosity may be testified not just by a degradation process but a transformation of the polymer supramolecular structure in the solution. This transformation may be caused by disintegration of chitosan macromolecule associates [7]. Therefore, it is impossible to definitely name the reason for viscosity reduction basing on the viscosimetric method only. For revealing the reasons for viscosity changes of polymer solutions there is proposed an absolute method for determining the molecular weight, namely a method combining sedimentation and diffusion [8].

8.2 EXPERIMENTAL PART

Three chitosan samples, CHT-1 (Chimmed Ltd., Russia), CHT-2 and CHT-3 (Bioprogress CJSC.) obtained by alkaline deacetylation of crab chitin and enzyme agents "Collagenase," "Liraza" ("Immunopreparat" SUE, Ufa) and "Tripsin" ("Microgen" FSUE SPA, Omsk) were chosen as objects for investigation. The chitosan 2% (mass.) solution was prepared by 24 h dissolution at ambient temperature. Acetic acid of 1 g/dl concentration was used as a solvent. 5% chitosan mass of the enzyme agent predissolved in a small amount of water was introduced in the polymer solution. Enzymatic destruction was carried out for 20 days in 25 °C. Sodium azide (0.04% from the chitosan mass) was added to the chitosan solution to prevent microbial contamination [9].

The intrinsic viscosity of chitosan was determined in a buffer solution of 0.3 m. of acetic acid and 0.2 m. of sodium acetate, whereas the relative viscosity was established for polymer solutions of 1% acetic acid by the standard method [8] in 25 °C.

After exposition with the enzyme, the chitosan was precipitated, washed by distilled water and dried to a constant weight. The sedimentation S_c and diffusion D_c coefficients were calculated from the sedimentation data received by the analytical ultracentrifuge MOM 3180 (Hungary) in 546 nm wave length and 25+ 0.1°C temperature with Philpot-Svensson optics. The angle of the phase plate (Philpot-Svensson's inclination angle) in both cases amounted to 30°. The rotor speed varied depending on the task set. The diffusion coefficients D_c for 4–5 concentrations in the range of 0.15–0.40 g/dL were found by the erosion rate of the solvent-solution boundary in time. The experiments were carried out in the two-sector cell in a rotor speed of 6000 rpm. The diffusion coefficients D_c were calculated in relation to the square Q under the gradient curve to the maximum ordinate H in the moment of time t:

$$D_c = (Q/H)^2 /4\pi t.$$

The sedimentation coefficient S_c within the same concentration range was in time determined by the movement rate of the solvent-solution boundary. The experiments were conducted in a two-sector cell in a rotor speed of 45,000 rpm. The S_c value was calculated by the formula:

$$S_c = (dx/dt)/\omega^2 t,$$

where x – a maximum coordinate of the curve of the gradient in the sedimentation boundary segment (sm); t – time (sec); $\omega = 2\pi n/60$ – angular velocity of rotation of the rotor; t – rotor speed per minute.

M_{SD} value was calculated by the found S_0 and D using the first Swedberg formula [8]:

$$M_{SD} = (S_0 D) [(RT/(1—v\rho_0)].$$

where R – gas constant; T – absolute temperature, K; v – partial specific gravity of the polymer in a solution, sm^3/g; ρ_0 _ density of the solvent g $/sm^3$. S_0 and D are sedimentation and diffusion constants received by S_c and D_c extrapolation on the zero concentration.

All concentration S_c and D_c dependences were linear which allowed carrying out the S_c and D_c value extrapolation on the zero concentration, finding true values of these characteristics (S_0 and D) and calculating the true value of M_{SD}, accordingly [8].

X-ray phase investigations were fulfilled on the X-ray diffractometer "Shimad-zu XRD 6000." CuKαradiation wave length 0.154 nm, initial angle 5.00 deg., measuring step 0.05 deg., the final angle 40.00 deg,was used.

8.3   RESULTS AND DISCUSSION

In the earlier researches [10–11], a considerate decrease in relative viscosity of chitosan solutions in acetic acids in the presence of enzyme agents "Liraza," "Collagenase" and "Tripsin" was established (Fig. 8.1, curves 1–3). Taking into account that acids are chitosan solvents, the reason for viscosity reduction in chitosan solutions may consist in the process of acid degradation rather than the influence of a nonspecific enzyme. Some decrease in relative viscosity of chitosan really occurs even in the absence of enzyme agents (Fig. 8.1, curve 4).

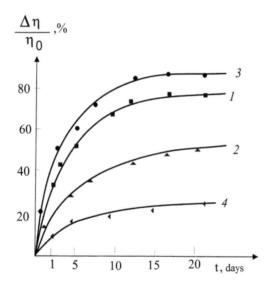

FIGURE 8.1   The dependence of relative viscosity changes of 2% chitosan solution from the exposure time with the enzyme agents "Liraza" (1), "Collagenase"(2) and "Tripsin" (3) and in the absence of enzyme agents (4); 1% acetic acid concentration. The chitosan:enzyme weight ratio (%) is 95:5.

Moreover, exposure of chitosan solutions with enzyme agents is accompanied by the decrease in characteristic viscosity values of its solutions (Fig. 8.2, curves 1–3). While no enzymes are used, the exposure of chitosan in the acetic acid also leads to drastic reduction in single values of its characteristic viscosity (Fig. 8.2, curve 4).

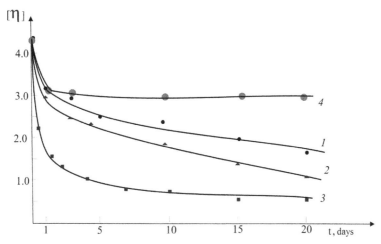

FIGURE 8.2   The dependence of characteristic viscosity changes from the time of chitosan exposure in the solution of CHT-1 separated from the 1% acetic acid solution in the absence of enzyme agents (4) and with enzyme agents "Tripsin" (1), "Liraza" (2), and "Collagenase"(3).

So the characteristic viscosity of CHT-1 failing to pass through the dissolving stage in the acetic acid was determined in an acetate buffer (pH=4.5, [η] = 4.45 dL/g). If this chitosan sample is sustained in a 1% acetic acid solution for 24 h and separated from the solution, its characteristic viscosity is determined in an acetate buffer and the received value [η] amounts to 3.1 dL/g which is considerably less than the initial value. If the polymer is sustained in the acetic acid solution for a longer period of time, it does not lead to further reduction of characteristic viscosity. Similar regularities are observed in all studied samples of chitosan. As seen from the Table 8.1, after chitosan is precipitated from the acetic acid solution, its characteristic viscosity always decreases. Additional heating of the sample is also accompanied by some reduction in viscosity values. The received experimental data may be described from the positions of the acid degradation process. However, the application of the method combining high-speed sedimentation and diffusion for investigating chitosan solutions in the acetic acid allows to register the following results (see Table 8.1). In the absence of enzyme agents the values of the sedimentation constant and the diffusion coefficient coincide with the values of the source chitosan samples both for the precipitated and heated ones. This definitely indicates that the value of the chitosan molecular weight during the process of its reprecipitation and heating does not change. Thus, dissolution of chitosan in the acetic acid solution is not accompanied by polymer degradation.

TABLE 8.1   Some Physical and Chemical Characteristics of Chitosan Samples in an Acetic Buffer (pH=4.5)

Chitosan sample	Used enzyme agent	$[\eta]$, dL/g	$S_0 \times 10^{13}$, s	$D \times 10^7$, sm²/s	$M_{SD} \times 10^{-5}$
CHT-1	–	4.45	3.00	1.30	1.60
CHT-1[2]	–	3.69	3.10	1.35	1.57
CHT-1[1]	–	3.50	3.15	1.38	1.59
CHT-1[3]	"Liraza"	1.10	1.84	4.00	0.25
CHT-1[3]	"Collagenase"	0.60	1.79	5.70	0.21
CHT-1[3]	"Tripsin"	1.80	1.97	3.50	0.42
CHT-2	–	6.10	7.13	1.02	3.34
CHT-2[1]	–	4.90	7.20	1.10	3.12
CHT-3	–	3.45	2.15	1.49	0.99
CHT-3[1]	–	2.82	2.20	1.56	0.97

[1] a chitosan sample precipitated from the solution in the acetic acid;
[2] a chitosan sample precipitated from the solution in the acetic acid and subjected to additional heating by 70–80°C in the solution of the acetic buffer;
[3] a chitosan sample separated from the enzyme-containing solution of 1% acetic acid, holding time of chitosan in the solution is 20 days and the enzyme content is 5% from the chitosan mass.

On the contrary, chitosan holding in an acetic acid solution in the presence of the enzymes leads to considerable reduction in the sedimentation constant and increase of the diffusion coefficient (see Table 8.1). Thus, it results in significant reduction of the chitosan molecular weight taking place during the biodegradation process.

The observed changes in the values of characteristic and intrinsic viscosity in the absence of enzyme agents are probably connected with structural transformations such as destruction of chitosan associates in the acetic acid solution. As a XRD analysis shows, the source chitosan and the one dissolved in the acetic acid have different supramolecular organization (Fig. 8.3).

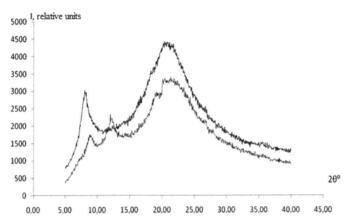

FIGURE 8.3   X-ray pattern of the source CHT-3 (1) and a film sample received from the chitosan solution in the 1% acetic acid (2).

8.4   CONCLUSIONS

The usage of the method of speed sedimentation and diffusion allows to state that any changes of chitosan molecular weight do not occur in the absence of enzyme agents. So the observed reduction in intrinsic viscosity of chitosan after the dissolving in the acetic acid is determined by transformation of the initial supramolecular structure of the polymer. On the contrary, the decrease of the molecular weight of the polymer takes place under the influence of the degradation enzymes of the chitosan main chain in the presence of the enzyme agents.

KEYWORDS

- **Chitosan**
- **Enzyme degradation**
- **Sedimentation**

REFERENCES

1. Skryabina, K. G., Vikhoreva, G. A. & Varlamova, V. P. (2002). *Chitin and Chitosan: Production, Properties and Application*; Nauka, 365.
2. Ilyina, A. V., Tkatcheva, Yu.V. & Varlamov, V. P. (2002). *Applied Biochemistry and Microbiology*, *38(2)*, 132.
3. Mullagaliev, I. R., Artuganov, G. E. & Melentiev, A.I. (2006). *In Modern Perspectives in Chitin and Chitosan Researches*, 305.

4. Vikhoreva, G. A., Rogovina, S. Z., Pchelko, O. M. & Galbrakh, L. S. (2001). *Polym. Sci., B ser, 43(6),* 1079.
5. Fedoseeva, E. N., Semtchikov, Yu.D. & Smirnova, L. A. (2006). *Polym. Sci., B ser, 48(10),* 1930.
6. Budtov, V. P. (1992). *Physical Chemistry of Polymer Solutions*; S. Petersburg, 384.
7. Kulish, E. I., Chernova, V. V., Vildanova, R. F., Bolodina, V. P. & Kolesov, S. V. (2011). *Bulletin of Bashkir State university, 16(3),* 681.
8. Rafikov, S. R., Budtov, V. P. & Monakov, Yu. B. (1978). *Introduction to Physics and Chemistry of Polymer Solutions*. Moscow, 320.
9. Martirosova, E. I., Plaschina, I. G., Feoktistova, N. A., Ozhimkova, E. V. & Sidorov, A. I. (2009). *All-Russia Scientific Conference of Topical Problems of Modern Science and Education.* URL: http://e-conf.nkras.ru/konferencii/2009/Martirosova.pdf
10. Chernova, V. V., Kulish, E. I., Volodina, V. P. & Kolesov, S. V. (2008). *Modern Perspectives in Chitin and Chitosan Researches: 9th International Conference*, Moscow, 234.
11. Kulish, E. I., Chernova, V. V., Volodina, V. P., Torlopov, M. A. & Kolesov, S. V. (2010). *Modern Perspectives in Chitin and Chitosan Researches: 10th International Conference*, Moscow, 274.

CHAPTER 9

DOUBLE-DECKER PHTHALOCYANINES: SPECTRAL REDOX PROPERTIES AND SENSOR ELEMENTS BASED ON THEM

G. A. GROMOVA and A. V. LOBANOV

CONTENTS

ABSTRACT

In this chapter, a correlation between the ionic radius of the metal-complexing in double-decker phthalocyanines and positions of maxima Q-bands in the electronic absorption spectra were determined in dimethylformamide and chloroform. The increasing of ionic radius from holmium to lutetium caused a regular change in the position of the maxima Q-bands in the absorption spectra. The behavior of metal diphthalocyaninates in supramolecular systems was also investigated. It was revealed that a new band shifted to the red region appeared in absorption spectra of sandwich phthalocyaninates of lutetium, erbium and ytterbium in albumin solution. This particular behavior allows us to consider phthalocyanines as prototype of sensitive biosensor system.

Aim and Background. The aim of this chapter is to study the dependence of the spectral and photochemical properties of the double-decker bis-phthalocyanines on ionic radius of lanthanides, which determines the distance between the planes of two phthalocyanine ligands.

9.1 INTRODUCTION

Double-decker phthalocyanine lanthanide complexes have a high stability [1], intense absorption and can be used in various technological [2] and biomedical applications [3]. Therefore, it is necessary to study the nature of the bands in the electronic absorption spectra, as well as the possibility of changing the spectral properties depending on the composition of multicomponent systems.

However, phthalocyanines soluble in extremely poor range of solvents and insoluble in water [1]. This problem can be solved by using various materials as solubilizers. They may be various macromolecular compounds (proteins, hydrophilic polymers, polycarbohydrates, etc.), supramolecular structures (micelles or liposomes) and nanosized carriers (nanoparticles of clay, silica, etc.). Phthalocyanines often represented as their H-aggregates in such supramolecular complexes. It is known that no fluorescence and generation of long-lived excited triplet states for H-aggregates phthalocyanines are observed [4, 5]. Organization of H-associates as well as their tendency to spontaneous formation depends strongly on the nature of the complexing internal metal ion. Systematic study of the relation of the "structure-property" for this process has not been carried out.

A double-decker complex of bis-phthalocyanines of lanthanides is an interesting model of H-aggregates of phthalocyanines. Therefore, it is important to study their properties.

9.2 EXPERIMENTAL PART

The objects of study were double-decker phthalocyanine lanthanide complexes with the lutetium (Lu), ytterbium (Yb), holmium (Ho) and erbium (Er) as metal-complexing agent. All compounds were synthesized at the Frumkin Institute of Physical Chemistry and Electrochemistry. Dimethylformamide (DMF) and chloroform (CHCl$_3$) for preparation of solutions were produced by LABTEH (Russia). Water was purified by distillation. Cetyltrimethylammonium bromide (CTABr) and sodium dodecyl sulfate (SDS) were used as detergents for micelles. Polyvinylpyrrolidone – 26500 (PVP), polyethylene glycol 10000 (PEG), poly N, N-dimethyldiallylammonium chloride (PDDA), polyvinyl alcohol (PVA) were polymers involved. Bovine serum albumin (BSA), bovine hemoglobin (Hem) and peroxidase (HRP) were macromolecular compound used.

To prepare the micelle solution of sodium dodecyl (SDS) at concentration of 0.0082 mol/L (M = 288.4 g/mol), a sample of dry SDS (0.0021 g) was dissolved in 10 mL distilled water. Solution of cetyltrimethylammonium bromide (CTABr) with concentrations of 0.0012 mol/L (M = 364.46 g/mol) were prepared by dissolving the dry sample of CTABr (0.0044 g) in 10 mL of distilled water. For preparing 2% solutions of polymers samples of 0.2 g of polyvinylpyrrolidone (PVP) (M = 26,500 g/mol), polyethylene glycol (PEG) (M = 10,000 g/mol) and poly N, N-dimethyldiallylammonium chloride (PDDA) (M = 20,000 g/mol) were dissolved each in 10 mL of distilled water. And for preparation of 2% alcohol solutions samples of 0.1 g of polyvinylpyrrolidone (PVP) and 0.002 g of polyvinyl alcohol (PVA) (M = 200,000 g/mol) were dissolved in 5 mL.

For preparation of 5% solutions of macromolecular compounds samples of 0.05 g of albumin (BSA) (M = 67,000 g/mol), peroxidase (HRP) (M = 44,100 g/mol) and hemoglobin (Hem) were taken. Nanosized dioxo silica with particle diameter 60 nm was taken as a store 1% solution. Mixtures of MPc$_2$/PVP, MPc$_2$/PVA, BSA/HRP and BSA/Hem were prepared from initial solutions in ratio 1:1.

For the formation of thin films quartz plate (size 30h10h1 mm) were used. Immediately before the experiment, the plate was washed successively with acetone, chloroform and distilled water. The mixtures of MPc$_2$/PVP and MPc$_2$/PVA were applied dropwise to the surface of the substrate using automatic pipette and allowed to dry completely under standard conditions. Thin films were examined under the binocular optical microscope BS-702B at a magnification of 200 μm and 20 μm.

Absorption spectra were recorded in thin film and in solutions of dimethylformamide and chloroform on spectrophotometer HACH DV 4000 V (USA). To record spectrum 2 mL of detergent solution or a polymer and 0.02 mL of 10^{-4} M phthalocyanine metal complex in DMF was placed in quartz cuvette (length1 cm). The analysis was performed at six different concentrations (1×10^{-6}; 2×10^{-6}; 3×10^{-6}; 4×10^{-6}; 5×10^{-6}; 6×10^{-6} M) of metal complexes with adding of 0.02 mL of phthalocyanine metal complex to the solution with a carrier.

9.3   RESULTS AND DISCUSSION

9.3.1   SOLUTIONS

Figure 9.1 shows the absorption spectra of solutions in chloroform for various metal phthalocyanines, which are characterized by λ_{max} = 660–664 nm absorption and the presence of low intensity at λ_{max} = 458–466 nm [6–8]. That indicates the predominance of neutral (green) form $[Pc^{2-}-M^{III}-Pc^{-}]^{0}$. A broad band of low intensity in the region of 920 nm corresponds to the radical part bis-phthalocyanine complex.

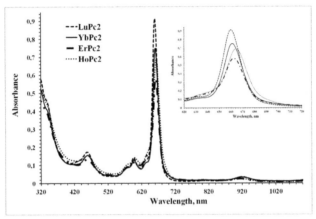

FIGURE 9.1   Electronic spectra of double-decker complexes bis-phthalocyanines lutetium, ytterbium, erbium and holmium in CHCl3.

In the transition from lutetium bis-phthalocyanine complex to complex with holmium light bathochromic effect is observed, in accordance with the change in the radius of the metal-complexing agent.

Complexes in solutions of DMF unlike solutions of CHCl$_3$ exist in the negative (blue) form $[Pc^{2-}-M^{III}-Pc^{2-}]^{-}$ (Fig. 9.2). Electronic absorption spectra of solutions of complexes exhibit a number of bands: the Soret band (332 nm), Qx-band (616–620 nm) and Qy-band (671–693 nm). Shifts of Qx-band to longer wavelengths are seen with increasing of radius of metal complexing. While Qy-band is shifted to shorter wavelengths.

Double-decker phthalocyanines are soluble in a narrow range of solvents and are insoluble in water, however, we have developed methods for preparing aqueous multicomponent systems based on lanthanide complexes phthalocyanines double-decker. Macromolecular structures (proteins, polymers), supramolecular structures (micelles) and nanosized silica were used as solubilizers.

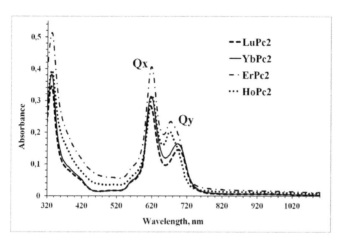

FIGURE 9.2   Electronic spectra of double-decker complexes bis-phthalocyanines lutetium, ytterbium, erbium and holmium in DMF.

With increasing concentration of phthalocyanine complexes in solutions, a regular increase in intensity were observed in accordance with the Bouguer-Lambert-Ber's law at the same time the position of the bands is not changed.

In solutions of polymers (PVP, PEG, PDDA) complete or partial neutralization of the negative charge of the anionic form of the initial metal complex was seen. Similar behavior was observed for metal complexes comprising micelles (SDS). In the case of micellar system CTABr lanthanide bis-phthalocyanine predominantly existed in the anionic form.

The analysis of the absorption spectra in solvents and in supramolecular systems showed correlation of the ionic radius of the metal complexing in double-decker phthalocyanines and the maxima Q-bands of the electronic spectra in dimethylformamide and chloroform. For other solutions methyl phthalocyanine complexes revealed no linear relationship between the position of the maxima Q-bands and the ionic radius of the metal-complexing agent. Table 9.1 shows a comparison of the spectral characteristics of the test solutions.

TABLE 9.1   Dependence of the Position of Q-Bands from Different Radius of the Metal

MPc_2	R иона (нм)	λ, нм								
		DMF		$CHCl_3$	CTABr		SDS	PVP		
		Qx	Qy		Qx	Qy			PEG	PDDA
$LuPc_2$	0.1	616	693	660	616	708	660	662	664	661
$YbPc_2$	0.101	618	692	662	620	706	660	663	664	662
$ErPc_2$	0.103	619	674	663	624	688	662	662	662	-
$HoPc_2$	0.104	620	671	664	626	674	672	632/676	636/670	673

The emergence of a new band with λ_{max} = 660 nm was seen in albumin solution (Fig. 9.3). The presence of oxygen in aqueous solution led to the formation of a complex with bis-phthalocyanine [9, 10] and the transition from the initial form to a neutral anion. Albumin presented in the solution prevented the oxidation process of phthalocyanine. Due to the hydrophobic nature of the lanthanide complexes, which bind to a hydrophobic pocket presented in albumin and thereby altering the spectral properties [11].

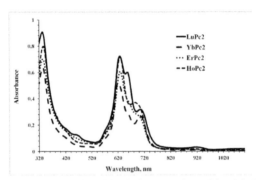

FIGURE 9.3   Electronic spectra of double-decker complexes bis-phthalocyanines lutetium, ytterbium, erbium and holmium in solution BSA.

This specific behavior in the presence of phthalocyanine complexes of albumin is of interest for the diagnosis of inflammatory processes, diabetes, liver disease [11, 12].

High sensitivity is very important for detection of albumin. In this paper we studied the sensitivity of the double-decker phthalocyanine complex. Isotherm (Fig. 9.4) shows that the absorption increases with the concentration of albumin in solution.

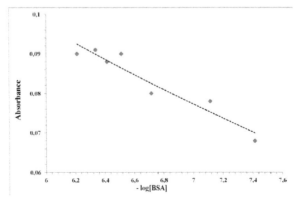

FIGURE 9.4   Dependence of BSA concentration from absorbance (λ = 704 nm) in the presence YbPc$_2$.

Selectivity is also important for determining the concentration of BSA in solutions. For this purpose, spectral measurements were carried out by adding peroxidase solutions (HRP) and bovine hemoglobin (Hem), as well as by mixing with a solution of BSA at a ratio of 1:1 (Figs. 9.5 and 9.6).

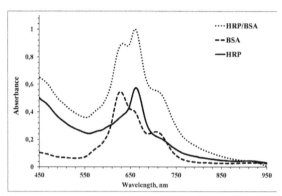

FIGURE 9.5   Electronic spectra of double-decker lutetium complex bis-phthalocyanines albumin solution, peroxidase solution and a solution of albumin/peroxidase in the ratio 1:1.

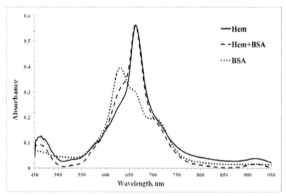

FIGURE 9.6   Electronic spectra of double-decker bis-phthalocyanines lutetium complex in a solution of albumin, hemoglobin and solution albumin/hemoglobin in ratio of 1:1.

Figure 9.5 shows that in the solution of water peroxidase phthalocyanine complex is in neutral form λ_{max} = 660 nm. In a solution of BSA/HRP we can observe the change in the shape of the spectrum. Presence of three absorption bands speaks about the difficulty of the transition from the blue to the green form.

In solution of Hem as well as HRP neutral (green) shape phthalocyanine is predominant, characterized by λ_{max} = 662 nm. In solution of BSA/Hem it is observed appearance of shoulder which corresponds to λ_{max} = 628 nm in the presence of BSA.

The inverse process of electron transfer is characteristic for lanthanide phthalocyanine complexes. Recovering green form to blue occurs in ethanol solution in the presence of nano-sized silica (Fig. 9.7).

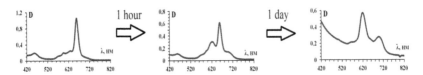

FIGURE 9.7   Scheme of changes in the shape of electronic absorption spectra bis-phthalocyanines ytterbium complex in alcoholic solution SiO_2 in time.

This can be explained by the fact that the double-decker lanthanide phthalocyanines enter into intermolecular interaction with SiO_2. Consequently sandwich complex molecules pull back partial negative charge from the surface of silica gel, passing to the anionic form.

9.3.2 THIN FILMS

Study of the spectral properties of double-decker phthalocyanine based on lanthanide complexes is one of the most promising area of research due to their attractiveness for applications primarily in the biomedical field [13] as a prototype sensitive model.

We have obtained thin film of a mixture of double-decker ytterbium and lutetium phthalocyanine with polymer solution PVP and PVA on four quartz substrates (Fig. 9.8). Next, a layer structure was examined under a microscope.

FIGURE 9.8   Image mixture $YbPc_2$/PVP (1) on a substrate; with the scale: (2) 200 µm; and (3) 20 µm.

Layer is heterogeneous and phthalocyanine crystals are formed on the glass substrate is irregular, forming a complex microstructure [14, 15].

It is important to keep the spectral properties of lanthanide complex in a thin layer. Because, specific interaction bis phthalocyanine with albumin plays an important role in the evaluation of protein. Band in the absorption spectrum (Fig. 9.9) phthalocyanine ytterbium in formed a thin layer is characterized by λ_{max} = 664 nm. Spectral shape and position of the absorption band corresponds to the neutral form [16].

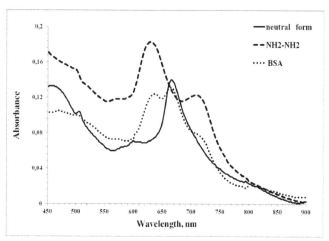

FIGURE 9.9   Electronic spectra of the YbPc$_2$ in a thin layer on the polymer matrix PVA. (1) neutral form; (2) hydrazine hydrate; (3) BSA.

Next, the obtained crystals were processed with 1% solution of hydrazine hydrate, resulting in change the spectral shape. Instead of one narrow, intense peak appeared two absorption bands (λ = 628 nm and 710 nm). This proves that the process of recovery of crystals occurred in the film (from the neutral to the anionic form, as shown in Fig. 9.9).

When applied to the film BSA solution have seen the emergence of a new band with λ = 660 nm, so that we observed in aqueous solutions of albumin (Fig. 9.3).

Thus, it can be concluded that a thin layer of double-decker phthalocyanines retain their spectral properties. This allows us to study these compounds as a sensing element for biosensors.

Nanoscale silicon oxide is widely used in microelectronics, optoelectronics and electronics are parts of the means of fire protection, coatings, high temperature adhesives, paints, various mortars. Medical and biotechnology applications comprising silica nanoparticles and sorbent molecular sieves of DNA delivery vehicles, proteins, anticancer agents [17].

A number of studies [18] submission of information on in vitro and in vivo toxicity of silica nanoparticles – both crystalline and amorphous. Most of the results on the toxicity in vitro is reduced to the analysis of size – and dose-dependent cytotoxicity, increased reactive oxygen species and proinflammatory stimulation. The data obtained from in vivo studies demonstrate nanoparticles induced lung inflammation and fibrosis, emphysema, and granuloma formation. It is therefore important to monitor the content of nano-sized silica in the body.

The recovery process a neutral form the phthalocyanine in the presence of SiO2 can also be consider as a prototype of the sensing system for a biosensor for the determination of the silica. Therefore, we investigated the influence of alcohol solution on the absorption spectra of thin films of phthalocyanine double-decker lutetium and ytterbium on polymeric matrices PVP and PVA 4 substrates (Fig. 9.10).

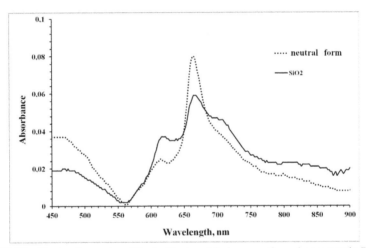

FIGURE 10   Electronic spectra of the LuPc$_2$ in a thin layer on the polymer matrix PVP. (1) neutral form; (2) SiO$_2$ in C$_2$H$_5$OH.

Band in the absorption spectrum of phthalocyanine lutetium form a thin layer is characterized by $\lambda = 662$ nm. This indicates that the sandwich complex is in neutral form. After treatment of the film an alcoholic solution happens recovery process. What does the reduction of the band intensity $\lambda = 662$ nm, an increase in absorption $\lambda = 618$ nm and the appearance of the shoulder $\lambda = 708$ nm. That the shape and position of the spectrum corresponds to phthalocyanine blue forms.

Important not only specific recognition, but also sensitivity to low concentrations of silicon dioxide in the solution. For this was carried out a series of measurements of the absorption spectra with different contents of SiO$_2$ ethanol solution. For obtained data were built kinetic curves (Fig. 9.11).

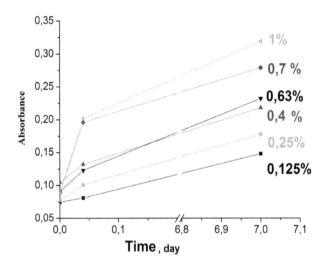

FIGURE 9.11   Kinetic curves showing the change of the absorption in time for various concentrations of SiO_2.

As seen from the graph, sandwich complex showing the sensitivity, even when lower concentrations silicon dioxide in the solution.

9.4   CONCLUSIONS

The nature of central atom has little effect on the differences in spectral properties of double-decker bis-phthalocyanine. Double-decker bis-phthalocyanine complexes of lanthanides $LuPc_2$, $YbPc_2$, $ErPc_2$, $HoPc_2$ are in monoradical neutral form in chloroform and anion form in dimethylformamide. In both cases the position of the bands at 660–668 nm and 462–466 nm (for the neutral form monoradical), 616–622 nm and 674–696 nm (for the anionic form) is linearly correlated with the value of the ionic radius of lanthanide. A broad band of low intensity in the region of 920 nm corresponds to the radical part of bis-phthalocyanine complex.

In supramolecular systems, in some cases (systems formed by addition of PVP, PEG, PDDA, SDS), partial or complete neutralization of the negative charge of the anionic form of the original bis-phthalocyanine is observed. In the case of micellar system CTABr lanthanide bis-phthalocyanines exist predominantly in the anionic form. The dependence of the optical density bis-phthalocyanine on their concentration in the composition of multicomponent systems revealed that in cases of different metal ions bis-phthalocyanine metal complex may be present in both molecular and associated forms.

The correlation between the ionic radius of the metal and the maxima Q-bands in the solvents DMF and chloroform is observed. All changes occur on the mechanisms of redox transitions between the two forms (blue and green). Coordination in the supramolecular system takes place with phthalocyanine ring, because the metal is not prone to further coordination. The systems studied can be considered as a prototype of sensitive element to BSA, SiO_2 to create biosensors.

KEYWORDS

- **Biosensor**
- **Lanthanides**
- **Phthalocyanine**
- **Supramolecular systems**

REFERENCES

1. Kılınc, N., Atilla, D., Gurek, A. G. & Ozturk, Z. (2009). Tetrakis(alkylthio)-substituted lutetium bisphthalocyanines for sensing NO_2 and O_3; *Sensors and Actuators B 142*, 73–81.
2. Bouvet, M., Parra, V., Locatelli, C. & Xiongb, H. (2009). Electrical transduction in phthalocyanine-based gas sensors: from classical chemiresistors to new functional structures; *J. Porphyrins Phthalocyanines, 13*, 84–91.
3. Giuliani, E., Fernandes, R., Brazaca, L. C., Rodriguez-Mendez, M. L., Antonio de Saja, J. & Zucolotto, V. (2011). Immobilization of lutetium bisphthalocyanine in nanostructured biomimetic sensors using the LBL technique for phenol detection; *Biosensors and Bioelectronics 26*, 4715–471.
4. Zimin, A. V., Field, Y., Yurre, T. A., Ramshaw, S. M., Mezdrogina, M. M. & Poletaev, N. K. (2010). Photoluminescence nitro phthalocyanines europium (III); *Semiconductor Physics and Technology, 44(8)*.
5. Belogorokhov, I. A., Martyshov, M. N., Gavriliouk, A. S., Drones, M. A. & Tomilova, L. G. (2008). Optical and electrical properties of semiconductor structures based on butylated phthalocyanines containing erbium ions; *Materials Science and Technology. Semiconductors. (3)*, 23–33.
6. Vivas, M., Fernandes, R. & Mendonca, R. (2012). Study of singlet excited state absorption spectrum of lutetium bisphthalocyanine using the femtosecond Z-scan technique; *Chemical Physics Letters 531*, 173–176.
7. Słota, R., Dyrda, G., Bloise, E. & Sole, R. (2012). Novel Lipophilic Lanthanide Bis-Phthalocyanines Functionalized by Pentadecylphenoxy Groups: Synthesis, Characterization and UV-Photostability; *Molecules, 17*, 10738–10753.
8. R. Słota, G. Dyrda, Bloise, E. & Sole, R. (2012). Novel Lipophilic Lanthanide Bis-Phthalocyanines Functionalized by Pentadecylphenoxy Groups: Synthesis, Characterization and UV-Photostability; *Molecules, 17*, 10738–10753.

9. Gumrukc, G., M. Ozgur, A. Altındal, Ozkayac, A., Salih, B. & Bekaro, O. (2011). Synthesis and electrochemical, electrical and gas sensing properties of novel mononuclear metal-free, Zn(II), Ni(II), Co(II), Cu(II), Lu(III) and double-decker Lu(III) phthalocyanines substituted with 2-(2H-1,2,3-benzotriazol-2-yl)- 4-(1,1,3,3-tetramethylbutyl) phenoxy; *Synthetic Metals 161*, 112–123.

10. Murdey, R., Bouvet, Sumimoto, M., M., Sakaki, S. & Sato, N. (2009). Direct observation of the energy gap in lutetium bisphthalocyanine thin films; *Synthetic Metals 159*, 1677–1681

11. Ishikawa, N. (2001). Electronic structures and spectral properties of double- and triple-decker phthalocyanine complexes in a localized molecular orbital view; *J. Porphyrins Phthalocyanines, 5*, 87–101.

12. Patra, D. (2010). Synchronous fluorescence based biosensor for albumin determination by cooperative binding of fluorescence probe in a supra-biomolecular host–protein assembly; *Biosensors and Bioelectronics 25*, 1149–1154.

13. I. Yahia, S., Abdulaziz, A., Khedhair, A., Musarrat, J. & Yakuphanoglu, F. (2011). Spectroscopy studies of the interaction between thiophanate methyl and human Optical serum albumin for biosensor applications; *Spectrochimica Acta Part A 79*, 1285–1290.

14. S. Smola, S., Snurnikova, O. V., Fadeyev, E. N., Sinelshchikova, A. A., Gorbunova, Y. G., Lapkina, L. A. & Tsivadze, A. Y. (2012). The First Example of Near-Infrared 4f Luminescence of Sandwich-Type Lanthanide Phthalocyaninates; *Macroheterocycles, 5(4–5)*, 343–349.

15. Bai, R., Shi, M., Ouyang, M., Cheng, Y., Zhou, H., Yang, L., Wang, M. & Chen, H. (2009). Erbium bisphthalocyanine nanowires by electrophoretic deposition: Morphology control and optical properties; *Thin Solid Films 517*, 2099–2105.

16. Gonidec, M., Biagi, R., Corradini, V., Moro, F. & Renzi, V. (2011). Surface Supramolecular Organization of a Terbium(III) Double-Decker Complex on Graphite and its Single Molecule Magnet Behavior; *J. Am. Chem. Soc, 133*, 6603–6612.

17. Basova, T., Gurek, A., Ahsen, V. & Ray, A. (2013). Electrochromic lutetium phthalocyanine films for in situ detection of NADH; *Optical Materials 35*, 634–637.

18. Gendrikson, O. D., Safenkova, I. V., Zherdev, A. V., Dzantiyev, B. B. & Popov, V. O. (2011). Methods of detection and identification of man-made nanoparticles; *J. Biophysics, 965–994*, 56–66.

CHAPTER 10

MICROWAVE SYNTHESIS AND ELECTRONIC STRUCTURE STUDIES OF LANTANIDE TETRAARYLPORPHYRIN COMPLEXES

A. S. GORSHKOVA, S. V. GORBACHEV, E. V. KOPYLOVA, V. D. RUMYANTSEVA, R. N. MOZHCHIL, and A. M. IONOV

CONTENTS

ABSTRACT

Yb and Er acetylacetonates were synthesized by ammoniac method. Influence of pH on the structure of formed compounds was studied. Data of UV-vis, IR, mass-spectra were given, the analysis of electronic structure and chemical composition was carried out by a method of photoelectron spectroscopy. On examples of TPP, symmetric 5,10,15,20-tetrakis(3-methoxyphenyl)porphyrin synthesis conditions with usage of MR and also conditions of lanthanides incorporation in a porphyrin macrocycle were optimized.

Aim and Background. Optimization of synthesis conditions of a number of lanthanide tetraarylporphyrin complexes including a use of microwave radiation; researches of their chemical composition, electronic structure, spectral characteristics.

10.1   INTRODUCTION

Interest in ytterbium porphyrins complexes is determined by a possibility of their use as effective markers in luminescent diagnostics of malignant tumors [1] since porphyrins are capable to accumulate in various types of cancer cells and tumors microvessels. However, drugs that are used at present for the cancer photodynamic therapy: Photofrin II, Foscan, Photolon, Radachlorin, Photosens, Photoditazin, etc. effectively generate the singlet oxygen that leads to undesirable side reactions at diagnostics. Unlike the free bases of porphyrins ytterbium complexes under the light irradiation do not create toxic concentration of singlet oxygen, O_2, retaining a high affinity to malignant tumors at the same time.

First works on synthesis of porphyrin lanthanide complexes, both hydrophobic tetraarylporphyrins [2, 3] and water-soluble [4], appeared at the end of the XX century. Classical techniques are based on boiling of porphyrin with lanthanide metal acetylacetonate in 1,2,4-trichlorobenzene in the inert atmosphere throughout 2.5÷3 h. In case of hydrophilic derivatives a short alloying in an imidazole was applied, then the imidazole sublimation was carried out and metallocomplexes were isolated by various chromatography methods.

Works that have recently appeared regarding the use of microwave radiation for synthesis of tetraarylporphyrins, phthalocyanines and their metallocomplexes with Er and Gd in the dimethylacetamide medium in presence of dry lithium chloride [5, 6], induced us to carry out similar syntheses on the example of TPP [7] and 5,10,15,20-tetrakis(3-methoxyphenyl)porphyrin. This porphyrin is interesting due to fact that on the basis of its dihydroderivative Bonnet [8] created a drug Foscan, which is used for the cancer photodynamic therapy. Octabromderivative TPP-Br$_8$ was synthesized from the TPP copper complex [9].

10.2 EXPERIMENTAL PART

Pyrrol, propionic acid and chloroform were refluxed at atmospheric pressure. DMF, 1,2-dichlorbenzene and 1,2,4-trichlorbenzene were refluxed in vacuum. Acetic acid and nitrobenzene were used without additional purification. Electronic absorption spectra (UV-vis) were registered on a spectrometer Jasco 7800 (Japan) in chloroform. Mass-spectra were measured on a serial mass-spectrometer MI-1201 at the energy of ionizing electrons 40 eV. IR-spectra were registered on a FT-spectrometer "Brucker EQUINO$_X$ 55" (Germany) in tablets with KBr. Microwave radiation (MR) synthesis experiments were carried out in microwave system ETHOS D by firm Milestone (Italy) in 50 mL fluoropolymer autoclaves. Silica gel 60 (Merck, Germany) was used for column chromatography. Melting point (M.p.) was received on a electrothermal "MEL-TEMP" (U.S.A.).

In this chapter, electronic structure and chemical bonding in tetrakis-porphyrins and precursors were studied experimentally by ultra-violet (UPS) and X-ray (XPS) photoemission spectroscopy. To obtain this information high resolution of N1s, C1s, valence band spectra were measured at Helmholtz-Zentrum Berlin synchrotron radiation source (the Russian-German beam-line at BESSY) in the photon energy range 100–1000 eV with total resolution (electron plus photons) about 120 meV and using Kratos AXIS Ultra DLD spectrometer (the photon energy 1486.69 eV, Al K$_\alpha$ mono) in the Institute of Solid State Physics of Russian Academy of Sciences (total resolution Ag 3d$_{5/2}$ was about 0.48 eV). Spectra were calibrated using Ag3d$_{5/2}$ and In 3d$_{5/2}$ lines. Samples for X-ray photoemission spectroscopy were prepared by chemical deposition (drop coating solution of the compounds) self-assembling techniques *ex situ* from the CHCl$_3$ or CCl$_4$ solution onto Ag and/or the sample was pressing in a In substrate *ex situ*. The surfaces of *ex situ* prepared samples were additionally cleaned *in situ* by resistive heating up to ~400 K in ultra high vacuum (UHV) and/ or cleaned by ion gun. The base pressure during measurements was in the range 5×10^{-10}–2×10^{-9} torr.

10.2.1 PRECURSORS AND PROCEDURE

10.2.1.1 YTTERBIUM ACETYLACETONATE (3)

A 0.215 g (5.6×10^{-4} mol) ytterbium nitrate was dissolved in 50 mL distilled water, when mixing 0.2 mL acetylacetone in 2 mL ethanol was added and then the diluted solution of ammonia was slowly dropped to pH 6–6.5. The white precipitation was isolated by centrifugation, three times washed by distilled water and dried in a vacuum to a constant weight. Yield 0.2 g (71%). M.p.132°C. The substance was recrystallized from a benzene and petroleum ether mix, M.p. 142–144°C. IR-spectrum, ν, cm⁻¹: 3391 (H$_2$O), 1616 и 1524 (acac). Mass-spectrum, m/z: 470 (M⁺,

29%), 370 (M$^+$ – acac, 100%), 270 (M$^+$ – 2xxacac, 35%). Found, %: C 34.38; H 4.71. C$_{15}$H$_{21}$O$_6$Yb×3H$_2$O. Accounted for, %: C 34.33; H 5.15.

10.2.1.2   ERBIUM ACETYLACETONATE (5)

A 0.29 g ($1.06×10^{-4}$ mol) erbium chloride was dissolved in 40 mL distilled water, 0.5 mL acetylacetone in 2 mL ethanol was added and then the diluted solution of ammonia was slowly dropped to this solution to pH 6.5. The light-pink precipitation was isolated by centrifugation, three times washed by distilled water and dried over KOH in a vacuum. Weight 0.263 g (52.6%). M.p. 120–122°C.

10.2.1.3   TETRAPHENYLPORPHYRIN

Tetraphenylporphyrin was synthesized in propionic acid from pyrrol and benzaldehyde by microwave irradiation technique [5] with yield 39%.

10.2.1.4   5,10,15,20-TETRAKIS(3-METHOXYPHENYL)PORPHYRIN (8A)

A 20 mL nitrobenzene, 40 mL propionic acid and 20 mL pure acetic acid were boiled for 30 min. 0.356 g (2.33 mmol) 3-methoxybenzaldehyde (7) was added and then 0.156 g (2.33 mmol) freshly distillated pyrrol (6) was slowly dropped. The mix was boiled for 2 h, then cooled to 40°C and blew with air for 30 min. 20 mL methanol was added and the mix was left in the freezer for the night. The precipitation was filtered, washed with isopropanol and recrystallized from chloroform with methanol. Weight 0.13 g (28.5%). UV-vis, λ_{max}: 418, 514, 549, 589, 644 nm.

10.2.1.5   YTTERBIUM COMPLEX OF 5,10,15,20-TETRAKIS(3-METHOXYPHENYL)PORPHYRIN (8B)

A 5 mg (0.0068 mmol) porphyrin (8a) was dissolved in 0.5 mL 1,2-dichlorbenzene, then 4.5 mL DMF, 7 mg (0.015 mmol) ytterbium acetylacetonate (3) and 5 mg lithium chloride were added. The mix was maintained for 15 min at 145°C and power 650 w in microwave oven. When reaction mass was cooled solvents were deleted at lowered pressure, and porphyrin complex was isolated by preparative chromatography on silica gel plates in chloroform. Weight 4 mg (58.4%). UV-vis λ_{max}: 417.6; 553.4; 590.4 nm. Luminescent spectrum, λ_{max}: 980 nm (DMSO).

10.2.1.6 ERBIUM COMPLEX OF 5,10,15,20-TETRAKIS(3-METHOXYPHENYL)PORPHYRIN (8C)

Synthesis was carried out similarly to ytterbium complex. Yield 4 mg (58.7%). UV-vis, λ_{max}: 418, 553.2, 590.4 nm.

10.2.1.7 YTTERBIUM COMPLEX OF TETRAPHENYLPORPHYRIN (10)

Synthesis was carried out similarly to ytterbium complex. Yielded 53%. UV-vis, λ_{max}: 420, 513, 549 nm.

10.2.1.8 YTTERBIUM COMPLEX OF OCTABROMTETRAPHENYLPORPHYRIN

A 21 mg (0.017 mmol) octabromtetraphenylporphyrin was dissolved in 15 mL 1,2.4-trichlorbenzene, and then 25 mg (0.053 mmol) ytterbium acetylacetonate (3) was added. The mix was boiled in inert gas flow within 2 h. The solvent was evaporated. Metallocomplex was isolated by column chromatography on a silica gel in the $CHCl_3-C_2H_5OH$ (9:1) system. Yield 15 mg (59%). UV-vis, λ_{max}: 472, 611, 662 nm.

10.3 RESULTS AND DISCUSSIONS

The first stage of our work was the synthesis of lanthanide elements salts (ytterbium and erbium) in a form of acetylacetonates. The rare-earth elements (REE) complexes in most cases have the coordination number (CN) more than six (7, 8, 9, 10 and even 12). CN of REE ions in complexes with organic polydentate ligands are high and variable [10]. The reason of this phenomenon lies in the big ionic radius, which decreases from 1.06 Å (La^{3+}) to 0.88 Å (Lu^{3+}) (the effect of "lanthanide compression"). The empty site of the coordination sphere is occupied by other ligands: water, hydroxyl ions, etc. In IR-spectrum the hydroxyl ion is characterized by a narrow strip at 3700–3600 cm^{-1}, it has higher frequency than water. Frequency v_{O-H} of water is located in a region of about 3600–3200 cm^{-1}.

β-Diketonates of rare-earth elements possess high stability [10] and maintain a high temperature (> 200°C). For a synthesis of acetylacetonates we used an ammoniac method on the basis of lanthanide nitrates and chlorides:

$$Yb(NO_3)_3 + 3\ CH_3COCH_2COCH_3 + 3\ NH_4OH \rightarrow Yb(acac)_3 + 3\ NH_4NO_3 + 3\ H_2O$$
(1)–(3)

When carrying out the reaction in water with REE salts it is required that a pH of reaction mass do not exceed a value at which ytterbium hydroxide is formed.

The value of a pH of ytterbium hydroxide sedimentation and its control during the course of the synthesis are the necessary conditions when receiving ytterbium complexes. Used oxygen-containing bidentate donors (acetylacetone) should be low basic. In this connection the reaction of interaction of ytterbium salts with such ligands can be carried out in neutral and very subacidic environments (pH 5–6).

For the synthesis of a REE complex connection, it is a matter of principle to have a pH-medium since the type of a lanthanides complex formation depends on it, and so does the character of a complex formation in water and in the water-organic mediums. When carrying out reaction in water with REE salts it is required that pH of a reaction mass doesn't exceed value at which ytterbium hydroxide is formed. Value of size pH sedimentation of hydroxide of ytterbium and its control in the course of synthesis are necessary conditions when receiving complexes of ytterbium.

Acetylacetone was dissolved in a small volume of ethanol, mixed with water solution of ytterbium nitrate and the diluted solution of ammonia when mixing was slowly dropped under a pH control by use of pH-meter. At pH ~ 5.7 the slight turbidity started, and at pH ~ 6.0 the white precipitate was formed. We tried not to raise a pH higher than 6.5 since there can be a replacement of the acetylacetonate residues with a hydroxyl:

$$Yb(acac)_3 + H\text{-}OH \rightarrow Yb(acac)_2OH + CH_3COCH_2COCH_3$$

The ytterbium acetylacetonate precipitate (3) was isolated on a centrifuge, carefully washed by a distilled water, dried and recrystallized from benzene–petroleum ether.

The accurate wide band is observed at 3391 cm^{-1} in IR-spectrum (Fig. 10.1), it is characteristic for crystallized water. The acetylacetonate residues possess bonds at 1616 and 1524 cm^{-1}. The substance melts at a temperature of 142–144°C.

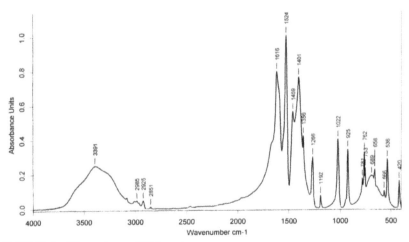

FIGURE 10.1   IR-spectrum of Yb(acac)$_3$.

In a mass spectrum (Fig. 10.2) there is a signal of a molecular ion 470 [M]$^+$, and also two fragments 370 and 270 corresponding to detachment of acetylacetonate ligands. The elemental analysis also shows existence of 3 molecules of crystallized water.

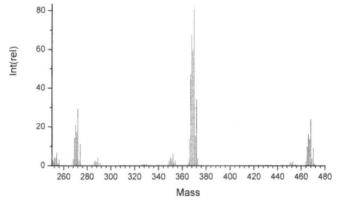

FIGURE 10.2   Mass-spectrum of Yb(acac)$_3$.

The X-ray photoemission data of all fabricated Yb salts (presented in Fig. 10.3–6) show the presence of all elements in spectra (except hydrogen) according to their states in the molecules.

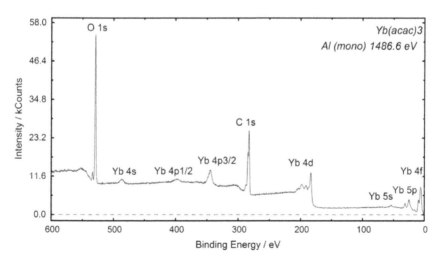

FIGURE 10.3   XPS raw spectra of Yb(acac)$_3$.

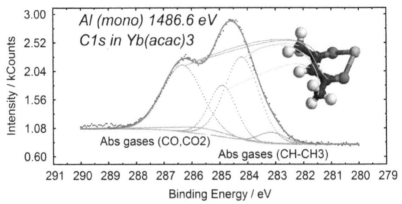

FIGURE 10.4   Spectra of C1s of Yb(acac)$_3$.

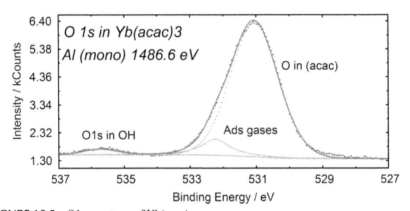

FIGURE 10.5   O1s spectrum of Yb(acac)$_3$.

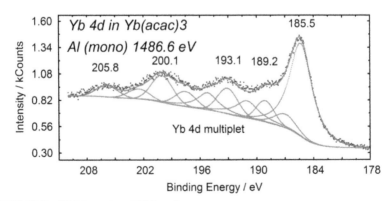

FIGURE 10.6   Yb4d spectra of Yb(acac)$_3$.

In core level spectra one can see peaks of C1s, O1s, Yb4d with relative composition close to chemical formula.

Wide peak of C1s core level is related with photoemission from three different bonding states of carbon in molecule (284.2 eV, 284.9 eV and 286.3 eV), as well as with small contaminations of adsorbed gases at 286 and 283 eV (see Fig. 10.4).

There are three peaks C1s in spectra with relative areas (CH_3 – 41.4%, CH – 20.4%, CO – 38.2%), which are proportional to carbon atoms in functional groups of salt (CH_3 – 40%, CH – 20%, CO – 40%).

In XPS spectra of O1s (Fig. 10.5) a peak with small shoulder at high binding energy side corresponding to two oxygen states is observed.

After peak analysis one can suppose that high binding energy shoulder at 532 eV is due to presence of small surface contamination of water and adsorbed gases, whereas main peak at 531.5 eV is related with acetylacetonate group.

Studies of core level Yb4d spectra exhibit a complex structure of Yb electronic states in acetylacetonate.

For careful analysis of Yb4d spectra in $Yb(acac)_3$, Yb4d spectra of Yb_2O_3 (Yb^{3+} – $4d^{10}4f^{13}$) and Yb metal (Yb^{2+} – $4d^{10}4f^{14}$) additionally were measured and decomposed. Divalent Yb has a $4f^{14}$ configuration, and the 4d spectra show the usual doublet with a 3:2 ratio, while for trivalent Yb, $4f^{13}$, the 4d peaks consist of a multiplet. Using the parameters of our analysis of metal Yb and sesquioxide of Yb and $Yb(acac)_3$ (see also Fig. 10.20) one can conclude that the multiplet 4d spectrum of $Yb(acac)_3$ is indicative of trivalent Yb in salt.

When carrying out the triple recrystallization of ytterbium acetylacetonate from acetone with water, the formation of a new compound takes place, with a higher temperature of melting (M.p. 179–181°C), with smaller percentage of carbon and hydrogen. The substance does not evaporate in a mass spectrometer under the conditions similar to ytterbium acetylacetonate (3).

In IR-spectrum (Fig. 10.7) along with the wide band relating to crystallized water–3396 cm^{-1}, two narrow bands appear with a frequency of 3598 and 3637 cm^{-1} which probably belong to the hydroxyl residues. Acetylacetonate groups appear at 1609 and 1521 cm^{-1}. Based on the analysis of physical and chemical data, the structure–$Yb_2(acac)_3xOH$ (Fig. 10.8), as described in Ref. [10], was attributed to high-melting compound.

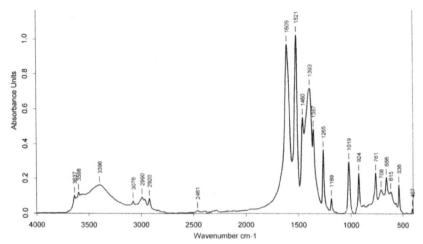

FIGURE 10.7   IR-spectrum of Yb$_2$(acac)$_3$xOH dimer.

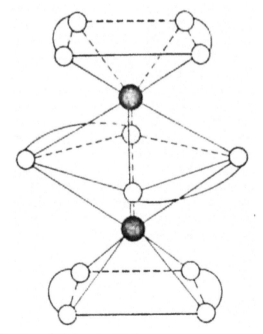

FIGURE 10.8   Structure of Yb$_2$(acac)$_3$xOH dimer.

X-ray photoelectron spectrum of Yb$_2$(acac)$_3$xOH is presented in Fig. 10.9 and quite similar to spectra of Yb(acac)$_3$.

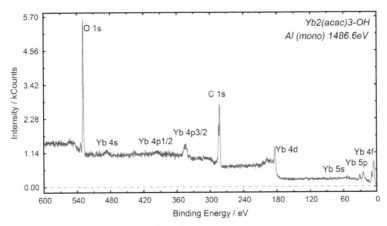

FIGURE 10.9   Raw spectrum of $Yb_2(acac)_3$ xOH.

In C1s core level spectra (Fig. 10.10) similar to the case of $Yb(acac)_3$ peaks of C1s core level are related with photoemission from three different bonding states of carbon at 284.4 eV, 284.9 and 286.5 eV (for $Yb(acac)_3$ 284.2 eV, 284.9 eV and 286.3 eV, respectively), as well as with contaminations of adsorbed gases at 286.5 and 283.5 eV.

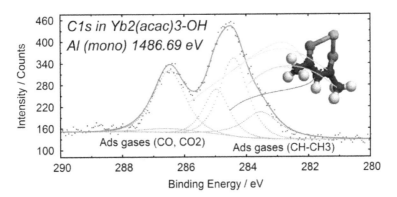

FIGURE 10.10   C1s spectra of $Yb_2(acac)_3$ xOH.

Three peaks in spectra C1s of $Yb_2(acac)_3$xOH with relative areas (CH_3 – 41.4%, CH – 20.4%, CO – 38.2%) are proportional to carbon atoms in functional groups of salt similar to $Yb(acac)_3$.

In XPS spectra of O1s (Fig. 10.11) a wide peak at about 531 eV and a peak at 535.7 are observed corresponding to two main oxygen states.

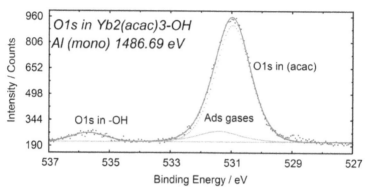

FIGURE 10.11   O1s spectra of Yb$_2$(acac)$_3$xOH.

The peak at 531 eV is related with acetylacetonate group like in Yb(acac)$_3$, peak at 535.7 is supposedly due to presence of hydroxyl group and small peak at 531.5 eV is related with adsorbed contamination.

Similar to Yb(acac)$_3$ core level Yb4d spectra (Fig. 10.12) exhibit a complex multiplet structure of Yb electronic states in acetylacetonate Yb$_2$(acac)$_3$xOH.

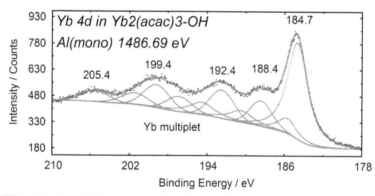

FIGURE 10.12   XPS Yb4d spectra and multiplet splitting analysis for Yb$_2$(acac)$_3$xOH

Using identical parameters of our analysis of Yb(acac)$_3$ (see also Fig. 10.20) one can also conclude the multiplet 4d spectrum typical for trivalent Yb in salt.

In case of erbium its acetylacetonate (5) was synthesized from metal chloride by a similar ammoniac method:

$$ErCl_3 + 3\ CH_3COCH_2COCH_3 + 3\ NH_4OH \rightarrow Er(acac)_3 + 3\ NH_4Cl + 3\ H_2O \quad (4)\,(5)$$

The received substance melted at a temperature of 120–122°C, and recrystallization from benzene–petroleum ether did not lead to a temperature change. For trihydrate Er(acac)$_3 \times 3H_2O$ in literature [11] it is given temperature of 129°C, and for the waterless–103°C. In IR-spectrum there is a characteristic wide water band at 3414 cm^{-1}, bands at 1610 and 1517 cm^{-1} belong to acetylacetonate residues.

Raw XPS spectrum of Er(acac)$_3$ is shown in Fig. 10.13.

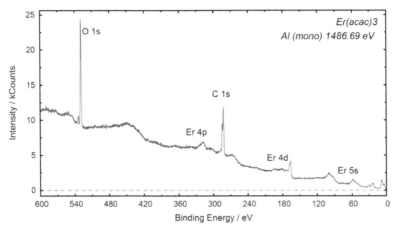

FIGURE 10.13   XPS spectrum of Er(acac)$_3$.

In core level spectra of Er(acac)$_3$ one can see peaks of C1s, O1s, Yb4d with relative composition close to chemical formula (see also Figs. 10.3 and 10.9 for Yb salts).

For Er(acac)$_3$ C1s spectra are characteristic for acetylacetonate group (Fig. 10.14), quite similar to Yb salts with main states at 284.5, 285 и 286.5eV and contaminations at 283.6eV. Three peaks in spectra C1s with relative areas (CH$_3$ – 38.7%, CH – 24%, CO – 37.3%) are proportional to carbon atoms number in functional groups of salt similar to Yb(acac)$_3$ (CH$_3$ – 40%, CH – 20%, CO – 40%).

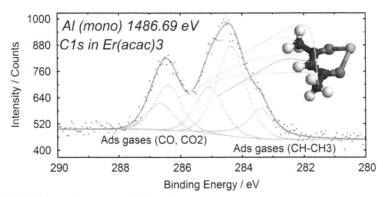

FIGURE 10.14   Cl s spectra of Er(acac)$_3$.

In XPS spectra of O1s of Er(acac)$_3$ OH (Fig. 10.15) a main peak at about 530.9 eV is related with acetylacetonate group like in Yb(acac)$_3$ and peak at 535.7 eV is due to presence of hydroxyl group and small peak at 531.6eV is related with adsorbed contaminations.

Similar to Yb compounds Er4d spectra show multiplet splitting, supporting REE^{+3} ion state in these salts.

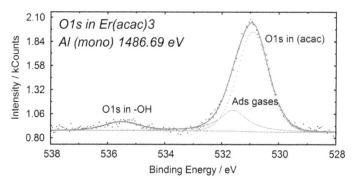

FIGURE 15   O1s spectra of Er(acac)$_3$OH.

The XPS spectra of valence band of Er(acac)$_3$ and acetylacetonate salt of Yb are shown in Fig. 10.16 (left and right panel).

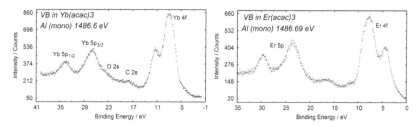

FIGURE 10.16   XPS valence band spectra of Er(acac)$_3$ and Yb$_2$(acac)$_3$.

The difference in electronic structure of VB Er(acac)$_3$ and acetylacetonate salt of Yb (Fig.10.16) is related mainly with energy position of semicore 5p and 4f states in Er и Yb salts.

10.3.1   SYNTHESIS OF PORPHYRINS AND REE METALLOPORPHYRINS

The next step was the synthesis of TPP, its octabromderivative and symmetric 3-methoxysubstituted derivative of tetraphenylporphyrin (8) (Fig. 10.17). The major difference between those porphyrins is electron-donor and electron-sink substitutes presence. The purpose was to find effective conditions for a synthesis of these substances by use of microwave radiation. Classical techniques of the synthesis for similar compounds are based on the method of Adler-Longo [12] using boiling of pyrrole and benzaldehyde in propionic acid, or on the technique improved by the Chinese scientists [13] with use of solvents threefold mix: $EtCOOH–AcOH–PhNO_2$ (2:1:1) where nitrobenzene performs as an oxidizer.

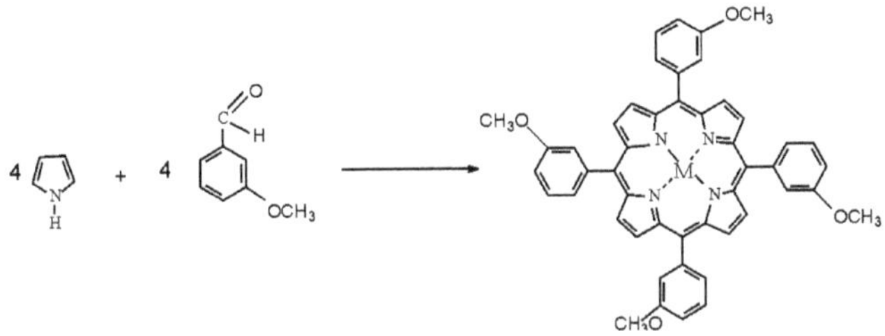

(6)   (7)   (8): (a) M = 2 H; (b) M = Yb(acac);
                                        (c) M = Er(acac)

FIGURE 10.17   Scheme of porphyrins synthesis. (a) M = 2 H; (b) M = Yb(acac); (c) M = Er(acac).

We carried out some experiments with use of MR, various solvents and their mixes (acetic and propionic acids, nitrobenzene); reaction time, temperature and MR power were also varied. It appeared that in absence of nitrobenzene porphyrin in the closed volume is formed only in trace quantities. In the conditions of classical synthesis when boiling in threefold mix and the subsequent blowing with air, porphyrin (8a) was prepared with a yield 28.5%.

Porphyrin molecules form stable complexes with lanthanide ions, these complexes have intensive absorption in a visible range of spectrum. Erbium, ytterbium and neodymium complexes are characterized by a 4f-luminescence in near IR-range of spectrum [1]. The most studied complexes with porphyrins are ytterbium complexes since Yb has smaller ionic radius in comparison with lanthanum (radius of Yb^{3+} ion is 1.01 Å), which determines higher stability of these metallocomplexes. Distinctive feature of Yb porphyrin complexes is a characteristic narrow and rather intensive luminescence band located in the IR-range at 975–985 nm, in so-called "therapeutic window of tissue transparency."

Nevertheless, ytterbium and erbium incorporation in a porphyrin macrocycle faces certain difficulties. For this purpose highly boiling solvent, long time of heating and use of inert gas are required. In Ref. [5] incorporation of erbium and gadolinium in a porphyrin cycle in the medium of dimethylacetamide {Boiling point (B.p.) 165°C} in a presence of dry lithium chloride is described.

At the first step we tried to incorporate Yb^{3+} ion in porphyrin by heating in a microwave oven a porphyrin and ytterbium acetylacetonate in a medium of 1,2,4-trichlorbenzene, however the result was negative. The best results were received using mix of solvents 1,2-dichlorbenzene (B.p. 180°C) and freshly distillated DMF (B.p. 153°C) in the ratio (1:9), and also additives of dry lithium chloride. Reaction time was 15–20 min., temperature in the oven was 145°C, and power was 650 w. After cooling solvents were deleted under a lowered pressure, and metallocomplexes were isolated by preparative chromatography on aluminum oxide of II activity degree or on 5/40 μm silica gel. The last adsorbent appears to be more preferable since lanthanide porphyrin metallocomplexes are less adsorbed on it and easily eluted by chloroform – isopropanol mix.

As a result of microwave synthesis metalloporphyrin Yb-complexes (8в) were obtained with an yield 58.4%, and Er-complexes (8c) with an yield 58.7%.

10.3.2 ELECTRONIC STRUCTURE OF REE METALLOPORPHYRINS: CORE LEVELS

All elements of pristine TPP and metalloporphyrins were found in XPS spectra (photon energy 1486.69 eV (Al K_a) and 100–600 eV of synchrotron radiation) after moderate annealing in UHV. Yb-TPPBr$_8$ apparently partially decomposed even after annealing at 100°C in UHV. XPS and UPS studies have also demonstrated higher thermal stability of Yb-TPP in UHV as compared with brominated compounds.

Let us discuss the details of electronic structure of REE (Yb, Er) metalloporphyrins – Yb(acac)TPPBr$_8$; Yb(acac)TPP; Er(acac)TPP and pristine TPP, TPPBr$_8$.

XPS spectra of N1s, O1s, C1s levels and VB spectra are shown in Fig. 10.18 (insets). The spectrum of the core levels of the N1s level in the TPP and TPPBr$_8$ there are two peaks of N1s with binding energies of 399.8 and 397.8 eV and were assigned to sp^3 and sp^2 nitrogen, respectively (pyrrol- and aza-states associated with the protonated and not protonated nitrogen [14, 15]). Central atom of Yb, replacing the two hydrogen atoms bound to the equivalent of all the nitrogen atoms. In metalloporphyrines charge distribution is more uniform for N1s spectra of Yb-TPP(OCH$_3$)$_4$ and other REE metalloporphyrins thus wide peak of N1s states reflecting small difference between pyrole- and aza-nitrogen in metalloporphyrines, in consequence of that formed a single N1s state. Note that there is imposing the peak of N1s and Yb4p, whose parameters are derived from the decomposition peak at Yb4p in Yb oxide.

Wide peak in the range 282 and 290 eV was seen in the wide C1s spectra (Fig. 10.19) that are related to the unequivalent C atoms in the molecules.

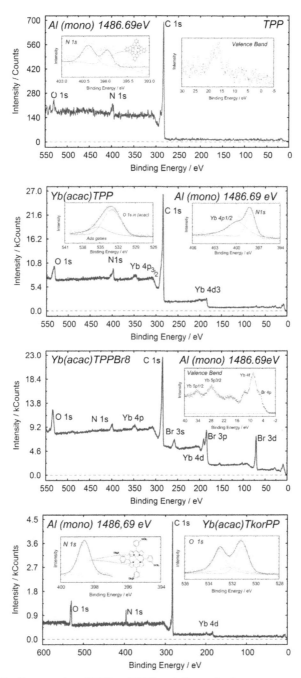

FIGURE 10.18 Raw spectra of TPP and REE metalloporphyrins.

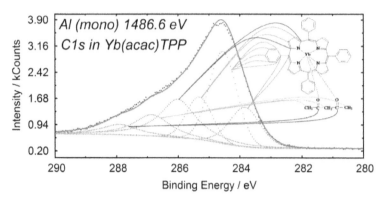

FIGURE 10.19   Typical C1s spectra of REE metalloporphyrins.

Different peaks observed in the C1s spectra after decomposition of spectra are related to the different C atoms in the porphyrin molecules (aromatic and C-N-C groups) as well as shake-up HOMO-LUMO satellite. This C1s peak in the XPS data has a number of subpeaks that correspond to nonequivalent positions of carbon atoms in the molecule. Calculated states of carbon atoms are situated very closely to each other, so they cannot be practically resolved experimentally. It is well agreed with results for thiolporphyrins and related to them phthalocyanine compounds [16]. Similar C1s state data for various porphyrins was received earlier by the XPS method and for related pyridyl-porphyrins.

The analysis of Yb4d shows that spectra do not consist of the usual spin-orbit split doublet, but instead are composed of asymmetric peak with multiplet splitting (Fig. 10.20).

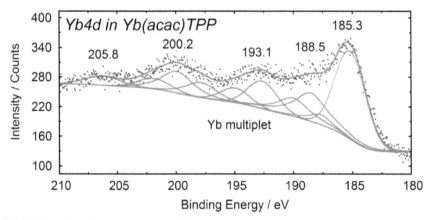

FIGURE 10.20   Yb4d spectra in YbTPP.

For Yb two cases are possible. Divalent Yb has a filled 4f shell, that is, a $4f^{14}$ configuration, and the 4d spectra show the usual doublet with a 3:2 ratio, while for trivalent Yb, $4f^{13}$, the 4d peaks consist of a multiplet. The transition from doublet to multiplet in the 4d spectra of Yb is illustrated nicely in the oxidation of metallic Yb, since metallic Yb is divalent and the sesquioxide is trivalent. Using the parameters of our analysis of metal Yb and sesquioxide of Yb one can conclude that the multiplet 4d spectrum of Yb(acac)TPP and Yb(acac)TPPBr$_8$ is clearly indicative of trivalent Yb (Fig. 10.21).

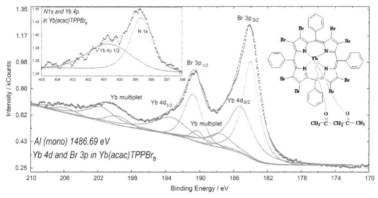

FIGURE 10.21   XPS Yb4d-Br3p spectra of Yb(acac) TPPBr$_8$. Inset- Yb4p-N1s region.

In spectrum Yb(acac)TPPBr$_8$ it is observed the imposing of lines Yb4d and Br3p that causes some difficulties in processing. But using preliminary processing of TP-PBr$_8$ (which spectra are not presented here) and Yb(acac)TPP this difficulty has been solved. We see that a complex multiplet Yb which testifies to a trivalent condition, is presented in both samples.

XPS spectra of other REE metalloporphyrin Er(acac)TPPBr$_8$ presented in Fig. 10.22. Like in the case of Yb metalloporphyrins in Er(acac)TPPBr8 porphyrins N1s spectra (Fig. 10.22 left inset) demonstrate single peak structure of N1s states with binding energy 398 eV similar to YbTPP (see Fig. 10.18).

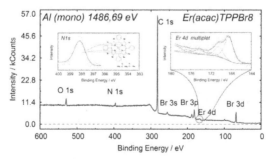

FIGURE 10.22   XPS spectra of Er(acac)TPPBr$_8$ porphyrin.

Similar results for C1s data were obtained for Er(acac)TPPBr$_8$ porphyrin. Wide peak observed in the C1s spectra are related to the unequivalent C atoms in the porphyrin molecules (aromatic and C-N-C groups) as well as with shake-up HOMO-LUMO satellite.

Evident multiplet structure in the 4d spectra of Er (Fig. 10.22 right inset) and comparison with our 4d spectra of metallic Er and oxide of Er clearly supports the trivalent state of Er in compound.

10.3.3   ELECTRONIC STRUCTURE: VALENCE BAND SPECTRA

We have examined the electronic structure of valence band of the pristine porphyrin and Yb metalloporphyrins using facility of Helmholtz-Zentrum Berlin synchrotron radiation source (the Russian-Germany beam-line at BESSY) in the photon energy range 100–1000 eV and Kratos AXIS Ultra DLD spectrometer (the photon energy 1486.69 eV, Al K$_\alpha$ mono). Typical spectra are presented in Figs. 10.22 and 10.24.

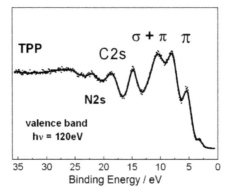

FIGURE 10.23   UPS VB spectra of TPP.

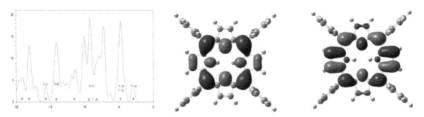

FIGURE 10.24   Calculated VB spectra and charge distribution HOMO (center) LUMO (right) of TPP porphyrin.

XPS and UPS studies of VB (performed at photon energy 120 eV) of TPP porphyrins show that the valence band is mainly formed by peaks corresponding to π (2–10 eV) and σ states (8–16 eV) of porphyrin macrocycles and apparently can be reproduced by superposition of aromatic benzene and pyrrole spectra. Peak of C2s states in TPP lied at about 18 eV in VB spectra. This spectrum is similar to the theoretical one (Fig. 10.24). Experimental photoemission spectra are in good agreement with quantum chemical calculation. HOMO states in TPP lied at 2 eV below Fermi edge and according to calculations are due to π (N2p) states of nitrogen in macrocycle (Fig. 10.24).

In present work *ab initio* quantum-chemical calculations were performed by Gaussian 03 using density-functional theory for tetraphenylporphyrin. 6–31G(d, p) basis was used for all atoms, core electrons of which were simulated with LanL2 pseudopotential with corresponding 2-exponent basis for valence electrons. Theoretical valence band spectra of the molecules were obtained from calculated molecular orbitals. Chemical shift has been modeled as a change of electrostatic potential of atoms and three well-resolved nitrogen states and nine states were obtained situated very closely to each other, so they cannot be resolved experimentally.

Valence band spectra of $Yb(acac)TPPBr_8$ (photon energy 1486.6 eV) are shown in Fig. 10.25. The analysis of experimental and literature data allow to suppose that the valence band of REE (Yb, Er) metalloprphyrins are similar to TPP, 3d- and Pt metalloporphyrins and is formed by π– (1–6 eV), π+σ– (5–10 eV) и σ-states (6–16 eV) of macrocycles with HOMO π (N2p) states at 2.5–2.7eV below E_F [14]. At present time careful *ab initio* quantum-chemical calculations using density-functional theory for $Yb(acac)TPPBr_8$, Er, Yb(acac)TPP and HOMO, LUMO states are in progress.

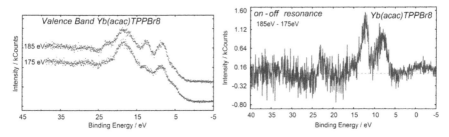

FIGURE 10.25 VB spectra and difference (on-resonance- off-resonance) spectra of $Yb(acac)TPPBr_8$.

For analysis of fine structure of Yb(acac)TPP VB additional experiments using resonance photoelectron spectroscopy were performed at Russian-Germany beamline (BESSY). Resonance UPS spectra for Yb(acac)TPP were measured at different photon energies including the values in the vicinity of 4d–4f resonance. As an example, Fig. 10.24 presents UPS spectra taken at hv = 175 eV and 185 eV (off- and on-resonance, respectively). Strong enhancement of the shallow core level photoemission at photon energies close to the $4d$–$4f$ photothreshold and resonant behavior of the photoemission spectra of rare earth compounds originates from the process of photoexcitation (near $4d$–$4f$ absorption region) followed by the decay of the excited configuration via direct and indirect processes. These processes produce the resonant enhancement of photoemission and can be described as shown below for Yb:

$$\text{direct: } 4d^{10}4f^{13} + hv \rightarrow! \ 4d^{10}4f^{12} + e^{-},$$

$$\text{indirect: } 4d^{10}4f^{13} + hv \rightarrow 4d^{9}4f^{14}! \ 4d^{10}4f^{12} + e^{-}.$$

Photoemission spectra of Yb(acac)TPPBr$_8$ clearly demonstrate (Fig. 10.25) a remarkable resonance of Yb $4f$ states and therefore the spectral intensity for Yb compound near 8 and 12 eV is strongly enhanced (Fig. 10.24, on-resonance: 185 eV), an effect which one tends to attribute to emission from Yb$4f$ states proving the position of Yb^{3+} ($4f^{13}$) states in metalloporphyrins.

10.4   CONCLUSION

Results of carried out researches demonstrate the possibility of receiving the lanthanide porphyrin complexes by use of microwave radiation technique. Thus, the time of process is considerably reduced, energy consumptions decreased, reactions are carried out at lower temperatures, there is no need for an inert atmosphere.

The XPS and UPS data show different atomic constituents in accordance with its states in the molecules, which can be related to the peaks appearing in the electronic spectra. Different peaks were seen in the C1s spectra are related to the chemically unequivalent C atoms in the molecules. The spectrum of the N1s level in the TPP and TPPBr$_8$ shows two peaks structure of N1s with binding energies of 399.8 and 397.8 eV which were assigned to sp^3 and sp^2 nitrogen, respectively (pyrrol- and aza-states associated with the protonated and not protonated nitrogen). In REE metalloporphyrins charge distribution for N is more uniform with single N1s peak at about 399 eV in spectra. Evident multiplet structure in the 4d spectra of Yb and Er metalloporphyrins clearly indicates the trivalent state of REE in compounds.

The UPS results give the position of the occupied states π and σ-states and REE in valence band of TPP and REE metalloporphyrins.

10.5   ACKNOWLEDGMENT

This chapter was supported by Ministry of Education and Science of the Russian Federation, Russian Academy of Sciences and the bilateral program "Russian-German Laboratory at BESSY." The calculations were provided at Freie Universität Berlin and Moscow State University by Dr. A. Volykhov. The financial and technical support from BESSY staff is gratefully acknowledged.

KEYWORDS

- Acetylacetone
- Electronic structure
- Erbium
- Metalloporphyrin
- Microwave synthesis
- MR
- XPS
- Ytterbium

REFERENCES

1. Romanova, N. N., Gravio, A. G. & Zyk, N. V. (2005). Microwave irradiation in organic synthesis. *Uspekhi khimii, 74(11),* 1059–1105.
2. Gaiduk, M. I., Grigoryants, V. V., Mironov, A. F., Chissov, V. I., Rumyantseva, V. D. & Sukhin, G. M. (1990). Fiber-laser IR luminescence diagnostics of malignant tumors using rare-earth porphyrins. *J. Photochem. Photobiol., B: Biology, 7(1),* 15–20.
3. Ivanov, A. V., Rumyantseva, V. D., Shchamkhalov, K. S. & Shilov, I. P. (2010). Luminescence diagnostics of malignant tumors in the IR spectral range using Yb-porphyrin metallocomplexes. *Laser Physics, 20(12),* 2056–2065.
4. Belogorokhov, A. I., Bozhko, S. I., Chaika, A. N., Ionov, A. M., Trophimov, S. A., Rumyantseva, V. D. & Vyalikh, D. (2009). Electronic Structure and Self-assembling Processes in Platinum Metalloporphyrines: Photoemission and AFM Studies Applied Physics A. *Materials Science and Processing, 94,* 473–476.
5. Belogorokhov, A. I., Bozhko, S. I., Ionov, A. M., Chaika, A. N., Trophimov, S. A., Rumyantseva, V. D. & Vyalikh, D. (2009). Synthesis of platinum metalloporphyrins and investigation of electron structure of complexes by photoelectron spectroscopy methods Poverkhnost.' *Rentgenovskie, sinkhotronnye i neironnye issledovaniya, (12),* 18–23.
6. Ivanov, A. V., Rumyantseva, V. D., Shchamkhalov, K. S. & Shilov, I. P. (2010). Luminescence diagnostics of malignant tumors in the IR spectral range using Yb-porphyrin metallocomplexes. *Laser Physics, 20(12),* 2056–2065.
7. Wong, C. P., Venteicher, R. F., Horrocks, W. & De, W. (1974). Lanthanide porphyrin complexes. A potential new class of nuclear magnetic resonance dipolar probe. *J. Am. Chem. Soc, 96(22),* 7149–7150.
8. Horrocks, W., De, W. & Wong, C. P. (1976). Lantanide porphyrin complexes evaluation of nuclear magnetic resonance dipolar probe and shift reagent capabilities. *J. Am. Chem. Soc, 98(23),* 7157–7162.
9. Horrocks, W., De, W. & Hove, E. G. (1978). Water-soluble lantahanide porphyrins: shift reagents for aqueous solution. *J. Am. Chem. Soc., 100(14),* 4386–4392.
10. Romanova, N. N., Gravio, A. G. & Zyk, N. V. (2005). Microwave irradiation in organic synthesis *Uspekhi khimii, 74(11),* 1059–1105.

11. Gürek, A. G. & Hirel, C. (2012). In. Photosensitizers in Medicine, Environment, and Security. T. Nyokong, V. B. Ahsen (Eds.): Springer Science+Business Media, 47–82.

12. Enikolopyan, N. S. (1985). Porphyrins: structure, properties, synthesis. Ed. Moscow. *Nauka*, 211–212.

13. Bonnet, R. (1989). *Eur. Patent № 337601*, Cl C 07D 487/22.

14. Bhyrappa, P. & Krishnan, V. (1991). Octabromotetraphenylporphyrin and its metal derivatives: Electronic structure and electrochemical properties. *Inorg. Chem., 30,* 239–245.

15. Spitsyn, V. I. & Martynenko, L. I. (1979). Coordination chemistry of rare-earth elements. Moscow, *MSU,* 248 p.

16. Przystal, J. K., Bos, W. G. & Liss, I. B. (1971). The preparation and characterization of some anhydrous rare earth trisacetylacetonates. *J. Inorg. Nucl. Chem., 33,* 679–689.

17. Adler, A. D., Longo, E. R. & Shergalis, W. (1964). Mechanistic investigations of porphyrin synthesis I Preliminary studies on ms-tetraphenylporphyrin. *J. Am. Chem. Soc, 86,* 3145–3149.

18. Sun Zh., She, Y. & Zhong, R. (2009). Synthesis of p-substituted tetraphenylporphyrins and corresponging ferric complexes with mixed-solvents method Front. *Chem. Eng. China, 3(4),* 457–461.

19. Belogorokhov, A. I., Bozhko, S. I., Chaika, A. N., Ionov, A. M., Trophimov, S. A., Rumyantseva, V. D. & Vyalikh, D. (2009). Electronic Structure and Self-assembling Processes in Platinum Metalloporphyrines: Photoemission and AFM Studies Applied Physics A. *Materials Science and Processing, 94,* 473–476.

20. Ghosh, A., Moulder, J., Bröring, M. & Vogel, E. (2001). X-Ray Photoelectron Spectroscopy of Porphycenes: Charge Asymmetry Across Low-Barrier Hydrogen Bonds *Angew. Chem. Int. Ed, 40(2),* 431–434. (1994). See also Ghosh, A., *Inog. Chem, 33,* 6057–6060.

21. Reisert, H., Knupfer, M. & Fink, J. (2002). Electronic structure of partially fluorinated copper phthalocyanine (CuPCF$_4$) and its interface to Au(1 0 0). *Surf. Sci, 515,* 491–498.

CHAPTER 11

SAME CONCEPTS OF COSMOLOGY IN THE MODERN NATURAL SCIENCE

M. D. GOLDFEIN and A. V. IVANOV

CONTENTS

Each molecule, therefore, throughout the Universe, bears impressed on it the stamp of a metric system as distinctly as does the meter of the Archives at Paris, or the double royal cubit of the Temple of Karnac.

J.C. Maxwell

11.1   INTRODUCTION

In article the modern concept of an origin of the Universe, feature of its structure and evolution, a structure of Solar System and parameters of its planets are considered. New data on cosmological characteristics of the Moon and methods of its study are provided.

11.2   ORIGIN OF THE UNIVERSE AND ITS STRUCTURE

The Universe represents the whole environment of the material world available to observation. Various types of objects differing in sizes and masses are contained into it, such as elementary particles, atoms and molecules, substances in various states of aggregation of matter, planets, stars, galaxies, and physical fields (gravitational, electromagnetic, etc.). In spite of the fact that the processes proceeding in the Universe cannot be researched in terrestrial conditions, the usual methodology accepted in natural sciences is used for their investigation. This is caused by that the Universe as a whole obeys the same universal laws which describe the behavior of its separate components. The section of physics and astronomy to study the Universe as a whole is called as *cosmology*. Gravitation has major importance for cosmological processes. Before Isaac Newton (Claudius Ptolemy, Nicolas Copernicus, Giordano Bruno, Galileo Galilei, and Johann Kepler) and after the creation of classical mechanics the Universe was considered as infinite. Now it is accepted that the Universe implies distances of 10^{28} m and time intervals of 10^{18} s of the order of magnitude (3×10^{14} h, 10^{13} days, 3×10^{10} years).

The initial stage of the Universe formation (in essence, its expansion) when the substance density, radiation energy, and temperature were infinitely high (the singular state) is usually called the *Big Bang*. This is the most developed concept with its heart containing the results of the works by Albert Einstein, Alexander Friedman (1888–1925), Edwin Hubble (1889–1953), George Gamow (1904–1968), and Brian Schmidt (b. 1967).

The Universe formation process as a result of the Big Bang is schematically seen as follows. At the most initial moment of explosion, the Universe's size was almost zero, and it itself was infinitely hot. In the process of expansion (the diameter of the Universe increased by 10^{50} for every 10^{-32} s) the radiation temperature sharply decreased and reached about 10 billion degrees in few seconds. Such fast expansion, subsequently called inflation, corresponds to an explosive process. At that time, the Universe consisted of electrons, protons, neutrons, neutrinos, and their antiparticles.

Its further expansion and temperature drop was accompanied by an increase in the electron—positron pair annihilation rate, which led to the formation of photons and to sharp reduction of the quantity of electrons. Approx. in hundred seconds after the Big Bang, the temperature decreased down to 1 billion degrees at which protons and neutrons started their merging to form deuterium nuclei. One fraction of deuterium nuclei then turned into hydrogen nuclei, and the other fraction captured protons and neutrons, turning into helium nuclei (containing two protons and two neutrons), and into heavier elements (lithium and beryllium) in small amounts. In few hours after the Big Bang, the formation of hydrogen, helium, and other elements stopped, but the Universe expansion still proceeded during ca. 1 million years, without being accompanied by special changes in the chemical composition. At last, when the temperature decreased down to few thousand degrees, the electrons and nuclei began to unite to form atoms. *The model of the hot Universe* was experimentally confirmed by the discovery of the relic radiation in 1965, that is, the microwave background radiation with a temperature of ca. 3 K, and by the existence of high concentrations of helium (everywhere) and deuterium in interstellar space. The Universe cooling, changes in its density, and the emergence of gravitational attraction between its separate areas 1 terminated its expansion and caused compression. Such processes proceeded long enough and approx. in billion years after the Big Bang, as a result of substance cooling, its compression and fluctuations of its density, the formation of galaxies to rather uniformly fill the whole space of the Universe began.

The development of modern cosmology actually begins with the creation of the special (at first) and general (subsequently) theories of relativity. It was Albert Einstein who offered *the first relativistic model of the Universe*, proceeding from the classical ideas of its stationary nature, that is, the invariance of properties in time. According to Einstein's hypothesis, the Universe is spatially finite but has no distinctly expressed borders. Naturally, such a concept is abstract, and it was subjected to criticism by Alexander Friedman who developed the bases of the nonstationary relativistic cosmology (1922). Proceeding from the possibility of changes in the curvature radius of the world space in time, he showed that the solutions of the equations of the general theory of relativity allowed considering the probability of three models of the Universe. In two of them the curvature radius of space monotonously grows, and the Universe infinitely expands (in one model – from a point, in another one – starting from some finite volume). The third model concerns the probability of existence of the pulsing Universe whose radius of curvature periodically varies. The reality of such models was confirmed by the effect of *redshift* (the galaxies move away far from each other) discovered by Edwin Hubble in 1929. All the galaxies (except for several ones most close to our galaxy) have been shown to go away from each other with certain speeds at expansion of the Universe. These speeds (v) can be found from the Christian Doppler shift of the spectral lines in the spectra of these galaxies, and at distances about 10 Mpc = 3×10^{23} m they obey Hubble's law: $v = H \times R$. The value of H depends on time only, and now its value (Hubble's constant) is within (50–100) km/s × Mpc = (1.6–3.2) ×10^{-18} s^{-1}). Calculations of these speeds allowed estimating the age of the Universe (counted from the singular state) as about 15 billion years. This value gives evidence of the finite nature of time passed from the instant of the Big Bang, which leads to the conclusion about the existence of *the cosmological horizon (Observable universe)*. This notion represents the border to separate the area of space observable at present from the area, which cannot be observed now. However, as light from the farther parts of the Universe reaches the observer, the cosmological horizon extends. Now, it is about 6000 Mpc, that is, covers more than a half of the space volume of the Universe available to observation.

Analysis of the density of the substance which the Universe consists of has shown that its value sharply falls upon transition from small objects to big ones: it is 10^{14} g/cm^3 in atomic nuclei and in neutron stars, about 1 g/cm^3 in planets and many stars, about 10^{-24} g/cm^3 in the Galaxy, and over the whole observable Universe the average density of substance is close to the so-called critical density estimated as $\sim 10^{-29}$ g/cm^3.

It is necessary to note that the concept of the nonstationary Universe does not contradict the concept about its uniformity. The uniformity of the Universe manifests itself, first, in essentially identical structural elements of stars and galaxies, the same physical laws and physical parameters, and, second, in the homogeneous distribution of substance over the whole Universe. This allows drawing a basic conclusion that our planet Earth does not occupy any exclusive position in the Universe.

The *Galaxies* are congestions of many stars and stellar remnants; they differ in size and appearance. They are conditionally divided into several types, namely: elliptical, spiral, irregular, interacting, radio galaxies, etc. The *elliptialc galaxies* are ellipsoids with different extents of compression; they are simplest by structure; the density of their stars monotonously decreases from the center. The *spiral galaxies* are considered as most numerous; our Galaxy (the Milky Way) and the Andromeda Galaxy belong to them. The *irregular galaxies* have no strict center and their structure features no certain regularities. The Large and Small Magellanic Clouds (which are satellites of our Galaxy and have, in comparison with it, considerably smaller masses and sizes) exemplify them. The *interacting galaxies* are usually located at small distances from each other, and some of them are overlapped. The *radio galaxies* possess very powerful radio emission; one of the known radio galaxies is in Cygnus, and another one is in the Centaurus constellation. Their radio emission has various causes. In particular, it is assumed that radio waves are emitted by the ionized hot gas (generally, hydrogen) in interstellar space, which is formed by hot stars and space rays. In recent years, it has been revealed that all galaxies are distributed nonuniformly and there are places in the Universe where they are absent at all.

Following points are to be noted:

1.   1 pc (parsec) is the unit of length equal to the distance at which the parallax is 1 s, which corresponds to 3.26 light years or 3.9×10^{13} km.

2. Parallax is the visible change of the location of a subject owing to moving of the observer's eye.
3. Light year (l.y.) is the way which light passes for one calendar year: 9.5×10^{12} km.

Our Galaxy includes about 150 billion stars, being a huge disk, which in turn, consists of a core and several spiral branches. The diameter of our Galaxy is 120,000 light years, its thickness is 10,000 light years; the mass is about 10^{11} of the Sun's mass; the speed of rotation has different values depending on the distance from its center, and no strict dependence exists. Therefore, various sites of the Galaxy have different periods of rotation. The stars and fogs within the Galaxy move in absolutely various directions. The Galaxy itself rotates round its axis quite slowly, spending 180 million years for a complete turn. The star density in the Galaxy is also nonuniform; for example, in the field of its core it reaches two thousand stars per parsec, which exceeds the average star density in the Sun vicinity almost by 20,000 times. Near the Sun, the average distance between stars is 10 million times higher than the average diameter of these stars; therefore, the stars almost never collide.

The following classification of stars is accepted in modern cosmology.

1. The variable stars are those whose shine, energy, and radiation spectrum are constantly changing. It is due to their periodic pulsation, that is, expansion or compression of the star, accompanied by temperature changes.
2. The double stars are systems of two stars rotating round their common center of masses.
3. The ecliptic-double stars are those whose total shine is constantly changing owing to periodic overlapping by each other.
4. The multiple stars are systems of three and more stars.
5. The red dwarfs are the stars whose diameter is less by 2–3 times, and whose density is higher by 2–3 times than in the Sun, and the luminosity is comparable with the Sun's.
6. The white dwarfs are the stars whose mass is close to the Sun's one, the diameter is 0.01 of its diameter, the density is ~ 10 t/cm^3, the luminosity is 10^{-4} of the Sun's.
7. The red giants are the stars whose diameter exceeds the diameter of the Sun by hundreds times, the density is lower by thousands times, and the luminosity exceeds that of the Sun by tens and hundreds times.
8. The neutron stars are huge congestions of neutrons; their mass is close to that of the Sun, the diameter is 2×10^{-5} of the diameter of the Sun, the density is $\sim 10^8$ t/cm^3.
9. The pulsars are the stars whose electromagnetic radiation periodically changes over the whole known range of wavelengths (from radio waves to gamma radiation).

10. The quasars are quasi-star (i.e., similar) sources of radio emission; their mass and diameter exceed those of the Sun by tens million times; the energy of radiation exceeds the total energy of radiation of several galaxies.

11.3   SOME CHARACTERISTICS OF THE SUN AND THE SOLAR SYSTEM PLANETS

Nicolas Copernicus' creation of the heliocentric system was a major event in our understanding of the Solar system structure, whose basis contains the following statements:
- the Sun rather than Earth is located at the center of the world;
- the spherical Earth rotates round its axis, and this causes the seeming daily motion of all stars;
- Earth, as well as all other planets, rotates round the Sun, and this explains the apparent motion of the Sun relative to stars.

Now the Solar system is known to include the Sun itself, planets and their satellites, and other space objects (asteroids, comets, meteorites, and even space dust) as well. According to modern representations, the formation of the Solar system about 5 billion years ago had the following stages:

1. The initial gas-dust cloud reached an appreciable density and started to compress under the influence of gravitational forces.
2. In the process of compression, the sizes of the gas-dust cloud decreased, and the speed of its rotation increased; the speeds of compression of the cloud parallel and perpendicular to its axis of rotation differed, which led to condensation of the cloud and the formation of a disk.
3. At achievement of some limiting density, the dust particles started to collide with each other and the released kinetic energy of the compressing gas-dust cloud led to temperature growth, with the central region of the disk to heat up most intensely.
4. At achievement of a temperature of few thousand degrees, the central region of the disk started to shine; it is a *protostar* (or a *protosun*) which falling of the cloud substance proceeded on, to increase the pressure and temperature at the center.
5. When the temperature at the center of the protostar reached millions degrees, thermonuclear reactions of hydrogen burning began.
6. The external areas of the disk remained relatively cold and, as a result of different cloud areas possessing different gravity, the largest condensations in these external areas formed planets.

The Sun has been established to move almost evenly on a big circle of the celestial sphere called as the *ecliptic*, from the west to the east (oppositely to rotation of the celestial sphere), making a complete turn for ca. 200 million years (a *galactic year*).

Depending on the nature of movement on the celestial sphere, the planets are divided into the lower ones (Mercury, Venus) and the higher ones (all the other planets, except Earth), or the internal and external planets relative to Earth's orbit. In their movement on the celestial sphere, Mercury and Venus never leave far from the Sun (Mercury – not farther than 18–28 degrees, Venus – not farther than 45–48 degrees) and can be either to the east or to the west from it. The instant of the longest angular distance of a planet to the east and to the west from the Sun is called as *the eastern or evening elongation* and *the western or morning elongation,* respectively. The movement of both higher and lower planets from the west to the east is called as forward movement, and that from the east to the west is the reverse one, and these movements alternate by certain cycles.

The question of where exactly the Solar system ends and the interstellar space begins has not still been finally solved, as its border depends on the influence of two various phenomena, namely, the solar wind and solar gravitation. Even outside of the *heliopause,* the Sun can keep other space objects by its gravitation (the heliopause is the border separating interstellar space from the Solar system substance).

The radius of the Sun is 7×10^8 m, the volume 1.4×10^{27} m³, the surface area 6×10^{18} m², the mass 2×10^{30} kg, the average density 1.4 mg/m³, the density at the center 160 mg/m³, the pressure at the center 3.4×10^{16} Pa (1 Pa is 10^{-5} atm), the total radiation (power) 3.8×10^{26} J/s, the average surface temperature is about 6,000 K, the temperature at the center ca. 10^7 K, the speed of movement relative to next stars 19.5 km/s, the distance from the Galaxy center about 30,000 l.y., the speed of rotation round the Galaxy center 250 km/s, the cycle time round the Galaxy center 200 million years, the age of the Sun is about 5 billion years. The chemical composition (by mass): ca. 70% of hydrogen atoms, 27% of helium atoms, 3% of other element atoms (Table 11.1).

TABLE 11.1 Some Characteristics of the Solar System Planets

Planet	Average distance from the Sun		Rotation Period	Cycle time on the equator	Equatorial diameter, km	Mass (in Earth's mass)	Average substance density, mg/m³	Number of satellites	Diameter (relative to the Earth)
	mln km	*a. u.*							
Mercury	57.9	0.39	87.97 days	58.6 days	4–878	0.055	5.7	–	0.38
Venus	108.2	0.72	227.70 days	243 days*	12–104	0.805	4.95	–	0.95
Earth	149.6	1.00	365.26 days	23 h 56 min	12–756	1.00	5.5	1	1.0
Mars	227.9	1.52	686.98 days	24 h 37 min	6–974	0.106	3.94	2	0.53
Jupiter	778.3	5.20	11.86 days	9 h 50 min	142,600	314.03	1.33	12	11.1
Saturn	1,427	9.54	29.46 years	10 h 14 min	120,200	94.01	0.70	16	9.4
Uranium	2,870	19.18	84.01 years	~20 h*	53,000	14.4	1.49	5	4.0
Neptune	4,496	30.06	164.81 years	~20 h	49,500	17.0	2.09	2	3.9
Pluto	5,946	39.75	247.7 years	6.39 days	4,000	0.002	0.4	1	0.47

*The rotation direction is opposite to the revolution direction.

11.4   COSMOLOGICAL CHARACTERISTICS OF THE MOON

The Moon is the sole natural satellite of Earth and the closest celestial body.

11.4.1   ORIGIN OF THE MOON

Three hypotheses of the Moon's origin are most known. At the end of the nineteenth century, J. Darwin put forward a hypothesis according to which the Moon and Earth originally constituted one common molten mass whose speed increased in the process of its cooling and compression; as a result, this mass became torn onto two parts: the big one (Earth) and the small one (the Moon). This hypothesis explains the small density of the Moon (formed of the external layers of the initial mass) but meets serious objections from both the viewpoint of the mechanism of this process and owing to essential geochemical distinctions between the terrestrial shell rocks and the lunar ones. According to the "capture" theory offered by Weitzekker, Alfvén, and Urey, the Moon originally was a small planet which, when passing near Earth, due to the terrestrial gravitation, turned into a satellite of Earth. Special calculations have shown that the probability of such an event is very low. According to the third theory developed by Schmidt in the mid-twentieth century, the Moon and Earth were formed simultaneously by association and consolidation of a large number of small particles. But, as the Moon has, on the average, a smaller density than Earth, the substance of the protoplanetary cloud should have been divided to form a considerably higher concentration of heavy elements in Earth. From whence it follows that Earth surrounded with a powerful atmosphere started to be formed first, and, at subsequent cooling, the substance of this atmosphere condensed into a ring of *planetesimals*, which the Moon was formed of. This theory is most developed scientifically.

11.4.2   SOME PHYSICAL CHARACTERISTICS OF THE MOON

The shape of the Moon is very close to a sphere with a radius of 1,737 km, which is 0.2724 of the equatorial radius of Earth. The surface area of the Moon is 3.8×10^7 sq. km, the volume 2.2×10^{25} cm^3. More detailed estimation of the geometrical shape of the Moon is complicated by the absence of oceans on it and there being no obviously expressed level surface to be a reference for measurements of heights and depths. Besides, as the Moon is always turned to Earth by one side, it is obviously possible to measure the radiuses of points on the visible hemisphere surface of the Moon (except the points on the very edge of the lunar disk) from Earth on the basis of the weak three-dimensional effect caused by *libration* only (see below). Studying of librations has allowed estimating the difference of the principal semiaxes of the Moon ellipsoid. Under the influence of tidal forces, the Moon is a little extended towards Earth. The polar axis is shorter than the equatorial ones directed towards Earth and perpendicular to the direction to Earth, by approx. 700 and 400 m, respectively. The mass of the Moon was more precisely determined from observation from artificial Earth satellites; it is 81 times less than the mass of Earth, that is, 7.35×10^{22} kg. The average density of the Moon is 3.34 g/cm^3. The gravity acceleration on the Moon surface is 6 times less than on Earth, that is, 162.3 cm/s^2 and decreases by 0.187 cm/s^2 when lifting by 1 km.

11.4.3 MOVEMENT FEATURES OF THE MOON

The Moon moves round Earth with an average speed of 1.02 km/s on an approximately elliptic orbit in the same direction in which the vast majority of other bodies of the Solar system do, that is, counterclockwise if taking a detached view of the Moon orbit from the North Pole of the world. Owing to its elliptic orbit and perturbations, the distance from the Moon varies between 356,400 and 406,800 km. The Moon cycle time round Earth, the so-called *sidereal* (stellar) month, is 27.32166 days, but is also subject to small variations. Studying of the Moon movement is one of the most difficult problems of celestial mechanics. Elliptic movement represents a certain approximation only since some processes caused by the attraction of the Sun and planets (including the Earth) are imposed on it. The Sun's attraction of the Moon is 2.2 times stronger than Earth's one, so, strictly speaking, one should consider the Moon movement round the Sun and Earth's influence on it. However, of interest is usually the Moon movement as seen from Earth and, consequently, the principal gravitational theory (offered by J. Hill) describes the Moon movement exactly round Earth.

The plane of the Moon orbit is inclined to the ecliptic at an angle of 5°8'43," subjected to small variations. The *ecliptic is* the big circle of the celestial sphere, which the visible annual moving of the Sun center proceeds along. The ecliptic plane is inclined to the plane of the celestial equator at an angle of 23°27' and is crossed with the celestial equator at the points of the spring and autumn equinox. *The celestial equator* is the big circle of the celestial sphere representing the line of its crossing by the plane perpendicular to the so-called axis of the World round which the visible rotation of the celestial sphere proceeds. The points of intersection of the Moon orbit with the ecliptic are called as the ascending and descending knots, have nonuniform movement, and make a complete turn on the ecliptic in 6,794 days (about 18 years). Thereof, the Moon comes back to the same knot through a time interval, a little shorter than the sidereal month (27.21222 days on the average). The periodicity of solar and lunar eclipses is related with this period. Besides, as the Moon rotates round the axis inclined to the ecliptic plane at an angle of 88°28,' and with the period precisely equal to a sidereal month, it is always turned to Earth by the same side. The combination of uniform rotation with nonuniform movement along the orbit leads to small periodic deviations from the invariable direction to Earth. So, it is possible to see only 59% and less of the whole lunar surface from

Earth at different times; such deviations are called the *libration* of the Moon. The planes of the lunar equator, of the ecliptics, and of the lunar orbit are always crossed in one straight line.

11.4.4   PHASES OF THE MOON

Without being self-shining, the Moon is only visible where either sunshine or the rays reflected by Earth fall. This explains the Moon phases. Every month the Moon, moving along its orbit, passes between Earth and the Sun and is turned to us by its dark side; at this time there is a new moon. After this, in 1–2 days a narrow bright sickle of the young Moon appears on the western part of the sky. The other part of the lunar disk at this time is poorly illuminated by Earth turned to the Moon by its day hemisphere. In 7 days, the Moon departs from the Sun by 90°, the first quarter comes, when exactly half of the lunar disk is illuminated and the *terminator* (the line between the light and dark sides) becomes a straight line, that is, a diameter of the lunar disk. In the subsequent days, the terminator becomes convex, the look of the Moon comes nearer to a light circle, and in 14–15 days a full moon comes. On the 22nd day the last quarter is observed. The angular Moon-to-Sun distance decreases, it again gets the shape of a sickle, and in 29.5 days a new moon comes again. The interval between two consecutive new moons is called as the synodic month, whose duration (29.5 days) exceeds a sidereal month, since Earth during this time passes about 1/13 of its orbit and the Moon, to pass between Earth and the Sun again, must additively pass 1/13 of its orbit, which requires some more than two days. If the new moon occurs near one of the corners of the lunar orbit, a solar eclipse is observed, and the full moon near the knot is accompanied by a lunar eclipse. It should be noted that the specified system of the Moon phases was the basis for a number of calendar systems.

11.4.5   MOON'S SURFACE AND ITS RELIEF

The Moon surface is quite dark, its *albedo* being 0.073, that is, it reflects only 7.3% of the light rays of the Sun on the average. The *albedo* is the ratio of the quantity of the radiant energy reflected by a body to the quantity of the energy falling on this body, to characterize the reflective ability of the body's surface. A day on the Moon lasts nearly 1.5 terrestrial days and a night proceeds as much. Without being protected by an atmosphere, the Moon surface heats up to +110°C in the afternoon, and cools down to –120°C at night. However, owing to the very weak heat conductivity of the surface layers, such quite high temperature variations penetrate into few decimeters only. Owing to the same cause, the warm surface is quickly cooled during full lunar eclipses. Visually, some irregularly shaped darkish extended spots are visible on the Moon, which were first mistaken for seas but finally appeared to be plains and ring-shaped mountains (craters). At the end of the 19th – the beginning of the twentieth century, a large atlas of the Moon was published according to the photos obtained by the Parisian observatory; later, a photographic album of the Moon was published by the Lick Observatory (the USA), and in the mid-twentieth century Kuiper compiled some detailed photo atlases of the Moon obtained by means of modern telescopes in various observatories of the world. The relief of the lunar surface was also established as a result of long-term telescopic observations. The "lunar seas" occupying about 40% of the visible Moon surface have, as a rule, a simple cap-like shape of a diameter from 15 to 200 km. The mountain heights have been estimated by the length of their shadows on the surface or photometrically. The compiled hypsometric cards for the most part of the visible Moon side (the 1:1,000,000 scale) are also used to account for the roughness of the edge Moon with the purpose to locate it. The absolute age of the lunar formations (mountains and craters) is precisely known at some points only. However, using some indirect methods, it is possible to establish that the age of the youngest large craters is tens and hundreds millions years, and the majority of large craters appeared 3–4 billion years ago.

Special calculations related with the thermal history of the Moon show that soon after formation its subsoil was warmed up by radioactive heat and considerably melted, which was accompanied by intense volcanic phenomena to form craters and cracks. Besides, at early stages, a very large number of meteorites and asteroids dropped onto the surface of the Moon, at whose explosions craters of different sizes and structures appeared as well. Now, meteorites drop onto the Moon much less often, and the volcanism degree has considerably decreased.

11.4.6 INTERNAL STRUCTURE OF THE MOON

The internal structure of the Moon has been explored as well as Earth's, that is, by the distribution nature of longitudinal and cross-seismic waves. From analysis of the gravitational field of the Moon it has been concluded that its density slightly changes with depth; unlike Earth, no high mass concentration at the center is observed. The upper layer of the visible side of the Moon represents a crust whose thickness is about 60 m. It is very probable that on the reverse side the crust is approx. 1.5 times thicker. In both cases, the crust is made of erupted crystal basalt rocks. However, the basalts of the continental and sea areas have appreciable differences by their mineralogical composition. The most ancient continental regions of the Moon are mainly formed by a light rock (anorthosite), almost completely consisting of plagioclase minerals with small impurities of pyroxene, olivine, magnetite, titan magnetite, etc. The crystal rocks of the lunar seas, like terrestrial basalts, are generally made of plagioclases and monoclinic pyroxenes (augites) formed when cooling the magmatic melt on the surface or near it. As the lunar basalts are less oxidized than the terrestrial ones, this means that they were crystallized with lower oxygen—metal interaction. Besides, in comparison with the terrestrial rocks, a lower content of some volatile elements and the richness by many refractory elements are observed. Due to the impurity of olivines and, especially, ilmenites, the areas of the seas look darker, and the density of the constituent rocks is higher than on the continents. Under the crust there is the mantle in which, similarly to the terrestrial one, it is possible to resolve the upper, middle, and lower ones. The thickness of the upper mantle is about 250 km, that of the middle one is 500 km, and its border with the lower mantle is located at a depth of approx. 1,000 km. To this level, the speeds of cross waves are almost constant; this means that the substance of the subsoil is in a solid state, being a powerful and relatively cold lithosphere, in which seismic vibrations do not fade for long. The composition of the upper mantle is assumingly olivine–pyroxenic, and at deeper depths, spinel and the mineral melilite (occurring in metamorphic and igneous rocks) are present. At the border with the lower mantle, the temperature is close to those intervals of temperatures where strong absorption of seismic waves begins. This area is a lunar asthenosphere. At the center there is a small liquid core of a radius less than 350 km which no cross-waves pass through. The core may be either iron-sulfide or iron; in the latter case it should be less, which agrees with the estimates of the density distribution by depth. Its mass does not possibly exceed 2% of the total Moon mass. The temperature in the core depends on its composition and varies from 1,300 to 1,900 K.

11.5   ANTHROPOCOSMISM CONCEPT

The idea of interrelation of the man and the Universe, put forward in an extreme antiquity, has an old tradition in both western and eastern philosophy. For the early representations of anthropocosmism (Constantine Tsiolkovsky, Vladimir Vernadsky, Alexander Chizhevsky, et al.) a naturalistic aspect is characteristic. The modern general scientific concept of anthropocosmism, being development and generalization of the previous versions, should be constructed on the basis of the sciences of not only space but also the man and his reason. A special contribution can be brought by theoretic and methodological innovations, characteristic of not only the era of cosmos but also the entire period of modernization of such fundamental ideas, which initiate civilization development in connection with the problem of transition to sustainable development. In the most general view, anthropocosmism is understood as the world outlook concept to display both the influence of space on the man and impact of the man on space. In other words, anthropocosmism should coincide in many respects with sociocosmism, the concept describing the relations in the society—Universe system. Wide space exploration is predicted as a number of certain stages of extraterrestrial space industrialization. The man's advance into the Universe will be accompanied by development of the production of goods out of Earth, transition from the "two-dimensional" infrastructure of economic activity to a "three-dimensional" one. Just as since the beginning of Neolithic revolution the production of goods became the basis of accelerated social and economic development, now the production activity on Earth and out of its limits becomes the base of the human penetration into cosmos.

At the initial stage of implementation of the ideas of astronautics (cosmonautics—in Russian) its founders and followers dreamed of the man's exit out of the limits of his planet, of massive resettlement of people into the Universe space. For example, Constantine Tsiolkovsky represented the mankind moving into space as a kind of nomads who has left its native planet and, then, the Solar system as well, wandering over the Universe in searching for new sources of power and negentropy. The scientist treated cosmos as that ecological environment, those processes and forces which could accelerate social progress in the conditions of our planet, would eliminate the threat of geological and, subsequently, cosmic disasters.

At the present stage of space exploration, the use of space means for scientific-technical, social-economic and ecological development on our planet appears most important. The concept showing that mankind must be in the focus of cumulative space activity as the process of space exploration has been named as "anthropogeo-cosmism" or "sociogeocosmism." From this concept it follows that our planet will long be the center of the ecosystem consisting of Earth and its surrounding space. However, further, along with Earth, there will appear other centers of space activity, settlements of people on artificial constructions and natural celestial bodies suitable for life, for example, on the Moon, Mars, the satellites of the Jupiter group planets.

At the same time, in the anthropocosmism concept, not only space exploration by means of the modern tools of astronomy is of great importance but also solving such problems as determination of the man's place in the system of the Universe and its reason in the Universe, finding a genetic and structural link between the man and space. Concretizing the ideas of our Universe as of manned system and space ecomedium of human development (including the noosphere formation), just the anthropic cosmological principle reveals the relation between the global properties and characteristics of the Universe; first of all, this concerns the relation between the processes of self-organizing and emergence of an hierarchy of structural levels of matter whose evolution has led to the appearance of the man. Besides, the modern anthropocosmism concept is closely linked with the problem of extraterrestrial civilizations.

11.6   PROBLEM OF EXTRATERRESTRIAL CIVILIZATIONS

The conclusion of the life and reason existence out of our planet is known to be hypothetical. These questions were discussed by the thinkers of antique Greece and New Era. Nowadays the problem of extraterrestrial civilizations is developed by special sciences, first of all, astronomy. A characteristic feature of modern exploring the problem of plurality (habitability) of cosmic worlds, unlike the post-Copernicus period, is the emergence of special scientific works rather than physiophilosophical ones. It is interesting that this research is focused on the problem of communication with extraterrestrial civilizations (this problem is abbreviated as CETI, Communication with ExtraTerrestrial Intelligence). However, approximately since the late 1970s, a new term SETI (Searching for ExtraTerrestrial Intelligence) appeared. The replacement of "communication" by "search" is quite justified, since it is necessary to find extraterrestrial civilizations first.

Since the time of Giordano Bruno, essential changes have occurred in the development of his philosophical ideas of the Universe and mankind. However, the experts engaged in the problem of extraterrestrial civilizations are poorly informed in the field of philosophy and social knowledge. In their search they, as a rule, are guided by a technoscience approach, paying much attention to the natural-scientific, astronomical conditions of the possible existence of extraterrestrial civilizations and technical aspects of communication with them. The role of economic and ecological factors in civilizational processes in the Universe is almost not considered.

The problem of extraterrestrial civilizations and communication with them is complex, interdisciplinary, and general-scientific. Development of the perspective of extraterrestrial civilizations essentially depends on the possibility of revealing regularities and tendencies of development inherent not only in our terrestrial civilization but also in other expected civilizations in space. During the development of this problem, two main problems are to be solved. The first problem consists in assistance to the progress of those sciences and solving those "terrestrial" problems of the mankind development which proceed from the ideas of our terrestrial civilization as about a systematic and complete developing object transforming into the cosmo-noosphere. The second problem is searching for extraterrestrial civilizations, their detection, and contact establishment. If the first problem is urgent just now, the second one appears more fundamental and its solution seems more distant in time. The conceptual character of this problem is caused not only by the basic possibility of contact establishment to other worlds but also by the need of deeper studying the prospects of the survival and development of the terrestrial mankind, revealing the regularities and prospects of noospherogenesis on our planet and beyond its limits. Just this globoterrestrial aspect prevails in the modern research concerning general characteristics and social regularities of the development of expected extraterrestrial civilizations.

Some scientists have an opinion that the solution of the problem of communication with extraterrestrial civilizations can give the chance of using their experience and knowledge. However, the situation is absolutely different: the terrestrial civilization is an information model, which the man gets hypothetical knowledge of extraterrestrial civilizations from. Until this knowledge gains a corresponding "space interpretation," it would be fondly to rely on the help of other civilizations. At solving any questions of this problem, the mankind surely proceeds (obviously or implicitly) from the life and civilization in space resembling the terrestrial ones. This is the basic methodological and epistemic-logical principle of every research, and any deviation from it would mean transfer of scientific research to the area of unscientific fantasy. Therefore, it is now the sole effective principle of the probabilistic-theoretical increment of our knowledge of extraterrestrial civilizations. When studying the logic and semiotics problems with the purpose to design a language for interstellar transfers, further enrichment and generalization of sign systems happens. Besides, it has been established that any originally created information means are of substantially anthropomorphous character and expect the communicant having the concepts developed by the people.

The ideas being developed in the works investigating the development regularities of extraterrestrial civilizations (physical, technical, social-economic, ecological, etc.) very much remind the ideas of global problems of the scientific and technical revolution, and especially global modeling. At treatment of the development problem of space civilizations, such quantitative characteristics are analyzed as settling space, the total number of population, the area and volume population density, the total consumed energy, the energy per consumer, the volume of accumulated information, its quantity per consumer etc. The results obtained at the development of global problems, "cleared" from terrestrial specifics and separate tactical nuances, can be used in research of the social aspect of the extraterrestrial civilization problem. The principal breakage with geocentrism, understanding of that the future of our civilization, positive solutions of global problems cannot anymore do without space exploration and the formation of the cosmo-noosphere is indicative in this regard. But until extraterrestrial civilizations are found, we can take information from purely theoretical contact to them as hypothetical objects, space copies of our civilization, which have made "great transition" to the cosmo-noosphere. Analysis of this problem allows us to deepen and expand our ideas of the *noospherogenesis* process at the cosmic development stage. It is a feature of the noospheric-methodological orientation: not only the outlook development of the world's scientific picture but also a certain social efficiency, urgency, the humanistic orientation of the search to comprehensive studying of the man and society.

If one proceeds from the possibility of the existence of extraterrestrial civilizations, each of them cannot be considered as an especially independent civilizational process covered with only general regularities together with other representatives of the social development stage. They can form the basis for a new community in the sense of integrity. It is a question of such association of civilizations into a uniform system (*the astroinfonoosphere*) in which the social stage of matter development will fully show its space nature. The prospective social evolution of matter will represent not only the purely planetary independently proceeding noospheric-civilizational process independent of other social space. Such autonomy, if any, takes place only till a certain historical period of the development of the noospheric civilization exploring cosmos in broad scales. In any case, civilizations of the noospheric development level steadily exist in space, sooner or later they will necessarily come into contact and form an information-noospheric system.

11.7   DARK MATTER AND UNIVERSAL EVOLUTION

At the end of the twentieth century and the beginning of the twenty-first century, scientific discoveries occurred which allow us to look at evolution processes in a new way. Our Universe is supposed not to be lonely and not to any more represent a Universum containing all the real. What was earlier called the Universum is now considered one of *miniuniverses*. Expansion of our representations of the evolution problem and an essentially new concept of universal evolution are substantially caused by the discovery of new and mysterious substances, namely: *dark energy* and *dark matter*. Before the discovery of dark energy, *hidden substance* (or *dark mass*) was ascribed to dark matter. If this second form of matter occupies about 20% of the mass and energy of the Universe, then the whole modern science occurs to study only few percent's of the whole material content of the Universe. Therefore, the modern scientific picture of the world must represent a systematic and complete image of both inanimate and animate nature, the man and mankind, obtained on the basis of synthesis and generalization of the whole scientific knowledge. In spite of the fact that since the second half of the twentieth century the ideas of a universal evolutionism lie at the heart of the general scientific picture of the world, from time to time this or that science unexpectedly makes a contribution which appears very principal for understanding of the Universe and the human role therein.

11.7.1   BASIC FORMS OF BEING AS WAYS OF MATTER SELF-PRESERVATION

As was told above, philosophy has always aspired to investigate the universe in its integrity and at various levels, to reveal the most fundamental and limiting primary origins and life bases therein. At the physiophilosophical stage, it aspired to get into nature's secrets, to find eternal and passing, steady and changeable, universal and specific in it. The interest to these questions began to be lost with the development of separate sciences, and the philosophy of science focused its attention on the philosophical and methodological analysis of topical problems and on the logic aspects of the philosophy—science relationship.

At the end of the twentieth century, the so-called dark energy was discovered and active search for experimental (generally observant) confirmations of its existence began. This unusual form of space matter being still poorly blends with the "material and evolutionary" scientific picture of the universe being created recently. Science accepts the hypothesis (though in a debatable form) about the existence of a very stable part of the Universe which, in the generalized look, we will call *dark matter* consisting of two main forms, namely, *dark energy* and *dark mass* (Table 11.2). The definition of the notion of "dark matter" is ambiguous. Some researchers believe that this notion, in general, is hardly applicable since there is already "dark substance" (dark mass) which, as well as the term "dark energy," is really dark because radiates no light.

TABLE 11.2 Fundamental Forms of Matter Being in the Universe

Material (observable) part of the Universe —baryonic matter: 3% of the world energy density;	Dark energy (space vacuum): 67% of the world energy density;	Dark mass (hidden substance): 30% of the world energy density.
Average density of mass (stars, molecular hydrogen clouds, etc.) 2×10^{-31} g/cm³	Average density of mass 7×10^{-30} g/cm³ (identical over the whole universe)	Average density of mass 2×10^{-30} g/cm³
Characteristics of development	Characteristics of development	Characteristics of development
1. Evolves (remains through evolution).	1. Does not change and does not evolve on the Big Bang, evenly fills the whole universe.	1. Changes but does not evolve as normal substance.
2. Obeys the law of universal gravitation.	2. Possesses the property of antigravitation, causing accelerated expansion of the Universe.	2. Obeys the law of universal gravitation.
3. Expands for about 7 billion years with acceleration.	3. The composition and structure are unknown; it is supposed as a homogeneous formation.	3. Consists of very heavy particles of unknown nature which poorly interact among themselves and with usual substance.
	4. Dark energy is influenced by neither substance nor dark mass, but a back influence has been established.	

They prefer a different name for dark energy, namely, "space vacuum." By different estimates, the dark components constitute 96–97% of the total material content of the Universe. They are invisible unlike the shining (or seen) Universe and remain, to some extent, in a poorly or generally nonevolutionary form against the distinctly expressed evolution of substance. Dark energy resists to gravitation and allows, owing to the "world antigravitation" property, the Universe to extend with acceleration. It is supposed that space vacuum, influencing the Universe expansion (antigravitation), is a rather stable and not changing form of matter influenced by nothing. By modern representations, the property of self-preservation over evolution prevails in this form of matter being. Many scientists assume that space vacuum could have undergone several phase transitions before the event called the Big Bang occurred. Therefore, a certain transformation of vacuum passed through several phases could precede the Big Bang.

So, dark matter, which is unobservable and does not radiate, consists of two parts essentially different in their nature, namely, the antigravitating dark energy with a density of about 67–75% of the total energy density of the Universe, and the "hidden substance" whose density is about 23–25% of the total energy density of the Universe, exceeding the density of usual observable substance (i.e., the "shining" Universe) by 5–6 times. On the share of the latter there is only 3–4% of the universal energy density. Dark mass, being exposed to gravity forces, interacts with neither substance nor radiation; it absorbs nothing and does not shine, however, some change processes distinct from substance evolution are quite distinctly traced therein. The development of gravitational inhomogeneities in dark mass during the period before electron recombination speaks for this. There was also a decrease in the dark mass density due to which universal gravitation ceased to dominate during a certain period, having conceded to the forces of "universal antigravitation." Probably, "hidden substance" consists of very long-living components of a very high density (from neutron stars, cooled-down white and brown dwarfs, black holes, including relic ones, and other components arisen as long ago as at the Big Bang).

Thus, the modern cosmological (still mainly hypothetical) picture of the world gives us unknown or poorly studied forms of matter self-preservation which is the cause and source of the existence of stars in a galaxy, congestions of galaxies, super congestions, and other forms of baryonic matter (quarks, bosons, leptons) which constitutes the observable Universe.

The nature of self-preservation of large-scale structures of the material world is related with the stability of our Universe during many billions years. This self-preservation as the dominating and fundamental component of matter being assumes some asymmetry between the dark and baryonic forms of matter. From the viewpoint of modern natural sciences it is important to answer the question: how can matter self-preserve without evolution and, the more so, without any changes? To answer now this question is essentially difficult because neither thermodynamics nor synergetic representations allow constructing a model of such matter self-preservation during nearly one and a half tens billions years. The more so that it is the dominating part of the Universe (nearly three quarters of its mass and power contents). If dark energy remains throughout many billions years, either still unknown laws of preservation must act therein or it is necessary to assume the total absence of movement there, characteristic of material matter. This means that, according to modern synergetic representations (applicable to the material part of the Universe), there exists an unusual way of matter preservation in space vacuum (the more so that dark energy is not self-organizing substance).

11.7.2 DARK MATTER INFLUENCE ON GLOBAL EVOLUTION

Approx. in the first 300,000 years after the Big Bang, dark mass generated gravitational heterogeneity of the substance distribution in the Universe which galactic congestions and galaxies were further formed in. Without its influence, global evolution would be impossible, since galaxies could not be formed in the absence or insufficient quantity of dark mass. In the first minutes and hours after the Big Bang, the distribution of common substance in the Universe was uniform and remained such till proton—electron recombination at the age of the Universe of about 270,000 years. The gravitational condensation of common substance was interfered by the pressure of radiation, which this substance intensely interacted with. Dark mass did not interact with radiation and nothing interfered with the formation of gravitational condensation. A certain structure of inhomogeneities was already formed in the Universe at that time, consisting of gravitating dark mass only. After electron recombination, common substance simply fell into the gravitational potential holes prepared by dark mass. If the dark mass did not have time to form "dark" *protogalaxies*, then galaxies could not be further formed from common substance, and this substance would dissipate in the Universe. Moreover, both modern galaxies and their congestions cannot exist out of the potential holes formed by dark mass. It is not excluded that dark mass can interact with common substance not only through gravitation. Interaction of this part of dark matter with common substance through weak interaction (as in the case with neutrino) is also probable.

Until recently, at discussion of the global evolution problems no questions arose concerning the role of dark matter in this process. Generally, it was spoken that some global characteristics, first of all, the main physical constants corresponding to the four types of fundamental interactions, their fine tuning, and some already known parameters of the Universe (the dimension of space and time, topology, etc.) are such which allow the evolution process, including the global evolution with the man on its top now. At the same time, the existing dark energy as space vacuum with a constant and not changing energy density makes very essential impact on the processes of evolution of the Universe's material part. Predominating in our Universe, dark energy surpasses three times all the other forms of space matter taken together by energy density, creating powerful world antigravitation. With the Universe age of 6–8 billion years, the era of cosmological expansion with acceleration began because the dark mass density gradually decreased and became lower than the vacuum density. This antigravitational expansion of the Universe replaced the cosmological era of gravitation domination over antigravitation and the material forms of matter over vacuum (dark energy).

The question of the world invariance was discussed since Einstein who believed the Universe to be invariable and static. But in 1917, having applied his general relativity theory to cosmology, Einstein unexpectedly found that the cosmological model created by him did not confirm the eternity, invariance, and static character of the Universe. To save his ideas of the static character and invariance of the Universe, he therefore introduced a so-called cosmological constant as one of the fundamental physical constants. The first cosmological model of the world offered by Einstein represented an ideally symmetric model of the Universe in space and time. Such an idea remained till Edwin Hubble's 1929 discovery of space expansion, the galaxies moving off from each other.

Thus, the idea of evolution and global evolution in the general scientific picture of the Universe, in connection with the discovery of dark sides of the Universe, is essentially transformed and an essentially new world outlook is formed. That type of evolution, which was often called as universal evolution, appears not so universal as considered quite recently.

11.7.3   COGNITION PROBLEM OF DARK MATTER

The cognition problem of the dark forms of matter has its own specific features. The *black hole* is one of such components of dark mass, which is a closed sphere formed as a result of collapse of a massive star and the trapped substance having got into it cannot leave it even as radiation because of enormous compression (gravitational attraction). Hence, the black hole is understood as a "space—time" area where the second space speed is equal to the velocity of light and the gravitational field does not let out even photons. The spatial border of a black hole was named as *the event horizon* outside of which getting no information of any events and conditions in this black hole is possible. Any space body, substance, and radiation are capable to get in, fall into a black hole, but cannot leave it. This means that the cognizing subject cannot get any information of the internal state of a black hole when being out of it. Especially it applies to the central part of a black hole, to singularity as a super-dense state where, as is supposed, no known laws of nature are obeyed. However, the space matter making a black hole does not lose the whole information since it can be characterized by the mass, electric charge, and the own moment of momen-tum. Other characteristics and properties of the matter constituting a black hole can further be found if it really contains information. Therefore, no matter what a black hole has been formed of, inside it the vast majority of the earlier available distinc-tions disappear in the superdense homogeneous medium ($\sim 10^8$ t/cm^3).

While the interior of a black hole cannot be perceived by the external observer, the existence of this latent space object can be revealed, in particular, owing to the enormous gravitational fields representing powerful potential energy sources which, in principle, can be released when substance gets onto the event horizon of the black hole. At this loss (accretion) of substance on the black hole, very large amount of energy can be radiated before it crosses the border (the event horizon) of the black hole (the existence of an X-ray aura round it is possible). Certainly, if there is no substance and radiation in the vicinity of the black hole, it will remain unfound. If in the vicinity of a black hole there is substance and it falls on the black hole, the effect of radiation emission (as though particles take off from this hidden space object) is simulated for the external observer. At getting of the substance of stars and gas clouds into supermassive black holes, their brightness will appear the most intense in our Universe as large energy quantities (by two orders of magnitude higher than those in nuclear reactions) are released in this case. Such supermassive black holes with most powerful gravitation can serve as "dark" (hidden) power sources in a number of space processes with huge power yield and luminosity (quasars, the cores of active galaxies). A black hole with its mass of about 3 million solar masses is apparently in the core of our Galaxy.

The cognition problem of dark matter and other hidden exotic superdense space objects is a special gnoseological problem because the reliability of the results of scientific search is estimated from indirect, side signs of their influence on the usual shining matter. However, analogies between the physics of black holes and thermodynamics, and between them and the gravitation theory are used in cosmological modeling. The cognition of space vacuum is also complicated by its interacting with nothing, including the observer, though its influence on substance on the scale of the Universe not simply takes place but is determining for the future universe. The evolution processes and related information processes play an important role not only in the Universe but also in its cognition by the man. The relation between the Universe and its properties has found reflection in the anthropic cosmological principle, which is realized in the observable Universe. The main idea of this principle is that the man and the material Universe assume their mutual existence.

KEYWORDS

- **Cosmology**
- **Evolution**
- **Galaxy**
- **Moon**
- **Planet**
- **Properties**
- **Star**
- **Structure**
- **Sun**

REFERENCES

1. Arkhagelskaya, I. V., Rozental, I. L. & Chernin, A. D. (2007). *Cosmology and Physical Vacuum*. Moscow, [Russ].

2. Cherepashcuk, A. M. & Chernin, A. D. (2005). Universe, Life, Black Holes. Friasino. [Russ].
3. Chernin, A. D. (2005). Cosmology: *the Big Bang. Friasino*. [Russ].
4. Khvan, N. P. (2006). *Furious Universe*: from the Big Bang to Accelerated Axpensions, from Quarks to Superstrings. Moscow. [Russ].
5. Uspensky, G. R. (1994). *Cosmonauties of the twenty-first century*. Moscow. [Russ].
6. Goldfein, M. D., Ursul, A. D., Ivanov, A. V. & Malikov, A. N. (2011). *Fundamentals of Natural Science World View*. Saratov. [Russ].

CHAPTER 12

A COMMENTARY ON THE CONCEPTS OF THE PHYSICAL CHEMISTRY OF POLYMERS IN TECHNOLOGIES AND ENVIRONMENT PROTECTION

M. D. GOLDFEIN and N. V. KOZHEVNIKOV

CONTENTS

ABSTRACT

In this chapter, results of researches of kinetics and the mechanism of the radical polymerization proceeding on air and in oxygen-free conditions, in mass, solution and an emulsion, in the presence of components of salts of metals of variable valence, aromatic amines, cupferronates and stable radicals are provided. The specified substances can be used as effective stabilizers of monomers. Scientific bases of technology of synthesis high-molecular flocculent for cleaning natural and waste water, technology of receiving rigid polyurethane foam heat insulation and technology of synthesis of the polymeric latex which isn't containing the surface active agents are developed. The key kinetic parameters of the studied responses are determined.

Physical chemistry is a most important fundamental science in the modern natural sciences. The concepts of physical chemistry qualitatively and quantitatively explain the mechanisms of various processes, such as the reactions of oxidation, burning and explosion, obtaining food products and drugs, oil hydrocarbon cracking, making polymeric materials, biochemical reactions underlying metabolism and genetic information transfer, etc. Physical chemistry comprises notions of the structure, properties, and reactivity of various substances, free radicals as active centers of chain processes proceeding in both mineral and organic nature, the scientific foundations of low-waste and resource-saving technologies. The design and wide usage of synthetic polymeric materials is a lead in chemistry. This results in the emergence of new environmental problems due to pollution of the environment with these materials and the wastes of their production and monomer synthesis. The presented below results of our studies of the kinetics and mechanism of polymerization of vinyl monomers which can proceed in quite various conditions point to their usability in the scientific justification of optimization of technological modes of monomer and polymer synthesis, and at solving both local and global environmental problems.

To obtain reliable experimental data and to correctly interpret them, we used such physicochemical and analytical techniques as dilatometry, viscometry, UV and IR spectroscopy, electronic paramagnetic resonance, light scattering spectroscopy, electron microscopy, and gas-liquid chromatography. To analyze the properties of polymeric dispersions, the turbidity spectrum method was used, and the efficiency of flocculants was estimated gravimetrically and by the sedimentation speed of special suspended imitators (e.g., copper oxide).

12.1   EFFECT OF SALTS OF METALS OF VARYING VALENCY

Additives of the stearates of iron (IS), copper (CpS), cobalt (CbS), zinc (ZS), and lead (LS) within a certain concentration range were found to increase the polymerization rate of styrene and methylmethacrylate (MMA) in comparison with their thermopolymerization. By initiating activity, they can be arranged as LS < CbS < ZS

< IS < CpS. The decrease in the values of the effective activation energy, the activation energy of reaction initiation, and the kinetic reaction order by monomer point to the active participation of the monomer in chain initiation. IR spectroscopy shows that an intermediate monomer–stearate complex is formed which then decomposes into active radicals to initiate polymerization. The benzoyl peroxide (BP)–IS (or CpS) systems can be used for effective polymerization initiation. A concentration inversion of the catalytic properties of stearates has been found, which depends on salt concentration and conversion degree. The efficiency of the accelerating influence decreases with increasing temperature, the BP–IS system possesses the highest initiating activity. The initiating mechanism for these systems is principally different that the redox one. It follows from experimental data (color changes before and in the course of polymerization, the absorption and IR spectra of reactive mixtures, electron microscopy observations, etc.) that initiation occurs due to stearate radicals formed at decomposition of the complex consisting of one BP molecule and two stearate ones.

The phenomenon of concentration and temperature inversion of the catalytic properties of gold, platinum, osmium and palladium chlorides at thermal and initiated polymerization of styrene and MMA has been discovered. The mechanism of ambiguous action of noble metal salts is caused by the competition of the initiating influence of monomer complexes with colloidal metal particles and the inhibition reaction proceeding by ligand transfer.

12.2 EFFECT OF SOME ORGANIC SOLVENTS

The influence of acetonitrile (ACN) and dimethyl formamide (DMFA) on the radical polymerization of styrene and (met)acrylic esters initiated with azo isobutyronitrile (AIBN) or BP was studied. Basic characteristics of the gross kinetics of polymerization in solution have been found. The reaction order by initiator concentrations is always 0.5, which points to bimolecular chain termination. The reaction order by monomer concentration varies within rather wide limits (above or below unity) depending on the chemical nature of the solvent, monomer, and initiator. The lowest (0.83) and highest (1.6) values were found for BP-initiated polymerization of MMA in DMFA solution and for AIBN-initiated polymerization of styrene in CAN solution, respectively. Such a high value of the order by monomer is caused by the abnormally low rate of AIBN decomposition in ACN, which, in turn, is explained by the donor-acceptor interaction of the alkyl and nitrile group of the initiator with the nitrile and alkyl group of the same solvent molecule. The influence of DMFA and CAN on all the elementary stages of polymerization has also been ascertained, which manifests itself in the dependences of the rate constants of the reactions of initiation, propagation, and termination on the monomer concentration. This is caused by such factors as changes in the initiator decomposition rate, macroradical solvation with the molecules of an electron-donor solvent, diffusional-controlled chain termination, and conformational changes of macromolecules in solution.

12.3   HOMOPOLYMERIZATION AND COPOLYMERIZATION OF ACRYLONITRILE IN AN AQUEOUS SOLUTION OF SODIUM SULFOCYANIDE

When acrylonitrile (AN)-based fiber-forming polymers are obtained, a spinning solution ready for fiber formation appears as a result of polymerization in some solvent. Organic solvents or solutions of inorganic salts are used as solvents in these cases. First, a comparative study was made of the kinetics and mechanism of polymerization of AN in DMFA and in an aqueous solution of sodium sulfocyanide (ASSSC) initiated with AIBN and some newly synthesized azo nitriles (azobiscyanopentanol, azobiscyanovalerial acid, azobisdimethylethylamidoxime). The polymerization rate in ASSSC turns out to be significantly higher in comparison with the reaction in DMFA, in spite of the initiation rate in the presence of the said azo nitriles being more than 1.5 times lower than that in DMFA. The lower initiation rate in ASSSC is associated with stronger manifestation of the "cell effect" due to the higher viscosity of the water–salt solvent and the ability of water to form H-bonds (which hinders initiation). The ratio of the rate constants of chain propagation and termination ($K_p/K_t^{0.5}$) was found to be ca. tenfold higher in ASSSC than in DMFA. The molecular mass of the polymer formed is correspondingly higher. These differences are caused by the influence of medium viscosity on K_t and formation of H-bonds with the nitrile groups of the end chain of a macroradical and the added monomer. But sodium sulfocyanide (which forms charge-transfer complexes with a molecule of AN or its radical to activate them) mainly contributes into increasing K_p.

The kinetics of AN copolymerization with methylacrylate (MA) or vinylacetate (VA) in ASSSC is qualitatively analogous to AN homopolymerization in identical conditions. At the same time, the initial reaction rate, copolymer molecular mass, effective activation energy (E_{ef}), orders by initiator and total monomer concentration differ. For example, the decrease in E_{ef} is due to the presence of a more reactive monomer (MA), and in the AN–VA system the nonend monomer chains in macroradicals influence the rate constant of cross chain termination. For these binary systems, the copolymerization constants were estimated, whose values point to a certain mechanism of chain growth, which leads to the MA concentration in the copolymer being significantly higher than in the source mixture. The same is observed in the case of AN with VA polymerization (naturally, the absolute amount of AN in the final product is much higher than MA or VA due to its higher initial concentration in the mixture).

Obtaining synthetic PAN fiber of a nitrone type is preceded by the formation of a spinning solution by means of copolymerization of AN with MA (or VA) and itaconic acid (IA) or acrylic acid (AA) or methacrylic acid (MAA) or methallyl sulfonate (MAS). Usually, the mixtures contain 15% of AN, 5–6% of the second monomer, 1–2% of the third one, and ca. 80 wt. % of 51.5% ASSSC. When a third monomer is introduced into the reaction mixture, the rate of the process and the molecular mass of the copolymer decrease, which makes these comonomers be peculiar low-effective inhibitors. In such a case, it becomes possible to estimate the inhibition (retardation) constant which is, in essence, the rate constant of one of the reactions of chain propagation. At copolymerization of ternary monomeric systems based on AN in ASSSC, the chain origination rate increases with the total monomer concentration. The initiation reaction order relative to the total monomer concentration varies from 0.5 to 0.8 depending on the degree of the retarding effect of the third monomer. Besides, AN in the three-component system is shown to participate less actively in the chain origination reaction than at its homopolymerization and copolymerization with MA or VA.

Thus, the obtained results enable regulating the copolymerization kinetics and the structure of the copolymer formed, which, finally, is a way of chemical modification of synthetic fibers.

12.4   STABLE RADICALS IN THE POLYMERIZATION KINETICS OF VINYL MONOMERS

Free radicals are neutral or charged particles with one or more uncoupled electrons. Unlike usual (short-living) radicals, stable ones (long-living) are characteristic of paramagnetic substances whose chemical particles possess strong delocalized uncoupled electrons and sterically screened reactivity centers. This is the very cause of the high stability of many classes of nitroxyl radicals of aromatic, fatty-aromatic and heterocyclic series, and ion radicals and their complexes.

Peculiarities of thermal and initiated polymerization of vinyl monomers in the presence of anion radicals of tetracyanoquinodimethane (TCQM) were investigated. TCQM anion radicals are shown to effectively inhibit both thermal and AIBN-initiated polymerization of styrene, MMA, and methylacrylate in acetonitrile and dimethylformamide solutions. Inhibition is accomplished by the recombination mechanism and by electron transfer to the primary (relative to the initiator) or polymeric radical. The electron-transfer reaction leads to the appearance of a neutral TCQM, which regenerates the inhibitor in the medium of electron-donor solvents. Our calculation of the corresponding radical-chain scheme has allowed us to derive an equation to describe the dependence of the induction period duration on the initiator and inhibitor concentrations, and how the polymerization rate changes with time. The mechanism of the initiating effect of the peroxide—TCQM system has been ascertained, according to which a single-electron transfer reaction proceeds between an anion radical and a BP molecule, with subsequent reactions between the formed neutral TCQM and benzoate anion, and a benzoate radical and one more anion radical TCQM. Free radicals initiating polymerization are formed at the redox interaction between the products of the said processes and peroxide molecules. The TCQM anion radical interacts with peroxide only at a rather high affinity of this peroxide to electron (BP, lauryl peroxide); in the presence of cumyl peroxide, the anion radical inhibits polymerization only.

Iminoxyl radicals which are stable in air and are easily synthesized in chemically pure state (mainly, crystalline brightly colored substances) present a principally new type of nitroxyl paramagnets. Organic paramagnets are used to intensify chemical processes, to increase the selectivity of catalytic systems, to improve the quality of production (anaerobic hermetics, epoxy resins, and polyolefins). They have found application in biophysical and molecular-biological studies as spin labels and probes, in forensic medical diagnostics, analytical chemistry, to improve the adhesion of polymeric coatings, at making cinema and photo materials, in device building, in oil-extracting geophysics and defectoscopy of solids, as effective inhibitors of polymerization, thermal and light oxidation of various materials, including polymers.

In this connection, systematic studies were made of the inhibiting effect of many stable mono- and polyradicals on the kinetics and mechanism of vinyl monomer polymerization. The efficiency of nitroxyls as free radical acceptors has promoted their usage to explore the mechanism of polymerization by inhibition. Usually, nitroxyls have time to react only with a part of the radicals formed at azonitrile decomposition, and they do not react at all with primary radicals at peroxide initiation. Iminoxyls have been found to terminate chains by both recombination and disproportionation in the presence of azo nitriles. Inhibitor regeneration proceeds as a result of detachment of a hydrogen atom from an iminoxyl by an active radical to form the corresponding nitroso compound. The mechanism of inhibition by a nitroso compound is addition of a growing chain to a $-N=O$ fragment to form a stable radical again. The interaction of iminoxyls with peroxides depends on the solvent type. For example, in vinyl monomers, induced decomposition of benzoyl peroxide occurs to form a heterocyclic oxide (nitrone) and benzoic acid. In contrast to iminoxyls, aromatic nitroxyls in a monomeric medium interact with peroxides to form nonradical products. Imidazoline-based nitroxyl radicals possess advantages over common azotoxides, which are their stability in acidic media (owing to the presence of an imin or nitrone functional groups) and the possibility of complex formation and cyclometalling with no radical center involved.

12.5   MONOMER STABILIZATION

The practical importance of inhibitors is often associated with their usage for monomer stabilization and preventing various spontaneous and undesirable polymerization processes. In industrial conditions, polymerization may proceed in the presence of air oxygen and, hence, peroxide radicals MOO· serve active centers of this chain reaction. In such cases, compounds with mobile hydrogen atoms, for example, phenols and aromatic amines, are used for monomer stabilization. They inhibit polymerization in the presence of oxygen only, that is, they are antioxidants. As inhibitors of polymerization of (met)acrylates proceeding in the atmosphere of air, some aromatic amines known as polymer stabilizers were studied, namely, dimethyldi-(n-phenyl-aminophenoxy)silane, dimethyldi-(n-β-naphthylaminophenoxy)silane, 2-oxy-1,3-di-(n-phenylaminophenoxy)propane, 2-oxy-1,3-di-(n-β-naphthylaminophenoxy)propane. These compounds have proven to be much more effective stabilizers in comparison with the widely used hydroquinone (HQ), which is evidenced by high values of the stoichiometric inhibition coefficients (by 3–5 times higher than that of HQ). It has been found that inhibition of thermal polymerization of the esters of acrylic and methacrylic acids at relatively high temperatures (100°C and higher) is characterized by a sharp increase in the induction periods when some critical concentration of the inhibitor $[X]_{cr}$ is exceeded. This is caused by that at polymerization in air, the formation of polymeric peroxides as a result of copolymerization of the monomers with oxygen should be taken into account. Decomposition of polyperoxides occurs during the induction period as well and can be regarded as degenerated branching. The presence of critical phenomena is characteristic of chain branched reactions. However, in early works describing inhibition of thermooxidative polymerization, no degenerated chain branching on polymeric peroxides was taken into account. It follows from the results obtained that the value of critical inhibitor concentration $[X]_{cr}$ can be one of the basic characteristics of its efficacy.

Inhibition of spontaneous polymerization of (meth)acrylates is necessary not only at their storage but also in the conditions of their synthesis which proceeds in the presence of sulfuric acid. In this case, monomer stabilization is more urgent, since sulfuric acid not only deactivates many inhibitors but also capable of intensifying the process of polymer formation. The concentration dependence of induction periods in these conditions has a brightly expressed nonlinear character. Unlike polymerization in bulk, in the presence of sulfuric acid, decomposition of polymeric peroxides is observed at relatively low temperatures, and the values $[X]_{cr}$ for the amines studied are by ca. 10 times lower than $[HQ]_{cr}$.

Synthesis of MMA from acetonecyanhydrine is a widely spread technique of its industrial synthesis. The process proceeds in the presence of sulfuric acid in several stages, when various monomers are formed and interconverted. Separate stages of this synthesis were modeled with reaction systems containing, along with MMA, methacrylamide and methacrylic acid, and water and sulfuric acid in various ratios. As heterogeneous and homogeneous systems appeared at this, inhibition was studied in both static and dynamic conditions. The aforesaid aromatic amines appear to effectively suppress polymerization at different stages of the synthesis and purification of MMA. Their advantages over hydroquinone are strongly exhibited in the presence of sulfuric acid in homogeneous conditions, or under stirring in biphasic reaction systems. Besides, application of polymerization inhibitors is highly needed in dynamic conditions at the stage of esterification.

The usage of monomer stabilizers to prevent various spontaneous polymerization processes implies further release of the monomer from the inhibitor prior to its processing into a polymer. It is usually achieved by monomer rectification, often with preliminary extraction or chemical deactivation of the inhibitor, which requires high-energy expenditures and entails large monomer losses and extra pollution of the environment. It would be optimal to develop such a way of stabilization where the inhibitor would effectively suppress polymerization at monomer storage but would almost not affect it at polymer synthesis. The usage of inhibitors low soluble in the monomer is one of possible variants. When the monomer is stored and the rate of polymerization initiation is low, the quantity of the inhibitor dissolved could be enough for stabilization. Besides, as the inhibitor is spent, its permanent replenishment is possible due to additional dissolution of the earlier unsolved substance. The ammonium salt of N-nitroso-N-phenylhydroxylamine (cupferon), and some cupferonates were studied as such low-soluble inhibitors. The solubility of these compounds in acrylates, its dependence on the monomer moisture degree, the influence of the quantity of the inhibitor and the duration of its dissolution on subsequent polymerization were studied. Differences in the action of cupferonates are due to their solubility in monomers, their various stability in solution and the ability of deactivation; all this results in poorer influence of the inhibitor on monomer polymerization at producing polymer.

12.6   SOME PECULIARITIES OF THE KINETICS AND MECHANISM OF EMULSION POLYMERIZATION

Emulsion polymerization, being one of the methods of polymer synthesis, enables the process to proceed with a high rate to form a polymer with a high molecular mass, high-concentrated latexes with a relatively low viscosity to be obtained, polymeric dispersions to be used at their processing without separation of the polymer from the reaction mixture, and the fire-resistance of the product to be significantly raised. At the same time, the kinetics and mechanism of polymerization in emulsion feature ambiguity, which is caused by such specific factors as the multiphasity of the reaction system and the variety of kinetic parameters whose values depend not so much on the reagent reactivity as on the character of their distribution over phases, reaction topochemistry, the way and mechanism of nucleation and stabilization of particles. The obtained results pointing to the discrepancy with classical concepts, can be characterized by the following effects: (i) recombination of radicals in an aqueous phase leading to a reduction of the number of particles and to the formation of surfactant oligomers capable of acting as emulsifiers; (ii) the presence of several growing radicals in polymer-monomer particles, which causes the appearance of gel effect and the increase in the polymerization rate at high conversion degrees; (iii) a decrease in the number of latex particles with the growth of conversion degree, which is associated with their flocculation at various polymerization stages; (iv) an increase in the number of particles in the course of reaction when using monomer-soluble emulsifiers, and also due to the formation of an "own" emulsifier (oligomers).

Surfactants (emulsifiers of various chemical nature) are usually applied as stabilizers of disperse systems, they are rather stable, poorly destructed under the influence of natural factors, and contaminate the environment. The principal possibility to synthesize emulsifier-free latexes was shown. In the absence of emulsifier (but in emulsion polymerization conditions) with the usage of persulfate-type initiators (e.g., ammonium persulfate), the particles of acrylate latexes can be stabilized with ionized endgroups of macromolecules. The $M_nSO_4^-$ ion radicals appearing in the aqueous phase of the reaction medium, having reached a critical chain length, precipitate to form primary particles, which flocculate up to the formation of aggregates with a charge density providing their stability. Besides, due to recombination of radicals, oligomeric molecules are formed in the aqueous phase, which possess properties of surfactants and are able to form micelle-like structures. Then, the monomer and oligomeric radicals are absorbed by these "micelles," where chains grow. In the absence of a specially introduced emulsifier, all basic kinetic regularities of emulsion polymerization are observed, and differences are concerned only with the stage of particle generation and the mechanism of their stabilization, which can be strengthened at copolymerization of hydrophobic monomers with highly hydrophilic comonomers. Increasing temperature results in the growth of the polymerization rate and the number of latex particles in the dispersion formed, decreasing their sizes and the quantity of the formed coagulum, and in improved stability of the dispersion. At emulsifier-free polymerization of alkyl acrylates, the stability of emulsions and obtained dispersions rises in the monomer row: methylacrylate < ethylacrylate < butylacrylate, that is, the stability growth at lowering the polarity of the main monomer.

Our account of the aforesaid factors influencing the kinetics and mechanism of emulsion polymerization (in both presence and absence of an emulsifier) has enabled the influence of comonomers on the processes of formation of polymeric dispersions based on (meth)acrylates to be explained. Changes of some conditions of reaction have turned out to affect the character of influence of other ones. For example, increasing the concentration of MAA at its copolymerization with MA at a relatively low initiation rate leads to a decrease in the rate and particle number and to an increase in the coagulum amount. But at high initiation rates, the number of particles in the dispersion in the presence of MAA rises and their stability improves. The same effects were revealed for emulsifier-free polymerization of butylacrylate as well, when at high temperatures its partial replacement by MAA results in better stabilization of the dispersion, an increase in the reaction rate and the number of particles (whereas their decrease was observed in the presence of an emulsifier).

Similar effects were found for AN as well, which worsens the stability of dispersion at relatively low temperatures but improves it at high ones. Increasing in the AN concentration in the ternary monomeric system with a high MAA content leads to a higher number of particles and better stability of dispersion at relatively low temperatures as well.

With the aim to explore the possibility to synthesize dispersions whose particles would contain reactive polymeric molecules with free multiple C=C bonds, emulsion copolymerization of acrylic monomers and unconjugated dienes was studied. The usage of such latexes to finish fabrics and some other materials promotes getting strong indelible coatings. The kinetics and mechanism of emulsion copolymerization of ethylacrylate (EA) and butylacrylate (BA) with allylacrylate (AlA) (with ammonium persulfate (APS) or the APS—sodium thiosulfate system as initiators) were studied. The found constants of copolymerization of AlA with EA ($r_{AlA} = 1.05$, $r_{EA} = 0.8$) and AlA with BA ($r_{AlA} = 1.1$, $r_{BA} = 0.4$) point to different degrees of the copolymer's unsaturation with AlA units. Emulsion copolymerization of multicomponent monomeric BA-based systems (with AN, MAA, and the unconjugated diene acryloxyethyl maleate (AOEM) as comonomers) was also studied. The influence of AN and MAA is described above. Copolymerization with AOEM depends on reaction conditions. For example, in MAA-free systems, AOEM reduces the polymerization rate. In the presence of MAA, the rate of the process at high conversion degrees increases with the AOEM concentration, this monomer promoting the gel effect due to partial chain linking in polymer–monomer particles by the side groups with C=C bonds. The degree of unsaturation of diene units in the copolymer is subject to the composition of monomers, AOEM concentration, and temperature. In the BA—AOEM system, an increase of the diene concentration results in an increase in the unsaturation degree. This means that the diene radical is added to "its own" monomer with a higher rate than to BA ($r_{AOEM} = 7.7$). The higher unsaturation of diene units in MAA-containing systems points to that the AOEM radical interacts with MAA with a higher rate than with BA, and the probability of cyclization reduces, the degree of polymer unsaturation increases, and the diene's retarding action upon polymerization weakens.

12.7 COMPOSITIONS FOR PRODUCTION OF RIGID FOAMED POLYURETHANE

In the field of polymer physicochemistry, studies were made according to the requirements of the Montreal Protocol (1987), which demands drastically reduce the production and consumption of chlorofluorocarbons (CFC, Freon™) and even replace oxone-dangerous substances by ozone-safe ones. Our investigations dealt with the replacement of trichlorofluoromethane (Freon-11), which had been used as a foam-maker in the synthesis of rigid foamed polyurethane (FPU) over a long period of time, which was thermal insulator in freezing chambers and building constructions. On the basis of our experimental dependences of the kinetic parameters of the foaming process (the instant of start, the time of structuration, and the instant of foam ending rising), the values of density and heat conductivity of pilot foamed plastics samples on the concentration of the reactive mixture components and physicochemical conditions of FPU synthesis, optimal compositions (recipes) of mixtures with Freon-11 replaced by a ozone-safe (with an ozone-destruction potentials by an order of magnitude lower in comparison with Freon-11) azeotropic mixture of dichlorotrifluoroethane and dichlorofluoroethane were found. Their practical implementation requires no principal changes of known technological procedures and usage of new chemical reagents, which is an important merit of our developments.

12.8 SCIENTIFIC BASICS OF THE SYNTHESIS OF A HIGH-MOLECULAR-WEIGHT FLOCCULANT

There exists a problem of purification of natural water and industrial sewage from various pollutants, including suspended and colloid-disperse particles, associated with the growing consumption of water and its deteriorating quality (owing to anthropogenic influence). It is known that flocculants can be used for these purposes, which are high-molecular-weight compounds capable of adsorption on disperse particles to form quickly sedimenting aggregates. Polyacrylamide (PAA) is most active. In connection with that many countries (including Russian Federation) suffer from acrylamide (AA) deficit, we have developed modifications of the synthesis of PAA flocculent by means of the usage of AN and sulfuric acid to implement the reactions of hydrolysis and polymerization. It has turned out that in the presence of a radical initiator of polymerization and sulfuric acid, AN participates in both processes simultaneously, and, as AA is formed from AN, their joint polymerization begins. Desired polymer properties were achieved at AN polymerization in an aqueous solution of sulfuric acid up to a certain conversion degree with subsequent hydrolysis of the polymerizate (a two-stage synthetic scheme) or at achieving on optimal ratio of the rates of these reactions proceeding in one stage. The influence of the nature and concentration of initiator, the content of sulfuric acid, temperature and reaction duration on the quantity and molecular mass of the polymer contained in the final product, its solubility in water and flocculating properties were studied. The required conversion degree at the first stage of synthesis by the two-stage scheme is determined by the concentrations of AN and the aqueous solution of sulfuric acid. At one-stage synthesis, changes in temperature and monomer amount almost equally affect the reaction rates of hydrolysis and polymerization and do not strongly affect the copolymer composition. But changes in the concentration of either acid or initiator rather strongly influence the molecular mass and composition of macromolecules, which causes external dependences of the flocculating activity on these factors.

12.9   CONCLUSIONS

Thus, on the basis of our above results, the following conclusions can be made.

- New kinetic regularities have been revealed at polymerization of vinyl monomers in homophasic and heterophasic conditions in the presence of additives of transition metal salts, azo nitriles, peroxides, stable nitroxyl radicals and anion radicals (and their complexes), aromatic amines and their derivatives, emulsifiers and solvents of various nature.
- The mechanisms of the studied processes have been established in the whole and as elementary stages, their basic kinetic characteristics have been determined.
- Equations to describe the behavior of the studied chemical systems in the reactions of polymerization proceeding in various physicochemical conditions have been derived.
- Scientific principles of regulating polymer synthesis processes have been elaborated to allow optimization of some industrial technologies and solving most important problems of environment protection.

KEYWORDS

- **Environment**
- **Flocculent**
- **Inhibitor**
- **Initiator**
- **Kinetics**
- **Polymerization**
- **Radical**
- **Research**
- **Technology**

REFERENCES

1. Rozantsev, E. G., Goldfein, M. D., Pulin. V. F. (2000). *The Organic Paramagnetics.* Saratov: Saratov State University. 340 p. [Russ].
2. Goldfein, M. D., Ivanov, A. V. (2013). *Modern Concepts of Natural Science.* New York: Nova Science Publishers, Inc. 248 p.

CHAPTER 13

AGGREGATION BEHAVIOR OF UNSUBSTITUTED METAL PHTHALOCYANINES IN SUPRAMOLECULAR SYSTEMS

G. S. DMITRIEVA and A. V. LOBANOV

CONTENTS

ABSTRACT

To determine the photophysical and photochemical properties of unsubstituted phthalocyanines (MPc, M = Mg, Al, ZrL_2, $SiCl_2$, V=O, 2H) in their aggregates supramolecular systems have been developed on the basis of protein (albumin), micelles, biocompatible hydrophilic polymers, and nanoscale silica. The ability of MPc in supramolecular systems to form aggregates of different types depends on the nature of the central metal ion and microenvironment. The dependence of the type of MPc photoactivity method on aggregation was shown. H-aggregates exhibit selective photoactivity in electron transfer to O_2 to form ROS (triplet–triplet energy transfer and fluorescence are impossible). Monomers and J-aggregates of MPc fluoresce and may also participate in the triplet–triplet energy transfer with the formation of $^1O_2$. Spectral properties of J-correspond to the requirements of the photosensitizer.

Aim and Background. Using complexes of unsubstituted metal phthalocyanines are limited by their extremely low solubility in water. Creation of supramolecular systems with metal complexes based on a variety of media and detergents solves this problem. The resulting supramolecular systems have new photophysical and photochemical properties, and leads to the formation of phthalocyanine aggregates. On the other hand, for the effective functioning of MPc as photosensitizers their presence as monomer (isolated) form is important. In this case, no annihilation processes (self-quenching) the excited triplet state take place, which leads to efficient energy transfer to oxygen and produces photodynamic anticancer action. However, for other purposes, first of all in medical diagnosis, it is extremely important to use the MPc without any side phototoxic properties. It can be achieved by directional receiving MPc associated forms with additional solubilizers and macromolecular carriers. Therefore, it is necessary to study the photophysical and photochemical properties of MPc aggregates in various supramolecular systems.

13.1   INTRODUCTION

Phthalocyanines are remarkable macrocyclic compounds having magnificent physical and chemical properties. Metal phthalocyanines (Fig. 13.1) have been investigated in detail for many years due to their wide applications in many fields, mostly in terms of their uses as blue-green dyes and catalysts. They have also found different applications in many fields ranging from industrial and technological to medical [1].

Metal phthalocyanines are ideal building blocks because of their high versatility and exceptional thermal and chemical stability, as well as their intriguing electronic and optical properties [2–4]. It is well established that MPc can form two types of one-dimensional aggregates, namely, face-to-face H-aggregates and head-to-tail J-aggregates, depending on the orientation of the induced transition dipoles of their constituent monomers. The properties of H-aggregates and J-aggregates are remarkably different. For J-aggregates, the optical properties of the molecules are desirably maximized, which are much more advantageous for various technological applications such as spectral sensitization, optical storage and nonlinear optics than H-aggregates. However, while extensive studies have been devoted to the field of Pc assemblies, the methods to assemble the J-aggregates are still quite rare [5].

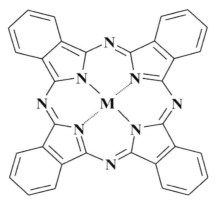

FIGURE 13.1 Structure of metal phthalocyanines.

Synthesis MPc with different substituents is possible to the formation of different of aggregates and their structures [6]. For this purpose, using deposition MPc on an inorganic carrier substrate by Langmuir-Blodgett method, for instance [7–10].

The tendency to form aggregates depends on the central metal ion and microenvironment. This work deals with the formation of aggregates of MPc (M = Mg, Al, ZrL_2, $SiCl_2$, V=O, 2H) in supramolecular systems using a variety of media.

13.2 EXPERIMENTAL PART

Initial solutions of metal phthalocyanine (MgPc, AlPc, ZrL_2Pc (L – $OOC_{14}H_{27}$), Si-Cl_2Pc, V=OPc, H_2Pc) for spectral measurements were prepared by adding the stock solutions of trace amounts of dye in pure solvent until an optical density of the solution was equal to 0.100. A series of base solutions of metal phthalocyanines AlPc (concentration of 0.00118 mol/L), ZrL_2Pc (concentration of 0.00126 mol/L), MgPc (concentration 0.001 mol/L), H_2Pc (concentration 0.001 mol/L), V=OPc (concentration 0.001 mol/L), $SiCl_2$Pc (concentration 0.001 mol/L) was prepared by dissolving dry dye of AlPc (0.0152 g) MgPc (0.0105 g) ZrL_2Pc (0.0135 g) in 20 mL DMF; H_2Pc (0.0103 g), V=OPc (0.0104 g) $SiCl_2$Pc (0.0112 g) in 20 mL H_2SO_4. The resulting concentration was determined by the absorption spectra. The solutions were stored in the dark at 4 °C. To prepare solutions of sodium dodecyl sulfate (SDS) to a concentration of 0.0082 mol/L (M = 288.4 g/mol) a sample of dry SDS (0.0021 g) was dissolved in 10 mL of distilled water. To obtain solutions of Triton (X-100) with a concentration of 0.00025 mol/L (M = 625 g/mol) sample of TX-100 (0.0016 g) was dissolved in 10 mL of distilled water. To prepare 2 wt.% polymer solution sample of 0.2 g of polyvinyl-N-pyrrolidone (PVP) (M = 26,500 g/mol) and poly-ethylene glycol (PEG) (M = 10,000 g/mol) taken were dissolved each in 10 mL of distilled water. A solutions of bovine serum albumin (BSA) (M = 67,000 g/mol) was prepared by dissolving 100 mg in 20 mL water (0.5 wt.%). Particles of nanoscale silica $(SiO_2)_n$ having a diameter of 60 nm were taken as finished 2 wt.% solution in ISPM RAS.

To record the absorption spectra the instrument HACH DV 4000 V (USA) was used. As a rule 2 mL of a detergent solution or polymer and 0,02 mL of phthalocya-nine metal complex in DMF were placed in a quartz cuvette with width of 1 cm. The analysis was performed at six different concentrations (1×10^{-6}, 2×10^{-6}, 3×10^{-6}, 4×10^{-6}, 5×10^{-6}, 6×10^{-6} mol/L) metal complexes adding 0.02 mL of MPc solution.

Formulas for the calculation:
1. Lifetime of singlet states τ_0, s

$$\tau_0 = \frac{3,5 * 10^8 * g}{v_{max}^2 * \varepsilon_{max} * \Delta v_{\frac{1}{2}}}$$

where, g = 1 – multiplicity of the quantum state; v_{max} – frequency position of the maximum of the absorption band (cm^{-1}); ε_{max} – extinc-tion coefficient (1 mol^{-1} cm^{-1}); $\Delta v_{1/2}$ – width of absorption band at its half-height.

2. Electronic transitions between the HOMO-LUMO isolated and associated states of metal phthalocyanine $E_{HCMO-B3MO}$, eV (HOMO – the highest occu-pied molecular orbital, LUMO – the lowest unoccupied molecular orbital)

$$E_{HCMO-B3MO} = h * \frac{c}{\lambda}$$

where, h = $4,136\times10^{-15}$ eV s – Planck's constant; c = 3×10^8 m s^{-1} λ – wavelength corresponding to the absorbance maximum of test metal complex of phthalocyanine solution (nm).

13.3   RESULTS AND DISCUSSION

Phthalocyanine metal complexes in the multicomponent systems can be in three states [5]: in the form of a monomer with narrow absorption band in the 665–675 nm [11–14], H-aggregates with their absorption band shifted relative to the monomer hypsochromically to ~630 nm, and J-aggregates with an absorption band batho-chromically shifted relative to the monomer to wavelengths of 730–850 nm. H- and

J-aggregates have a different structure: H-aggregates are stacking structure type and J-aggregate are brickwork type construction [15].

The absorption spectrum of the majority of metal phthalocyanines (e.g., MgPc in DMF, see Fig. 13.2) in solvents has a narrow absorption band in the range 665–675 nm (depending on the central metal ion) characteristic for a monomer state of metal complex [16–19].

As it was mentioned earlier, the aggregate state of metal phthalocyanines depends on the central metal ion. Figure 13.3 shows the absorption spectrum of metal-free phthalocyanine in different supramolecular systems. Both for polymers and micellar solutions a broad absorption band at 600–650 nm was observed. Hence metal-free phthalocyanine in these supramolecular systems is presented in the form of its H-aggregates.

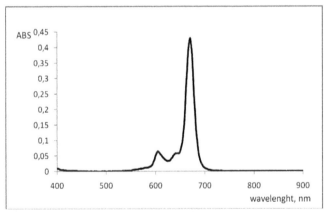

FIGURE 13.2   Absorption spectra of MgPc ($1'10^{-6}$ mol/L) in DMFA.

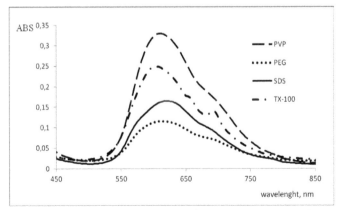

FIGURE 13.3   Absorption spectra of H$_2$Pc ($6'10^{-6}$ mol/L) in supramolecular systems.

Influence of environment on the formation of MgPc aggregates is presented in Fig. 13.4. For MgPc – TX-100 system narrow absorption band with wavelength of 668 nm is observed. Therefore, the metal complex is in a monomeric state. For systems based on SDS micelles absorption band corresponding to the monomer and a new absorption band at 830 nm are seen. The new absorption band corresponds to the presence of J-aggregates. Both PVP and PEG polymer solutions absorption lead to the appearance of new bands in the region 790–800 nm. The wavelength corresponding to the monomer does not change, so the metal complex is presented with monomer and J-aggregates. The difference in behavior of MgPc in TX-100 from other carriers can be probably explained by the structure of nonionic TX-100 micelles, which preventing convergence molecules of MgPc.

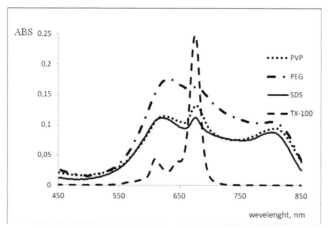

FIGURE 13.4   Absorption spectra of MgPc (6×10^{-6} mol/L) in supramolecular systems.

Behavior of metal complexes in supramolecular systems with different carriers also depends on the presence or absence extraligand in the central metal ion. Under consideration ZrL_2Pc, $SiCl_2Pc$, and V=OPc in carrier PVP complexes different results were obtained (Fig. 13.5). For ZrL_2Pc absorption maximum at 678 nm is observed. This behavior shows that ZrL_2Pc molecules are in an isolated state in the supramolecular system because of the presence near central metal ion long chain extraligand preventing the formation of aggregates. Absorption spectrum of $SiCl_2Pc$ has a narrow absorption band at 678 nm corresponding to the monomeric state, and despite the presence of two anions of Cl at central metal ion, there is a small peak at 850 nm, evidently corresponding to J-aggregate. For V=OPc on the absorption spectrum three absorption bands are observed: band at 630 nm, corresponding to H-aggregates and the absorption bands in the 730 and 830 nm. Both absorption bands correspond to J-aggregates that obviously have different structure. Monomer absorption band is absent.

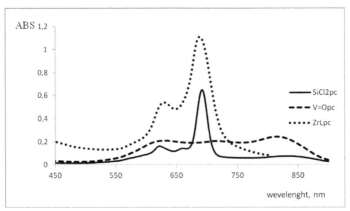

FIGURE 13.5   Absorption spectra of metal phthalocyanines (6×10^{-6} mol/L) in solution of PVP.

Consideration of ZrL_2Pc, $SiCl_2Pc$, $V=OPc$ in PEG solution, as well as in the case of PVP, shows different results were obtained (Fig. 13.6). Solution of ZrL_2Pc in PEG has absorption maximum at 678 nm, which corresponds to a monomeric state of the metal complex. The electronic spectrum of $SiCl_2Pc$ contain absorption band at 678 nm corresponding to the monomeric state. In addition new absorption maximum corresponding to J-aggregate bathochromicaly shifted to 850 nm appear. For $V=OPc$ three absorption bands in the band 630, 730 and 830 nm are observed in absorption spectrum, but no monomer absorption bands are seen in $V=OPc$ spectrum. Absorption band at 630 nm is characteristic for H-aggregates and absorption bands in the 730 and 830 nm can be attributed to J-aggregates. Thus, we can say that J-aggregates of $V=OPc$ in supramolecular systems differ in structure.

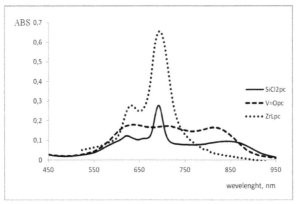

FIGURE 13.6   Absorption spectra of metal phthalocyanines (6×10^{-6} mol/L) in solution of PEG.

For media of SDS micelles (Fig. 13.7), we observed the formation of various forms metal complex. Molecules of $SiCl_2Pc$ have absorption bands at 678 nm corresponding to the monomeric state and band bathochromic shift of 200 nm relative to the monomer. This fact demonstrates a formation of $SiCl_2Pc$ J-aggregates stabilized by SDS micelles. For V=OPc a well-defined band corresponding to J-aggregates at wavelength of 830 nm is observed. Absorption band of H-aggregates at 630 nm is expressed badly and the absorption band of V=OPc monomer is absent. So we can say that the molecules of V=OPc form J-aggregates spontaneously in the majority of supramolecular complexes. Molecules of ZrL_2Pc have completely isolated state in SDS micelles.

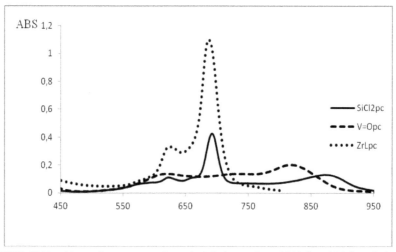

FIGURE 13.7   Absorption spectra of metal phthalocyanines ($6'10^{-6}$ mol/L) in solution of SDS.

Behavior of metal complexes of phthalocyanines in supramolecular systems with TX-100 (Fig. 13.8) is very different from other systems of micellar carriers. Tetra-substituted ZrL_2Pc, as in other supramolecular systems, has an absorption band with a wavelength of 678 nm corresponding to monomeric of molecules in the micelles. In micellar solutions of $SiCl_2Pc$ absorption maximum at wavelength of 676 nm is observed only, indicating a completely isolated metal complex in this conditions. Unlike other systems, V=OPc has a well-defined absorption band at 678 nm, which corresponds to the presence of monomers in the system, the band bathochromic shifted to 830 nm, which corresponds to J-aggregates and the absorption band hypsochromic shifted to 630 nm, which corresponds to H-aggregates. Hence, V=OPc molecules interaction with TX-100 micelles form three different type aggregation states of the molecules.

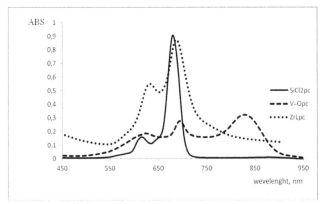

FIGURE 13.8 Absorption spectra of metal phthalocyanines (6×10^{-6} mol/L) in solution of TX-100.

On the absorption spectrum of ZnPc with nanoscale silica (Fig. 13.9) there is a peak at 750 nm, which corresponds to the presence of J-aggregates in the system, while at wavelengths corresponding to the monomeric state of the system, no absorption band maximum is not observed. The study of the optical absorption of MgPc at the nanoscale silica (Fig. 13.9) showed that, as well as in the case of ZnPc, MgPc is presented J-type aggregates with absorption band at 850 nm. Molecules of AlClPc in various supramolecular systems are characterized with isolated state [20–21]. However, in the case of nanoscale silica (Fig. 13.9) AlClPc has a broad absorption band at 630 nm, which corresponds to the presence of H-aggregates in the system. On the absorption spectra of MPc onto nanoscale silica particles there is absorbance throughout the visible region (Fig. 13.9), which can be attributed to the increase of nanoparticle size. Obviously it appears under the action of supported MPc.

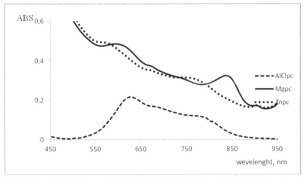

FIGURE 13.9 Absorption spectra of metal phthalocyanines (6×10^{-6} mol/L) in solution of nanoscale silica.

Behavior of metal phthalocyanines in the system with BSA (Fig. 13.10) is significantly different from their properties in other aforesaid supramolecular systems. Molecules of MgPc have an absorption band with wavelength of 667 nm characteristics for a monomeric state of the metal complexes, and the absorption band at 750 nm that indicates the presence of J-type aggregates. Behavior of AlClPc is reminiscent of feature of MgPc in BSA system and can be described by monomeric state (667 nm) and assembled J-aggregates with an absorption band in the region of 750 nm. Similar behavior of both metal complexes can be explained by the presence of hydrophobic pore in BSA globule, which attracts metal complex molecules to form aggregates. Molecules of ZnPc have one absorption band at 667 nm only in BSA solution, which indicates the completely isolated monomeric state in this system.

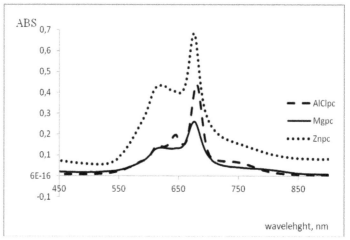

FIGURE 13.10   Absorption spectra of metal phthalocyanines (6×10^{-6} mol/L) in BSA.

On the base of the spectral characteristics of metal phthalocyanines data the energy of the electronic transitions for monomers and associates metal phthalocyanine, lifetime of singlet states of metal phthalocyanines in the systems studied were obtained. Magnitude of the electronic transitions for monomeric complexes do not change practically for ZnPc (Table 13.1), and for AlPc (Table 13.2), indicating the absence of influence of solvent molecules on the frontier molecular orbitals of molecules (the absence of outer coordination of phthalocyanine macroheterocycles). In associated molecules (H-aggregates), apparently different redistribution of electron density between the molecules of associates in different supramolecular complexes takes place, so there is a difference in the energy of the electronic transitions. Position of the absorption maxima of H-aggregates depending on the phthalocyanine supramolecular system varies from 580 to 630 nm.

TABLE 13.1 The Energy Values of Electronic Transitions Between the HOMO-LUMO for Monomeric and Associated States of ZnPc

The solvent (vehicle)	$E_{HOMO-LUMO}$, eV	
	ZnPc	$(ZnPc)_n$
DMFA	1.85	1.93
PEG	1.84	2.00
TX-100	1.84	2.03
PVP	1.84	2.02
SDS	1.84	2.01
$(SiO_2)_n$	1.86	1.94

TABLE 13.2 The Energy Values of Electronic Transitions Between the HOMO-LUMO for Monomeric and Associated Forms of AlClPc

The solvent	$E_{HOMO-LUMO}$, eV	
	AlClPc	$(AlClPc)_n$
DMFA	1.85	1.93
PEG	1.82	1.91
TX-100	1.82	1.91
PVP	1.82	1.91
SDS	1.81	2.03
$(SiO_2)_n$	1.85	1.96

Table 13.3 shows the lifetimes of singlet states in the studied metal phthalocyanine systems. This value is informative, since because of interactions of pigment-pigment and pigment-carrier (pigment-solvent) deactivation of the excited states with the transition into heat is observed, which leads to a reduction emission lifetimes. As it can be seen from values in Table 13.3, the lifetime varies by about an order confirming the assumption of a significant manifestation of the interaction of pigment-carrier and, in some cases, the existence of phthalocyanine in the associated state. Thus, for complexes with nanoscale silica, for which association Pc was the most clearly seen, lifetimes of excited states have a minimum value.

TABLE 13.3   Emission Lifetime of the Singlet State of Phthalocyanine Complexes of Zinc and Aluminum

The solvent	τ_0, s	
	ZnPc	**AlClPc**
DMFA	4.49×10^{-8}	3.82×10^{-8}
PEG	—	8.33×10^{-8}
TX-100	—	5.61×10^{-8}
PVP	—	1.09×10^{-7}
SDS	—	—
$(SiO_2)_n$	1.65×10^{-8}	2.88×10^{-8}

13.4 CONCLUSION

Study of the behavior of metal phthalocyanines in various supramolecular systems showed that aggregate formation depends on the nature of the central metal ion, the presence of extraligands, and nature of microenvironment. Metal-free phthalocyanine possesses the most tends to aggregation to form H-aggregates that appears in all supramolecular complexes studied. For ZrL_2Pc monomeric state is typical due to the central metal ion of additional long-chain ligand. Molecules of MgPc and $SiCl_2Pc$ form J-type aggregates simultaneously. An exception is the system with the carrier TX-100, both metal complexes have monomeric structure, which is explained by the carrier arrangement of the molecules. Complexes of V=OPc are presented as J-type aggregates, while their structure is different. Using of BSA as a carrier results in the formation of rare for AlPc J-aggregates that can be explained by the presence in protein of a hydrophobic pore wherein the molecules of metal complex form the aggregates.

Supramolecular systems with H-aggregates do not possess fluorescence [2], however they can be recommended for PDT of first type. Supramolecular systems of monomeric metal phthalocyanines possess fluorescence and are effective in the diagnosis as fluorescent labels. J-type aggregates are also fluoresce, but their absorption band is bathochromic shifted. This property increases the possibility of using of aggregated supramolecular systems in the diagnosis of several diseases.

KEYWORDS

- H- and J-aggregates
- Metal phthalocyanines
- Photophysical properties
- Photosensitizers
- Supramolecular systems

REFERENCES

1. Mahmut Durmus, Hanifi Yaman, Cem Göl, Vefa Ahsen, & Tebello Nyokong. (2011). Water-soluble quaternized mercaptopyridine-substituted zinc-phthalocyanines: Synthesis, photophysical, photochemical and bovine serum albumin binding properties. *Dyes and Pigments 91*, 153–163.
2. Mursel Arıcı, Duygu Arıcan, & Ahmet Lutfi Ugur. (2013) Electrochemical and spectroelectrochemical characterization of newly synthesized manganese, cobalt, iron and copper phthalocyanines. *Electrochimica Acta 87*, 554–566.
3. Aliye Aslı Esenpınar, Ali Rıza Ö zkaya, & Mustafa Bulut. (2011). Synthesis and electrochemical properties of crown ether functionalized coumarin substituted cobalt and copper phthalocyanines. *J Organometallic Chemistry 696*, 3873–3881.
4. Zafer Odabas, Ahmet Altındal, & Mustafa Bulut. (2011). Synthesis characterization and electrical properties of novel metal-free, Co(II), Cu(II), Fe(II), Mn(II), Sn(II) phthalocyanines peripherally tetra-substituted with 2,3-dihydro-1H-inden-5-yloxy moiety. *Synthetic Metals 161*, 1742–1752.
5. Jiaxiang Yi, Zihui Chen, Junhui Xiang, & Fushi Zhang. (2011). Photocontrollable J-Aggregation of a Diarylethene Phthalocyanine Hybrid and Its Aggregation-Stabilized Photochromic Behavior. *Langmuir, 27,* 8061–8066.
6. Joseph M. Fox, Thomas J. Katz, & Sven Van Elshocht. (1999). Synthesis, Self-Assembly, and Nonlinear Optical Properties of Conjugated Helical Metal Phthalocyanine Derivatives. *J. Am. Chem. Soc., 121,* 3453–3459.
7. Yunfeng Qiu, Penglei Chen, & Minghua Liu. (2008). Interfacial Assembly of an Achiral Zinc Phthalocyanine at the Air/Water Interface: A Surface Pressure Dependent Aggregation and Supramolecular Chirality. *Langmuir, 24,* 7200–7207.
8. Mehmet Kandaz, Meryem N. Yarasir, & Atıf Koca. (2009). Selective metal sensor phthalocyanines bearing nonperipheral functionalities: Synthesis, spectroscopy, electrochemistry and spectroelectrochemistry. *Polyhedron, 28,* 257–262.
9. Sampsa K. Hamalainen, Mariia Stepanova, & Robert Drost. (2012). Self-Assembly of Cobalt-Phthalocyanine Molecules on Epitaxial Graphene on Ir (111). *J. Phys. Chem. C, 116,* 20433–20437.
10. Maillard, P., Krausz, P. & Giannotti, C. (1980). Photoinduced activation of molecular oxygen by various porphyrins, bis-porphyrins, phthalocyanines, pyridinoporphyrazins, and their metal derivatives. *J Organometallic Chemistry, 197*–285–290.

11. Rıza Bayrak, Fatih Dumludag, & Hakkı Turker Akcay. (2013). Synthesis, characterization and electrical properties of peripherally tetra-aldazine substituted novel metal free phthalocyanine and its zinc(II) and nickel(II) complexes. *Molecular and Biomolecular Spectroscopy, 105,* 550–556.

12. Tebello Nyokong. (2007). Effects of substituents on the photochemical and photophysical properties of main group metal phthalocyanines. *Coordination Chemistry Reviews, 251,* 1707–1722.

13. Vongani Chauke, Mahmut Durmus, & Tebello Nyokong. (2007). Photochemistry, photophysics and nonlinear optical parameters of phenoxy and *tert*-butylphenoxy substituted indium(III) phthalocyanines. *Chemistry, 192,* 179–187.

14. Tamer Ezzat Youssef, & Hanan H. Mohamed. (2011). Synthesis and photophysicochemical properties of novel mononuclear rhodium(III) phthalocyanines. *Polyhedron, 30,* 2045–2050.

15. Andreas Gouloumis, David Gonzalez-Rodrıguez, & Purificacion Vazquez. (2006). Control Over Charge Separation in Phthalocyanine-Anthraquinone Conjugates as a Function of the Aggregation Status. *J. AM. CHEM. SOC, 128,* 12674–12684.

16. Hulya Yanıka, Duygu Aydına, & Mahmut Durmus. (2009). Peripheral and nonperipheral tetrasubstituted aluminum., gallium and indium phthalocyanines: Synthesis, photophysics and photochemistry. *J. Photochemistry and Photobiology A: Chemistry 206,* 18–26.

17. Victor Akpe, Hjalmar Brismar, Tebello Nyokong, & Osadebe, P. O. Photophysical and photochemical parameters of octakis (benzylthio)phthalocyaninato zinc, aluminum and tin: Red shift index concept in solvent effect on the ground state absorption of zinc phthalocyanine derivatives.

18. Staicu, A., Pascu, A. & Boni, M. (2013). Photophysical study of Zn phthalocyanine in binary solvent mixtures. *J. Molecular Structure 1044*–188–193.

19. Salih Agırtas, M., Metin Çelebi, & Selçuk Gümüs. (2013). New water soluble phenoxy phenyl diazenyl benzoic acid substituted phthalocyanine derivatives: Synthesis, antioxidant activities, atypical aggregation behavior and electronic properties. *Dyes and Pigments 99,* 423–431.

20. Raquel F. Correia, & Suzana M. Andrade. (2012). Aggregation and Disaggregation of Anionic Aluminium Phthalocyanines in Cationic Pre-Micelle and Micelle Media: a Fluorescence Study. *J. Photochemistry and Photobiology A: Chemistry.* 1–27.

21. Cesar A. T. Laia, Sılvia M. B. Costa, & David Phillips. (2004). Electron-Transfer Kinetics in Sulfonated Aluminum Phthalocyanines/Cytochrome *c* Complexes. *J. Phys. Chem.,* 7506–7514.

CHAPTER 14

THERMOPLASTIC COMPOSITES OF POLYPROPYLENE AND ETHYLENE-OCTENE RUBBER

MARIA RAJKIEWICZ, MARCIN ŚLĄCZKA, and JAKUB CZAKAJ

CONTENTS

ABSTRACT

The structure and physical properties of the thermoplastic vulcanisates (TPE-V) produced in the process of the reactive processing of polypropylene (PP) and ethylene-octene elastomer (EOE) in the form of alloy, using the cross-linking system was analyzed. With the DMTA, SEM and DSC it has been demonstrated that the dynamically produced vulcanisates constitute a typical dispersoid, where semicrystal PP produces a continuous phase, and the dispersed phase consists of molecules of the cross-linked ethylene-octene elastomer, which play a role of a modifier of the properties and a stabilizer of the two-phase structure. It has been found that the mechanical as well as the thermal properties depend on the content of the elastomer in the blends, exposed to mechanical strain and temperature. The best results have been achieved for grafted/cross-linked blends with the contents of iPP/EOE-55/45%.

Three units of ethene/n-octene copolymers (I) were blended with a unit of isotactic polypropylene (II) grafted/crosslinked with mixtures of unsaturated silanes and $(PhCMe_2O)_2O$ under conditions of reactive extrusion at 170–190 °C and studied for mechanical properties, thermal stability and microstructure. The composite materials consisted of a semicrystal II matrix and dispersed small pertides of cross-linked I. The best mechanical properties were when the II/I mass ratio was 55/45.

14.1   INTRODUCTION

The structure and physical properties of the thermoplastic vulcanisates (TPE-V) produced in the process of the reactive processing of polypropylene (PP) and ethylene-octene elastomer (EOE) in the form of alloy, using the cross-linking system was analyzed. With the DMTA, SEM and DSC it has been demonstrated that the dynamically produced vulcanisates constitute a typical dispersoid, where semicrystal PP produces a continuous phase, and the dispersed phase consists of molecules of the cross-linked ethylene-octene elastomer, which play a role of a modifier of the properties and a stabilizer of the two-phase structure. It has been found that the mechanical as well as the thermal properties depend on the content of the elastomer in the blends, exposed to mechanical strain and temperature. The best results have been achieved for grafted/cross-linked blends with the contents of iPP/EOE-55/45%.

Three units of ethene/n-octene copolymers (I) were blended with a unit of isotactic polypropylene (II) grafted/crosslinked with mixtures of unsaturated silanes and $(PhCMe_2O)_2O$ under conditions of reactive extrusion at 170–190 °C and studied for mechanical properties, thermal stability and microstructure. The composite materials consisted of a semicrystal II matrix and dispersed small pertides of cross-linked I. The best mechanical properties were when the II/I mass ratio was 55/45.

The thermoplastic elastomers (TPE) are a new class of the polymeric materials, which combine the properties of the chemically cross-linked rubbers and easiness of processing and recycling of the thermoplastics [1–8]. The characteristics of the TPE are phase micrononuniformity and specific domain morphology. Their properties are intermediate and are in the range between those, which characterize the polymers, which produce the rigid and elastic phase. These properties of TPE, regardless of its type and structure, are a function of its type, structure and content of both phases, nature and value of interphase actions and manner the phases are linked in the system.

The progress in the area of TPE is connected with the research oriented to improve thermal stability of the rigid phase (higher T_g) and to increase chemical resistance as well as thermal and thermo-oxidative stability of the elastic phase [2]. A specific group among the TPE described in the literature and used in the technology are microheterogeneous mixes of rubbers and plastomers, where the plastomer constitutes a continuous phase and the molecules of rubber dispersed in it are cross-linked during a dynamic vulcanizing process. The dynamic vulcanization is conducted during the reactive mixing of rubber and thermoplast in the smelted state, in conditions of action of variable coagulating and stretching stresses and of high coagulating speed, caused by operating unit of the equipment. Manner of producing the mixtures and their properties as well as morphological traits made them be called thermoplastic vulcanisates (TPE-V) [9, 10]. They are a group of "customizable materials" with configurable properties. To their advantage is that most of them can be produced in standard equipment for processing synthetics and rubbers, using the already available generations of rubbers as well as generations of rubbers newly introduced to the market with improved properties. A requirement of developing the system morphology (a dispersion of macromolecules of the cross-linked rubber with optimum size in the continuous phase of plastomer) and achieving an appropriate thermoplasticity necessary for TPE-V are the carefully selected conditions of preparing the mixture, (temperature, coagulation speed, type of equipment, type and amount of the cross-linking substance). When selecting type and content of elastomer, the properties of the newly created material can be adjusted toward the desirable direction. Presence of the cross-linked elastomer phase allows for avoiding glutinous flow under the load, what means better elasticity and less permanent distortion when squeezing and stretching the material produced in such a manner as compared to the traditional mixtures prepared from identical input materials, each of which produces its own continuous phase. With the dynamic vulcanization process many new materials with configured properties have been achieved and introduced to the market. The most important group of TPE-V, which has commercial significance, are products of the dynamic vulcanization of isotactic polypropylene (iPP) and ethylene-propylene-diene elastomer (EPDM). It is a result of properties of the PP and EPDM system, which, due to presence of the double bonds, may be cross-linked with conventional systems, which are of relatively low price, good contents miscibility and ability to be used within the temperature range 233–408 K [11–15].

Next level in the field of thermoplastic elastomers began with development of the metallocene catalysts and their use in stereoblock polymerization of ethylene and propylene and copolymerization of these olefins with other monomers, leading to macromolecules with a "customized" structure with a microstructure and stereo-regularity defined upfront. The catalysts enabled production of the homogeneous olefin copolymers, which have narrow distribution of molecular weights (RCC=M_w/ M_n<2.5), according to the developed technology called Insite and using the on-place catalyst [16, 17]. The Dow Chemical Company produces the olefin elastomers Engage™, which contain over 8% of octene. Co-polymers Engage are characterized by lack of relation between the traditional Mooney viscosity and the technological properties. Compared to other homogeneous polymers with the same flow index, they are characterized by higher dynamic viscosity at zero coagulation speed and decreasing viscosity at increasing coagulation speed. They have no fixed yield point. Saturated nature of elastomer, caused by absence of diene in the chain, results in some restrictions in choice of the cross-linking system. The ethylene-octene elastomers can be easily radiation cross-linked with peroxides or moisture, if they are formerly grafted with silanes. There is relatively not much description of the behavior of the ethylene-octene elastomer in the dynamic vulcanization process in the literature. The specific physical properties of such elastomers and possibility to process them within a periodic process as well as within a continuous process, due to convenient form of the commercial product (granulate), encouraged us to start the recognition works on development of the technology of producing thermoplastic vulcanisates from the mixture of elastomer Engage and iPP.

Use a silane-based cross-linking system in the dynamic vulcanization process seemed the most interesting. For the research works one of the known methods of cross-linking polyolefins with silanes was used, assuming that the cross-linking of EOE would proceed according to the analogous mechanism. In the seventies of the twentieth century the Dow Corning Company developed two methods of the hydrolytic cross-linking of polyolefins grafted with vinylosilanes according to the radical mechanism [18, 19]. Nowadays three polyolefin cross-linking methods are widely used in the industrial production. The grounds for distinguishing them are technological equipment and procedure. It is one-phase and two-phase method and a "dry silane method," available only under license [20]. The mechanism of cross-linking PE with the cross-linking system: silane/peroxide/moisture is shown schematically in Fig. 14.1.

The process of catalytic hydrolyzes of the alco-xylene groups of the grafted silane to the silane groups and, then, the catalytic condensation of the silane groups leads to production of the cross-linked structure through the siloxane groups. The hydrolyze and the condensation take place in an increased temperature with presence of the catalyst and water. Dibutyl tin dilaurate (DBTL) is most often used as a catalyst of the reaction. The catalyst may be added either to the polymeric blend (it constitutes an increased risk of the premature cross-linking), or in the form of a premixed reagent during the processing. The mechanism of action of DBTL, as a cross-linking catalyst, is complex and has not been sufficiently explained.

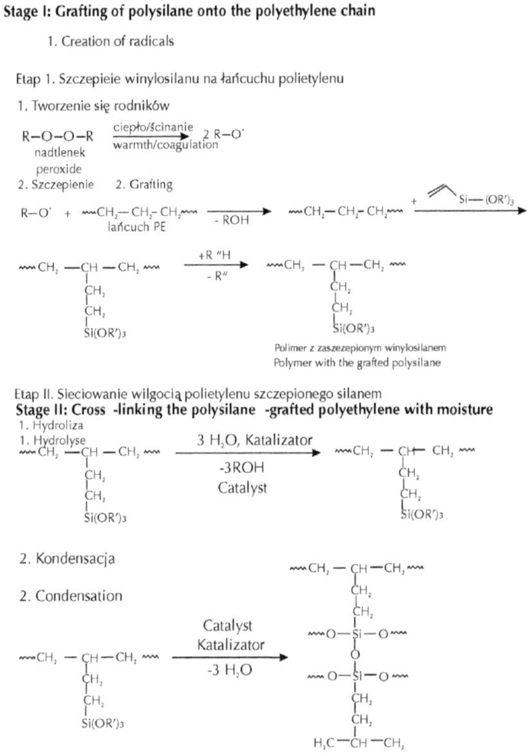

Stage I: Grafting of polysilane onto the polyethylene chain

1. Creation of radicals

Etap 1. Szczepieie winylosilanu na łańcuchu polietylenu

1. Tworzenie się rodników

$$R-O-O-R \xrightarrow[\text{warmth/coagulation}]{\text{ciepło/ścinanie}} 2\ R-O^{\cdot}$$

nadtlenek
peroxide

2. Szczepienie   2. Grafting

$$R-O^{\cdot} + \text{~~}CH_2-CH_2-CH_2\text{~~} \xrightarrow{-ROH} \text{~~}CH_2-CH_2-CH_2\text{~~} + \overset{}{Si}-(OR')_3$$

łańcuch PE

$$\text{~~}CH_2-CH-CH_2\text{~~} \xrightarrow[-R'']{+R''H} \text{~~}CH_2-CH-CH_2\text{~~}$$

Si(OR')₃   Si(OR')₃

Polimer z zaszczepionym winylosilanem
Polymer with the grafted polysilane

Etap II. Sieciowanie wilgocią polietylenu szczepionego silanem
Stage II: Cross -linking the polysilane -grafted polyethylene with moisture

1. Hydroliza
1. Hydrolyse

$$\text{~~}CH_2-CH-CH_2\text{~~} \xrightarrow[\text{Catalyst}]{3\ H_2O,\ \text{Katalizator}\\ -3ROH} \text{~~}CH_2-CH-CH_2\text{~~}$$

Si(OR')₃   Si(OR')₃

2. Kondensacja
2. Condensation

$$\text{~~}CH_2-CH-CH_2\text{~~} \xrightarrow[-3\ H_2O]{\text{Catalyst}\\ \text{Katalizator}}$$

Si(OR')₃

FIGURE 14.1 Crosslinking mechanism of polyolefins with a silane. (a) Rigid bond C-C;
(b) Elastic bond Si-O-Si.

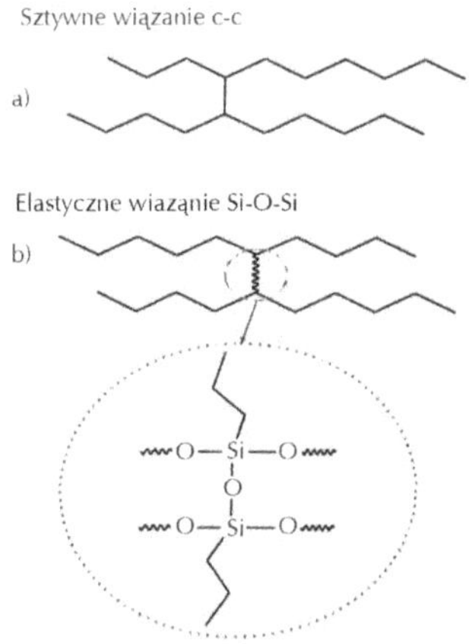

Sztywne wiązanie c-c

a)

Elastyczne wiazanie Si-O-Si

b)

~~O—Si—O~~
      O
~~O—Si—O~~

FIGURE 14.2 Structure of polyolefins crosslinked with a (a) peroxide or radiation, (b) with
a silane.

As a result of cross-linking the polyolefins with silanes the Si-O-Si bonds are produced, which are more elastic than the rigid bonds C-C created as a result of cross-linking of polymers induced by radiation and peroxides (Fig. 14.2). Use of silanes gives more elastic products and the cross-linking process is more cost-effective.

14.2 EXPERIMENTAL PART

14.2.1 RAW MATERIALS

- Isotactic polypropylene Malen P-F401 iPP, for extrusion, made by Orlen SA; flow index 2.4–33.2 g/10 min, yield point in stretching 28.4 MPa, crystallinity level 95%;
- The ethylene-n-octene elastomers EOE type Engage, synthesized according to the Insite technological process, manufactured by DuPont Dow Chemical Elastomers (Table 14.1);
- Silanes: Silquest A-172 vinylo-tris (2-methoxyethoxysilane), Silquest A-174–3-methacryloxypropyltrimethoxysilane, manufactured by Vitco SA;
- Dicumyl peroxide with 99% content of the neat peroxide, manufactured by ELF Atochem;
- Antioxidant tetra-kis (3,5-di-tetra-butyl-4-hydroxyphenyl) propionate, manufactured Ciba-Geigy.

TABLE 14.1 Properties of Ethylene-Octene Elastomers Engage

Elastomer type properties	Engage I	Engage II	Engage III
Co-monomer content, % of weight (^{13}C NMR/ FTR)	42	40	38
Density, g/cm^3, ASTM	0.863	0.868	0.870
Mooney viscosity, ML (1+4) 121 °C	35	35	8
MFR, deg/min, ASTM D-1238	0.5	0.5	5.0
Shore hardness A, ASTM D-2240	66	75	75

14.2.2 TEST METHOD

Three types of EOE from the wide range offered by the manufacturer were selected (Table 14.1). The general-purpose elastomers were selected with high content of octene and a defined characteristic.

The test were made aimed for determining a threshold value of content of elastomer in the iPP/EOE mixture, which was subject of the dynamic vulcanization process, considering the influence of these parameters on variable properties of iPP. A series of tests was made, in which the proportions of PP and elastomer Engage I were changed in the range 15–60%, with continuous addition of the cross-linking system (silane A-172/ dicumyl peroxide) 3/0.03% in relation to elastomer and antioxidant additive 0.2%.

For the tests of preparing dynamic vulcanisates in the continuous process of reactive extrusion a twin-screw mixer-extruder DSK 42/6D manufactured by Brabender was used. The vulcanisates were produced dynamically in the process of one-stage or two-stage extrusion process, setting the favorable operating parameters for the device, which had been determined based on multiple tests: distribution of temperatures in each heating area of the extruder: 170/180/190 °C, screw rotation: 40/min. In the one-stage process all the components provided in the formula (elastomer, iPP, antioxidant, silane initiating system/peroxide) was initially mixed in a fast-rotating mixer type Stephan in temperature of 50 °C, next a granulate was extruded. In the two-stage process in the first stage the iPP, elastomer and antioxidant mixture was extruded, next – after mixing the granulate with the cross-linking system – it was extruded again.

The profiles were formed from the granulates with an injection molding machine type ARBURG-420 M1000–25 all-rounder. For the tests the actual injection at speed of 10 cm^3/s was used with addition at speed of 15 cm^3/s, injection temperatures: 195/200/210/210 °C and blend injection time was slightly lower than for iPP itself.

14.2.2.1   METHODS OF ANALYZING THE BLEND

Hardness was marked according to the Shore method, scale D according to PN-ISO 868 or according to the ball insertion method according to the PN-ISO 868 (MPa). Flow speed index (MFR) was determined according to PN-ISO 1033. Resistance properties of the blend with static stretching were tested according to ISO-527, using a digital tester Instron 4505 (tear off speed: 50 mm/min). The bending properties were determined according to PN-EN ISO 178. In addition to the regular tests, the selected blends were subject to specialist examination, such as the thermogravimetric analysis (TGA), electron microscopy (SEM), differential scanning calorimetry (DSC) and dynamic thermal analysis of mechanical properties (DMTA).

The samples were heated in the ambient temperature in temperature range of 30–490 °C with speed of 5 deg/min. The test was conducted with thermobalance TGA manufactured by "Perkin Elmer." Turning points were made after freezing the samples in the liquid nitrogen for about 3 min. The surfaces of the turning points were concocted with gold with vacuum powdering. The scanning electron microscope JSM 6100 manufactured by JEOL was used to conduct the tests. The photographs have been made in magnification of 2000x

14.3   RESULTS OF THE TESTING

Influence of the content of elastomer Engage I on physical properties of TPE with PP and EOE modified (grafting/cross-linking) with a silane/peroxide cross-linking system has been shown in Fig. 14.3 and Table 14.2. The content of comonomer had significant influence on such properties of the elastomer as elasticity, modulus, density and hardness. Values of two last parameters decreased with the increase of content of n-octene in elastomer. It has been stated that properties of the dynamically vulcanized blends could be adjusted with content of the elastomer phase. With the increase of content of EOE in range 15–45% tensile strength increased (18–30 MPa), and, in the same time, relative elongation increased with tear off (300–700%). With elastomer content over 50% a visible decrease of both properties occurred, which came to 15 MPa and 600%, respectively. Whereas hardness expressed in Sh degrees or in MPa) systematically decreased with the increase of the content of EOE in the blend. The optimum content of EOE introduced to PP was 45% and therefore in most subsequent tests a blend was used, in which iPP/EOE ratio was 55/45%. Such contents had also the blends listed in Table 14.3, made of three types of EOE and two types of silane, with constant content of the cross-linking system (silane A–174/dicumyl peroxide 3.0/0.01%, irganox 1010–0.2%. The blends containing elastomers Engage I and Engage II, with difference of content of octene by 2% Shore hardness A (66 and 75, respectively) and with very similar Mooney viscosities, showed comparable resistance and rheological characteristic. The blends containing elastomer Engage III, with the lowest octene content, were characterized by slightly lower variables of tensile strength (tension at the tear off), elongation and hardness, but by a much higher tension at the yield point and high flow index.

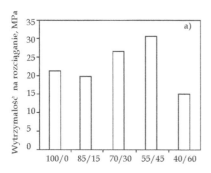 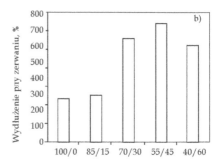

FIGURE 14.3   Mechanical properties of the dynamically cross-linked blends in relation to PP/EOE ratio, (a) tensile strength, (b) elongation at the tear off.

TABLE 14.2 Selected Properties of the Dynamically Crosslinked Blends in Relations to PP/EOE Ratio

Ratio PP/Engage I, % of weight	100/0	85/15	70/30	55/45	40/60
Hardness, °ShD	80	63	57	50	36
MFR (190 °C, 2, 16 kg), g/10 min	2.4	1.63	1.29	1.28	1.15
MFR (190 °C, 5), g/10 min	—	5.06	5.89	5.80	4.90
T_{A120}, °C*	152	143	130	106	~60
Hardness HK, MPa**	24.7	16.1	12.2	11.8	8.7
Solubility of elastomer in cyclohexane, %	—	—	13.9	12.03	14.2
Solubility of elastomer in boiling xylene, %	—	—	24.0	33.0	42.0

*T_{A120} – Vicat softening point;

HK – ball pan hardness method.

TABLE 14.3 Effect of Type of Elastomer Engage Modified with Silane A-174 on the Properties of the Dynamically Cross-Linked PP/EOE–55/45% Blends

Elastomer Blend properties	Engage I	Engage II	Engage III
MFR, g/10 min (2,16 kg, 190 °C)	1.86	1.80	4.07
Gardbess, °Sh, D	42/39	42/40	39/38
ϵ_B, %	720	752	660
σ_M, MPa	25.2	29.5	21.9
$\sigma_{100\%}$, MPa	11.1	12.6	11.0
σ_y, MPa	11.1	12.6	11.1
ϵ_y, %	24.0	27.9	39.9

Symbols: $\epsilon_{100\%}$ – tension at 100% elongation, σ_M – maximum tension, ϵ_B – relative elongation at the tear off, σ_y – yield point, ϵ_y – elongation on yield point.

Such behavior of elastomer Engage III blended with PP resulted probably from its different rheological characteristic, including its four times lower viscosity and very high flow index as compared to Engage I, which was recognized as the most suitable for production of nonsaturated blends using the dynamic vulcanization method.

Blend with the selected optimum contents iPP/Engage I 55/45% and the selected cross-linking system (silane/peroxide 3/0.03%) were characterized by high thermal stability, independent from type of the material employed to cross-linking silane. It has been confirmed with tests of TGA of blend containing silane A-172 and silane A-174 (samples PL-1 and PL-2, respectively), what is shown in Table 14.4, in the temperature of 230 °C the decrease of weight did not exceed 0.5%.

TABLE 14.4 Results of the Thermogravimetric Analysis of iPP/EOE-55/45% Blend (Engage I)

PL-1(Silane A-172)		PL-2 (Silane A-174)	
Temperature, °C	Decrease of weight, %	Temperature, °C	Decrease of weight, %
230	0.23	230	0.36
300	7.43	300	7.66
352	25.97	378	54.36
363	40.06	405	89.63
430	94.05	426	94.36

In temperature 300 °C came to as much as 7.5%, and the further increase of temperature caused the progressive degradation process. Analysis of the morphological structure of grafted/cross-linked iPP/Engage I blend using the SEM,

DSC and DMTA methods showed that the blends produced with dynamic vulcanization had a special two-phase structure. With scanning electron microscopy photographs of surface of turning points of iPP samples and iPP/Engage I 55/45% blends have been made (Fig. 14.4). The SEM analysis showed that the obtained blends were mixtures of two thermodynamically nonmiscible structures. The continuous phase of iPP had a visible semicrystal structure and the spherical and oval molecules of the dispersed phase of the cross-linked elastomer were not connected to the continuous phase. The viscoelastic properties, assessed with the DMTA Mk II equipment manufactured by Polymer Laboratories in the sinusoidally variable load conditions at bending with frequency of 1 Hz, in the temperature range between −100 and +100 °C also showed heterogeneous structure of the produced blends.

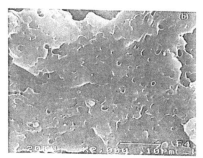

FIGURE 14.4 SEM microphotographs of: (a) neat iPP, (b) dynamically cross-linked PP/EOE blend – 55/45%; magnification 2000x.

In. Figure. 14.5 course of changes of the storage modulus E,' loss modulus E" and vibration damping factor gδ for iPP and iPP/Engage I blends with content of 85/15, 70/30 and 55/45% in relation to temperature has been shown. For iPP/EOE blends two, clear relaxation transitions in the range of glass transition are visible, near glass points of iPP and EOE.

Addition of elastomer slightly moved the glass point of iPP toward higher temperatures. PP showed higher values of the E' modulus as compared to the analyzed composites, whereas in the chart E" one maximum appeared corresponding to T_g PP.

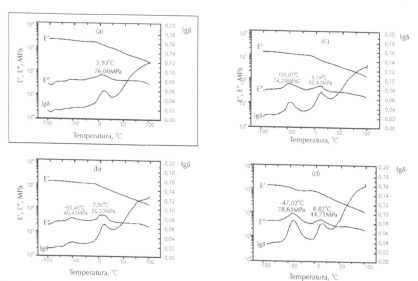

FIGURE 14.5 The dynamic mechanical properties of neat PP and dynamically cross-linked PP/EOE blends in relation to temperature: storage modulus (E'), loss modulus (E"), loss tangent (tanδ); (a) PP (b) PP/EOE 85/15, (c)-PP/EOE 70/30, (d)PP/EOE 55/45.

On the DSC thermal images made in positive temperatures (50–210 °C) a visible maximum appeared, which was connected with thermal transition corresponding to the melting point of iPP. Systematic decrease of melting point of the thermoplastic phase of iPP in iPP/Engage I blends related to the original polymer was observed (Fig. 14.6). Causes of these changes could not be unambiguously determined – it is supposed that here the phenomena such, as degradation of iPP in conditions of the high-temperature processing change of semicrystal structure of iPP may have occurred. In order to compare the properties of iPP/Engage I blend with content of 55/45%, produced with periodical method and in the one-stage or two-stage continuous process, a series of tests with use of the general formula was performed. Properties of the selected blends cross-linked with the silane A-174/dicumyl peroxide (TE-1) and silane A-172/dicumyl peroxide (TE-2) systems have been listed in table 14.5.

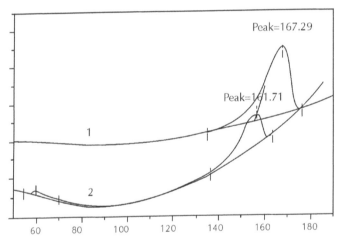

FIGURE 14.6   DSC curves of (a) PP, (b) PP/EOE dynamically cross-linked 55/45%.

TABLE 14.5   Selected Properties of PP and Dynamically Cross-Linked PP/EOE Blend PP/EOE (55/45%) in the Reactive Processing

Properties	PP	TE-1 (Silane A-174)	TE-2 (Silane A-172
Yield point, MPa	31.4	12.1	12.5
Relative elongation of yield point, %	9.4	25.1	33.0
Tensile strength, MPa	15.1	15.5	18.3
Elongation at the tear off, %	196	443	440
Tensile shear modulus, MPa	1455	476	430
Bending strength, MPa	39.6	13.0	10.4

TABLE 14.5 *(Continued)*

Properties	PP	TE-1 (Silane A-174)	TE-2 (Silane A-172
Tensile bending modulus, MPa	1499	504	415
Young modulus, MPa	1520	515	502
HDT, load 1.8 MPa, °C	50.5	38	38
Izod notched impact strength, kJ/m^2	3.16	46.7	43.0

14.4 CONCLUSIONS

- The conducted tests leaded to developing grounds for the technology for dynamic vulcanization of materials with thermo-elastoplastic properties, in which a thermoplastic polymer constitutes a continuous phase, whereas the dispersed phase consists of cross-linked elastomer particles. Basic elastomers are polyisoprene with isotactivity level of 85% or higher and copolymer EOE containing over 30% of *n*-octene.
- The blend properties can be adjusted with content of the elastomeric phase and with cross-linking silane/peroxide system. The achieved material has a heterogeneous structure and favorable set of resistance properties, including higher Izod notched impact strength as compared to PP.
- The developed technology allows for achieving new materials with precon-figured properties, competitive to the unmodified iPP and to physical PP/elas-tomer mixtures. They could be processed in the equipment for synthetics. These blends may be used as structural materials, characterized by higher thermal resistance as compared to the unmodified PP. They may also be a generation of modifiers for polyolefins and polymeric mixtures.

KEYWORDS

- **Ethylene-octane**
- **Polypropylene**
- **Thermoplastic composites**

REFERENCES

1. Rzymski, W. & Radusch, H. J. (2002). *Polimery, 47*, 229.
2. Spontak Richard, J. & Patel Nikunj, P. (2000). *Current Opinion in Colloids and Interface Sci., 5*, 333.
3. Rzymski, W. & Radusch, H. J. (2005). *Polimery, 50*, 247.

4. Radusch, H. J., Dosher, P. & Lohse, G. (2005). *Polimery, 50,* 279.
5. Rzymski, W. M. (1998). *Stosowanie i przetwórstwo materiałów polimerowych.* Wyd. Polit. Częstochowskiej, Częstochowa. 17–28.
6. Holden, G. (2000). *Understanding Thermoplastic Elastomers,* Hanser Publishers, Munich.
7. C.P. Rader. (1991). *Modern Plastic Encyclopedia.*
8. Rader, C. P. (1993). *Kuststoffe, 83,* 777.
9. Rzymski, W. & Radusch, H. J. (2001). *Elastomery, 5(2),* 19.
10. Rzymski, W. & Radusch, H. J. (2001). *Elastomery, 5(3),* 3.
11. Winters, R. (2001). *Polimery, 42,* 9745.
12. Trinh An Huy, T., Luepke, H. & Radusch, J. (2001). *App. Polym. Sci.,* 80, 148.
13. Jain, A. K., Nagpal, A. K., Singhal, R. & Gupta Neeraj, K. (2000). *J. Appl. Polym.Sci., 78,* 2089.
14. Gupta Neeraj, K., Janil Anil, K., Singhal, R. & Nagpal, A. K. (2000). *J. Appl. Polym. Sci., 78,* 2104.
15. Suresh Chandra Kumar, S., Alagar, M. & Anand Prabu, A. (2003). *Eur. Polym. Journal, 39,* 805.
16. Fanicher, L. & Clayfield, T. (1997). *Elastomery, 1(4).*
17. ENGAGE – polyolefin elastomers. A product of Du Pont Dow Elastomers, product Information, 2003.
18. Voight, H. U. (1981). *Kautsch. Gum. Kunstst., 34,* 197.
19. Toynbee, J. (1994). *Polymer, 35,* 428.
20. Special Chem. Crosslinking Agent Center, Dane techniczne, 2004.

CHAPTER 15

A LECTURE NOTE ON CULTIVATION OF LEGUME GRASSES FOR SEEDS IN MOUNTAIN ZONE

S. A. BEKUZAROVA, I. M. KHANIYEVA, and L. S. GHISHKAYEVA

CONTENTS

ABSTRACT

Red clover was sown solidly on slope lands with subsequent transition to wide-space sowing, where alfalfa was sown as well. Before sowing alfalfa seeds were treated by mixture of alanite clay and molybdenum-acid ammonium. On the stage of budding seed grass was fertilized additionally by extract of ragweed.

15.1 INTRODUCTION

Degradation of mountain ecosystems breaches natural processes of energy exchange. It diminishes soil fertility and impoverishes species number of plant organisms. One of the methods of restoration of disturbed landscapes is the undersow of perennial grasses, especially legumes, which have vigorous root system, high content of protein and capacity to accumulate biological nitrogen in soil [1].

In intensification of fodder production and making agriculture more friendly to environment special role belongs to field fodder grass cultivation. Perennial legumes are an important component of it. Due to their vigorous root system, perennial grasses restrain processes of erosion on slope lands [2, 3].

In modern conditions creation of seed grass is necessary, especially in mountain conditions on slope lands, where soil and micronutrients are wiped out. Known technologies of legume cultivation include additional fertilizing of grass by molybdenum [4]. Molybdenum facilitates increase of nitrogen fixation, development of plant and in the final result increase of seed yield of clover. A method is proposed to sprinkle the plants by molybdenum-acid ammonium [5]. But several other micronutrients are necessary to increase the activity of tubercles. Besides, molybdenum is sprayed on the stages of budding and flowering, that is, at the time when the grass has already formed, and it decreases the potential of clover plant. Treatment by micronutrients even in small doses on large areas increases costs.

Undersow of legume grasses such as clover, alfalfa, melilot, Coronilla and others is fulfilled for conservation of soil and restoration of fertility on degraded mountain hayfields [3]. This agro-technical process is accomplished on the second way of life, when seed grass is formed. Plants on the second year of life after the harvest of seeds fall out, and as a result biological diversity decreases on mountain hayfields and pastures. Usually two crops—clover (*Trifoliumpratense L.*) and alfalfa (*Medicago sativa L.*)—are sown at the same time [5]. These legumes oppress each other in the mixed crop, especially on the second year of life. In the first year after joint sowing those grasses are yet poorly developed. As a result soil erosion takes place. We have used wide-space method of undersow of alfalfa seeds under the grass of clover of the second year of life [6, 7].

Peculiar to legumes hardness of seeds has a significant influence on their germination rate, so it has been necessary to use special methods to increase their germination [8].

Clover plants develop well on slightly acid soils, while alfalfa prefers neutral soils, so for sowing of the second crop alfalfa seeds were mixed with zeolite-containing clay alanite.

Recently in Russia and abroad natural zeolites are used successfully to increase the productivity of agriculture and yield of various crops. Such interest in these minerals was evoked by their unique specific property. Carcass aluminosilicates have high resistance to acids and thermal stability. They are also highly active sorbents, cheap selective ionites and molecular sieves. Their selectivity, molecular-sieve effect and sorption capacity to cations of alkaline, alkali-earth, rare-earth and several heavy metals, the existence of large deposits, possibility of use before preliminary concentration and low cost make possible to use zeolites widely in agriculture, industry and environment protection [9].

In Northern Caucasus zeolite-containing clays (such as irlites, alanites, tereklites, leskenites) are found on flood-plain of river Terek.

15.2 MATERIALS AND METHODOLOGY

On the first year clover was sown solidly on the forecrop–winter wheat or winter barley. After the first hay harvest on the second year of life of clover the wide-space seed grass was formed with the following undersow of alfalfa in row-spacing with lowered norm of sowing. Before sowing alfalfa seeds were mixed with zeolite-containing clay, including alanite, (5–6 kg/ha) adding molybdenum-acid ammonium (30–40 g/ha). Hard seeds of alfalfa were treated for 6–8 h in 0.2–0.3% water solution of ragweed. Before sowing wet seeds were enveloped by bacteria from pounded tubercles taken from root system of alfalfa plants of second or third year of life on the stage of budding or flowering. Selection of tubercle bacteria was accomplished on old-age plants in past years. For hectare norm of alfalfa sowing in clover row-spacing (4 kg/ha) just 5 to 8 plants are needed, in which tubercle bacteria were taken without separation of roots from soil.

15.3 RESULTS AND DISCUSSION

Choice of winter wheat or winter barley as the forecrop for sowing of legume grasses is explained by the fact that those crops leave in soil unused microelements (boron, molybdenum, cobalt, copper, manganese), which promote increase of chlorophyll content in leaves, strengthening of plant's assimilation activity and process of photosynthesis. They also influence positively on seed development and their sowing qualities. Under influence of microelements plants become more resistant to various diseases and unfavorable environmental conditions (soil and aerial drought, excessive moisture, high and low temperatures). For legume grasses, clover and alfalfa substantial quantity of microelements is needed. Placement of legume grasses after winter cereals permits to cut costs on micronutrients and increase their fodder and seed productivity [5].

For the beginning we have sown clover solidly to cut the number of weeds in the crop. The next year on the stage of budding we have received fodder of good quality in the hay harvest. After the first hay harvest we formed from solid sowing of clover with the help of cultivator a wide-space sowing with row-spacing 60 cm wide, where alfalfa was undersown with lowered norm of sowing 4 kg/ha. This norm of sowing secures better lighting of grass, increase of quantity of generative organs and more active pollination by bees. Additional fodder is received in the year of alfalfa sowing, in the end of its vegetation. The following year (the third year of clover's life) the first legume crop (clover) falls out, and seed grass of alfalfa remains. It is harvested in the first hay-crop.

Due to the fact that clover plants develop well on slightly acid soils (pH 5–6), while alfalfa grows better on neutral soils (pH 7–7.5), special technique was applied for sowing the second crop. Seeds of alfalfa were mixed with zeolite-containing clay alanite, where molybdenum-acid ammonium was dissolved preliminarily. High content of calcium in alanite (30–40%) leads to decrease of soil acidity, and it has a favorable influence on alfalfa development.

Alfalfa grass develops a vigorous root system, securing yields of fodder and seeds for 6–8 years with simultaneous increase of soil fertility due to legume grasses' capacity to accumulate biological nitrogen.

Mixture of zeolite-containing clay alanite with molybdenum-acid ammonium leads to normalization of soil solution and decreases acidity in seed bed. So, joint treatment by molybdenum-acid ammonium and alanite is necessary for the development of alfalfa plants. Zeolite-containing clay prolongs the action of the mixture. It slowly releases elements containing in the mixture.

The above-mentioned agro-technical process–treatment of seeds before sowing in the water solution of ragweed and enveloping of wet seeds by tubercle bacteria from the root system of alfalfa–has significant importance. This process permits to increase germination rate of seeds and energy of sprouting as well as to decrease seeds' hardness.

Ragweed (*Ambrosia artemisiifolia L.*) contains the complex of chemical substances, including essential oil, glycosides, macroelements, which stimulate seed germination. Unlike other phyto-stimulators (nettle, wormwood, etc.), ragweed grows everywhere and is often found as a weed in various crops.

Results of the experiment showed that germination rate of alfalfa seeds grows from 72% (control–soaking in the water) to 92% in the optimal way of ragweed treatment. At the same time hardness of seeds decreased from 13% to 8% in the optimal variant. As a result erosion processes on slope lands weaken 2 to 3 times.

15.4 CONCLUSIONS

- Placement of legume grasses after winter cereals secures increase of seed yield of legume grasses.
- Treatment of seeds before sowing by mixture of zeolite-containing clay and molybdenum-acid ammonium by ragweed solution facilitates decrease of hardness of legume grasses' seeds.
- Additional fertilizing of seed grass by water solution of ragweed increase seed yield of clover and alfalfa on 18–25%.
- Using natural resources (zeolite-containing clay, weed plant ragweed), it's possible without additional costs to increase seed yield on slope lands and weaken erosion processes 2 to 3 times.

KEYWORDS

- **Additional fertilizing of seed plants**
- **Alanite**
- **Legume grasses**
- **Molybdenum**
- **Seeds**
- **Solid sowing**

REFERENCES

1. Gazdanov A. U., Bekuzarova, S. A. & Yefimova, V. A. (2006). Destructive processes on mountain hayfields and methods of their melioration. Vladikavkaz. Publication of Gorsky State Agrarian University, 96 p. (In Russian).
2. Vasin, V. G., Vasin, A. V. & Kiselyova, L. V. (2009). Perennial grasses in pure and mixed sowing in the system of green conveyor Kormoproizvodstvo (Fodder production). 2, 14–16 (In Russian).
3. Bekuzarova, S. A. & Gasieyev, V. I. (2012). Fodder crops in North Ossetia–Alania. A monograph. Vladikavkaz. Publication of Institute of raising the levels of instructors' skills 'Mavr,' 148 p. (In Russian).
4. Novosyolova, A. S. (1986). Selection and seed-growing of clover. Moscow. Agropromizdat, 174–175 (In Russian).
5. Bekuzarova, S. A., Aznaurova, Zh. U., Tsagarayeva, E. A. (2002). The method of nonroot treatment of seed grass of clover Patent for invention # 2189719, published on September 27, 2002. Bulletin # 27 (In Russian).
6. Bekuzarova, S. A., Bekmurzov, A. D., Basiyev, S. S., Shogenov, M. L. & Pliyev, Y. V. (2000). The method of sowing of legume grasses in crop rotation Patent for invention # 2155463, published on September 10, 2000. Bulletin # 25 (In Russian).

7. Adayev, N. L., Bekuzarova, S. A. & Abasov, S. M. (2011). The method of placement of legume seed grass on slope lands Patent for invention # 2425476, published on August 10, 2011 (In Russian).

8. Zherukov, B. H., Khaniyeva, I. M., Khaniyev, M. H., Magomedov, K. G., Bekuzarova, S. A. & Boziyev, A. L. (2013). The method of alfalfa seed treatment before sowing Patent for invention # 2479974, published on November 24, 2013 (In Russian).

9. Postnikov, A. V., Loboda, B. P. & Sokolov, A. V. (1992). Use of zeolites in agriculture Bulletin of Russian Academy of agricultural sciences, #5, 49–51 (In Russian).

CHAPTER 16

RED CLOVER (*Trifolium pratense* L.) FOR HAYFIELD-PASTURE USE

S. A. BEKUZAROVA and V. A. BELYAYEVA

CONTENTS

ABSTRACT

Results are represented of an experiment on introduction of wild samples of red clover as well as clover cultivars from geographically distant regions in the foothills of North Ossetia. Samples are chosen with the largest genetic value for creation of clover cultivars of hayfield-pasture type.

16.1 INTRODUCTION

Creation of highly productive, genetically valuable forms with high competitiveness in ecosystems of mountain hayfields and pastures is an important element of fodder base formation in North Caucasian region. Existing in the region clover cultivars of hayfield-pasture type has low productive longevity, and this fact creates substantial hardships for formation of agrocenoses in mountain and foothill conditions. The main obstacle on the way of growth of the biological potential of this species is the low adaptive capacity of recommended cultivars in conditions of vertical zoning. The specifics of environmental conditions in mountain regions with billowy relief, where more than half of all agricultural lands are situated in complex topographic conditions characterized by changes of soil-climatic gradients, demands the use of stress-tolerant cultivars. Highly productive cultivars and hybrids are usually less tolerant to such conditions, less effective in conditions of undersow due to low competitiveness with native species, and they have low survival rate of sprouts. Individuals, which have survived, don't live for long and soon fall out the grass. On practice it leads to unjustified costs of labor and funds [1]. Evaluation and use of genetic potential of local wild populations of red clover has a special importance due to their specific stress tolerance.

Wild mountain species of clover are known for their longevity, frost hardiness and high content of nutrients. They are more tolerant to high levels of ultraviolet radiation. These species are highly competitive in phytocenosis of mountain hayfields and pastures, and they have larger number of leaves and shoots per plant. But one shouldn't ignore the rich genetic material of the existing clover cultivars from other regions with wide number of undoubted merits such as high productivity and quality. However, it's necessary to take into consideration that geographic and climatic differences demand the careful study of introduced forms in conditions of any definite region [2].

Segregation of initial samples for selection is impossible without elaborate biochemical evaluation with establishment of rate of influence on the parameters under study of the complex of factors effecting plant organism, including stages of development [3]. With the knowledge of amplitude of chemical trait's variation within the limits of population difference it's possible to choose initial forms for hybridization [1, 4].

In connection with it the aim of our research consisted in the study of economic and biological traits of introduced forms of the red clover and in selection of samples promising for creation of hayfield-pasture cultivars in North Ossetia – Alania.

16.2   MATERIALS AND METHODOLOGY

We have fulfilled study and selection of red clover cultivars, introduced from different ecological-geographic natural habitats. Productivity and biochemical content were determined among Belarussian introduced cultivars and samples of red clover (Minskiy Mutant, Ustodlivy, Yaskravy, SL-38, T-46), an introduced cultivar from Siberian region (SibNIIK-10) as well as native wild forms growing in conditions of vertical zoning in North Ossetia – Alania: Dargavski (1800 m above sea level) and Gornaya Saniba (1200 m above sea level). Parameters of economic and biological traits of all samples were evaluated in comparison with standard cultivar Daryal, which had been recommended for North Ossetia – Alania from 1993. The authors of the cultivar are S. A. Bekuzarova and B. K. Mamsurov. Biochemical analysis of initial samples of red clover was fulfilled according to common methods [5]. The following parameters were evaluated: content of dry matter, protein, nitrogen, ash, calcium, phosphorus, cellulose, fat and sugar. Statistical analysis included descriptive statistics and analysis of variance (ANOVA).

16.3   RESULTS AND DISCUSSION

Biochemical analysis of samples in our study showed that on the stage of stooling percent content of dry matter was lower than in the following stages of ontogeny. On the contrary, content of raw protein, potassium, phosphorus, fat and sugar slightly exceeded similar indices in the stages of budding and beginning of flowering. High content of dry matter and raw protein on the stage of stooling distinguished introduced forms SL-38 (20.12 and 21.37%, respectively) and Dargavski (19.76 and 20.06%). The largest content of sugar was found in SL-38 (5.43%) and Gornaya Saniba (4.82%), while calcium and potassium were abundant in standard Daryal (2.08 and 2.87%, respectively). The largest content of ash was observed in SL-38 (12.27%) and standard Daryal (12.66%), phosphorus was abundant in SL-38 (1.09%) and wild introduced forms (0.98–1.02%), and cultivar Yaskravy exceeded all other samples on cellulose content (21.72%) (Fig. 16.1; Table 16.1).

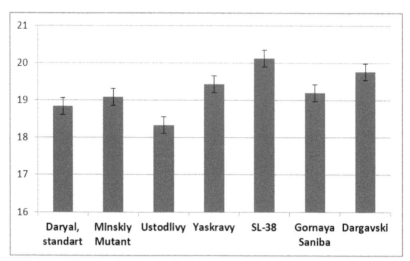

FIGURE 16.1   Content of dry matter in clover samples on the stage of stooling, %.

TABLE 16.1   Chemical Content of Red Clover Samples on the Stage of Stooling, % to Dry Matter Content

Cultivar, sample	Nitrogen	Protein	Ash	Calcium	Potassium	Phosphorus	Fat	Cellulose	Sugar
Daryal (standard)	2.87	17.94	12.66	2.08	2.87	0.86	4.78	17.46	4.55
Minskiy Mutant	2.74	17.13	11.84	1.37	2.81	0.84	3.69	17.53	4.34
Ustodlivy	2.25	14.06	11.05	1.41	2.61	0.87	5.71	17.31	4.72
Yaskravy	2.73	17.06	11.23	1.50	2.84	0.91	4.73	21.72	4.16
SL-38	3.42	21.37	12.67	1.53	2.07	1.09	4.85	16.29	5.43
Gornaya Saniba	3.10	19.37	12.13	1.46	2.55	1.02	4.92	16.42	4.82
Dargavski	3.21	20.06	11.24	1.41	2.68	0.98	5.02	15.81	4.69

Results of biochemical content studies of cultural cultivars and native populations of red clover on the stages of budding and beginning of flowering are evident of that individual indices have slight positive dynamics in the ontogeny process.

It's known that energetic component and nutrient value of fodder depend directly upon dry matter content in the yield. On the stage of budding the largest content of dry matter was observed in samples Minskiy Mutant (21.42%) and SL-38 (21.14%) (Fig. 16.2).

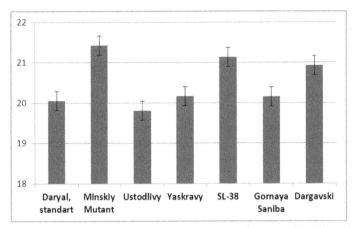

FIGURE 16.2   Content of dry matter in clover samples on the stage of budding, %.

Raw protein is the most valuable component of clover's green mass. It is characterized by good balance of amino acid content. Biological value of clover protein reaches 83–95% to the protein of hen's egg. The content of raw protein on the stage of budding was maximal in cultivar SL-38 (16.88%) and minimal in sample Ustodlivy (9.31%) (Table 16.2).

TABLE 16.2   Chemical Content of Red Clover Samples on the Stage of Budding, % to Content of Dry Matter

Cultivar, sample	Nitrogen	Protein	Ash	Calcium	Potassium	Phosphorus	Fat	Cellulose	Sugar
Daryal, standard	2.33	14.56	10.72	1.72	2.05	0.74	4.66	19.46	2.81
Minskiy Mutant	2.10	13.13	10.30	1.29	1.95	0.73	3.21	20.44	3.61
Ustodlivy	1.49	9.31	9.11	1.33	1.62	0.74	5.50	20.31	3.34
Yaskravy	2.31	14.43	10.28	1.38	1.95	0.82	4.55	26.44	3.98
SL-38	2.70	16.88	10.49	1.45	1.04	1.00	4.30	20.65	4.10
Gornaya Saniba	2.43	15.22	10.10	1.40	1.81	0.97	4.71	22.90	3.45
Dargavski	2.37	14.80	9.34	1.38	1.78	0.91	4.80	20.56	3.60

Results of the experiment are evident of the positive correlation in introduced cultivar SL-38 and some other samples between content of dry matter and raw protein.

Besides organic matters, the green mass of clover contains a lot of ash and different elements. In particular, the content of ash in studied samples varied on the stage of budding from 9.11 to 10.72%. Standard cultivar Daryal had the best index – 10.72%.

Potassium plays important role in physiological processes of plant organism. It encourages formation of protein substances in plants, increase of sugar synthesis in leaves and their further transportation to other organs, regulates proportion of free and combined water in tissues and increases hydrophilic of protoplasm. Calcium is no less important for vital functions of plants, and its lack leads to sliming and rot of the root system and loss of tolerance to fungal diseases [6].

According to our data, standard cultivar Daryal contained the largest amount of calcium and potassium – 1.72% and 2.05%, respectively. Other samples had nearly the same concentration of calcium, but differed on phosphorus accumulation. Cultivar SL-38 contained the largest quantity of phosphorus (1.0%), while in other samples this index varied from 0.73 to 0.93%. It is known that phosphorus is a part of nucleoproteins, and reproductive organs of plant and seed embryos are rich in this element. Lack of phosphorus leads to slow growth, depression of flowering and seed formation. According to our observations, ratio of calcium and phosphorus varied among different cultivars from 1.4 (SL-38, Gornaya Saniba) to 2.3 (Daryal). Unlike North Ossetia, in Leningrad Region the ratio of these elements averages 1.4, which can be considered as an optimal level. Such a ratio of calcium and phosphorus is more inherent in another clover species – *Trifolium repens* L. [7].

Fats are an important elemental part of fodder grasses, and it's necessary to note that their content in legumes is higher than in gramineous plants. Fat acids in grasses are represented mainly by linoleic and linoleic acids. They can't be produced in animal organism and should be received with fodder. In our study cultivar Ustodlivy contained on the stage of budding the maximal amount of fat (5.50%), while sample Minskiy Mutant had the minimal quantity (3.21%).

On the content of cellulose only cultivars Yaskravy (26.44%) and Daryal (19.46%) differed significantly from other samples, among which the amount of cellulose varied in limits 20.31…22.90%.

Cultivar SL-38 had the largest content of sugar (4.10%), while standard cultivar Daryal had the smallest content (2.81%).

The highest content of dry matter at the beginning of flowering stage was observed in samples Minskiy Mutant and SL-38–22.62 and 22.12%, respectively (Fig. 16.3).

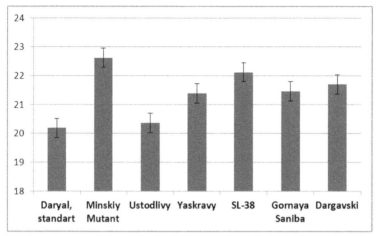

FIGURE 16.3   Content of dry matter in clover samples at the beginning of flowering, %.

Content of dry matter in cultivar Yaskravy, samples Gornaya Saniba and Dargavski exceeded on 1.20–1.52% the similar index of standard cultivar Daryal, while cultivar Ustodlivy didn't differ substantially from standard cultivar. From the stage of budding to the beginning of flowering the quantity of dry matter grew slightly in all samples of clover.

Insignificant increase of raw protein content at the beginning of flowering in comparison with budding was detected in all samples. Cultivar SL-38 had the highest level of protein (17.88%) and cultivar Ustodlivy had the lowest one (9.75%) (Table 16.3).

TABLE 16.3   Chemical Content of Red Clover Samples at the Beginning of Flowering (% to Content of Dry Matter)

Cultivar, sample	Nitrogen	Protein	Ash	Calcium	Potassium	Phosphorus	Fat	Cellulose	Sugar
Daryal, standard	2.38	14.89	11.86	1.82	1.92	0.70	4.80	22.54	3.10
Minskiy Mutant	2.12	13.25	12.30	1.56	1.80	0.71	3.30	26.68	4.10
Ustodlivy	1.56	9.75	12.26	1.36	1.76	0.68	4.90	28.62	3.80
Yaskravy	2.42	15.13	11.26	1.62	1.67	0.86	4.60	28.56	4.30
SL-38	2.86	17.88	11.12	1.48	1.24	0.95	3.80	22.30	4.50
Gornaya Saniba	2.60	16.22	11.18	1.46	1.90	0.95	4.80	26.77	4.28
Dargavski	2.48	15.53	10.20	1.42	1.81	0.88	4.83	25.94	4.17

Content of ash elements slightly increased in all samples on the stage of the beginning of flowering. Maximal content of ash (12.3%) was observed in cultivar Minskiy Mutant.

Dynamics of calcium content on the stages of budding and flowering was positive in all clover samples. Maximal content of calcium at the beginning of flowering was observed in cultivar Daryal (1.82%) and the minimal one in cultivar Ustodlivy (1.36%). Dynamics of potassium content was individual in different samples. Potassium content had positive trend in samples SL-38, Ustodlivy, Gornaya Saniba and Dargavski and negative trend in samples Minskiy Mutant, Daryal and Yaskravy. Cultivar Daryal contained the largest quantity of potassium at the beginning of flowering (1.92%).

Phosphorus plays an important role in formation of reproductive organs of plants. Analysis of its dynamics showed that its content at the beginning of flowering decreased in comparison with the stage of budding in all samples, except for Yaskravy. Minimal phosphorus content on the stage of flowering was observed in cultivar Ustodlivy – 0.68%, the maximal one in cultivar SL-38–0.95%. Later on the stage of ripening we have observed further decrease of phosphorus content in samples.

Clover samples differed substantially on fat content. For example, in cultivar Yaskravy this index remained at the beginning of flowering practically on the same level as on the stage of budding (change +0.05%). In cultivars Daryal and Minskiy Mutant fat content slightly increased (+0.11% and +0.09%, respectively). At the same time fat content substantially reduced in cultivars Ustodlivy (–0.35%) and SL-38 (–0.50%). Despite this reduction, Ustodlivy had the maximal quantity of fat (4.90%).

Quantity of cellulose at the beginning of flowering increased in all clover samples. This index was the highest in cultivar Ustodlivy (28.62%) and the lowest in cultivar SL-38 (22.30%).

It is known that sugar plays an important role in vital functions of plants, because there is a direct connection between sugar content and processes of amino acid production occurring in conditions of nitrogen assimilation. Besides, the level of frost hardiness depends upon the quantity of sugar on late stages of vegetation [8]. It was established in our study that dynamics of sugar accumulation was positive in all clover samples.

It's possible to state that on the stage of stooling the green mass of most clover samples has a good quality. On the stages of budding and beginning of flowering (which are especially important for evaluation of economic and biological traits) content of dry matter, protein, ash, calcium and sugar has an insignificant positive dynamics in all tested samples. At the same time content of cellulose increased significantly.

Standard cultivar Daryal exceeded all other samples on the stage of budding on content of ash, calcium and potassium, and at the beginning of flowering on content of calcium and potassium. But it yielded to other cultivars in content of dry matter, nitrogen, raw protein, phosphorus, fat, cellulose and sugar.

Belarussian cultivar SL-38 was in the lead on content of dry matter, protein, sugar and phosphorus. This cultivar along with sample Gornaya Saniba is characterized by optimal ratio of calcium and phosphorus.

Our studies showed that clover cultivar from Belarus SL-38 and T-46 after introduction to foothill zone displayed high indices of productivity, though they yielded in seed number to wild samples Dargavski and Gornaya Saniba.

Maximal height of plant was observed in standard cultivar Daryal (77.3 cm), the minimal one in SibNIIK-10 (54.0 cm) (Fig. 16.4).

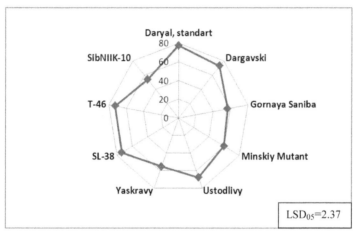

FIGURE 16.4 Height of red clover samples, cm.

Introduced cultivars of Belarussian selection Yaskravy, Minskiy Mutant and Ustodlivy had medium indices, while T-46 and SL-38 distinguished themselves by greater height – 72.7 and 74.0 cm, respectively.

Cultivar SL-38 had the highest productivity of the green mass – 3.0 kg/m² (125% to standard), while sample Gornaya Saniba had the minimal productivity – 1.1 kg/m² (45.8% to standard) (Fig. 16.5). Another native sample Dargavski also had the low productivity of the green mass – 1.5 kg/m² (62.5% to standard).

It's necessary to note that usually mountain clovers in their natural habitats have relatively small height. The explanation is that in the conditions of vertical zoning the existing soil-climatic gradients lead to formation of agrophytocenoses with maximal adaptation to sharp temperature drops, humidity, sun radiation and eating by animals.

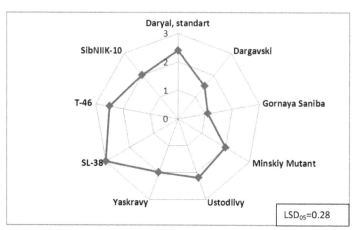

FIGURE 16.5   Yield of green mass in red clover samples, kg/m².

Evidently, natural selection among native clover populations went in the direction of decreasing plant height as an adaptive reaction to more economic consumption of soil nutrients and water resources.

In our study introduced wild samples of clover didn't have low height in all cases. In particular, sample Gornaya Saniba retained this trait in conditions of cultivation on 600 m above sea level. On the contrary, plants of sample Dargavski had larger height than sample Gornaya Saniba. The possible explanation of this difference is the individual type of reaction to meteo-climatic conditions of foothill zone according to adaptive capacities of the sample. Increase of height in sample Dargavski combined with slight decrease of leave number per shoot and bushiness. At the same time sample Gornaya Saniba was lower, but had greater bushiness.

Another direction of natural selection among red clover plants in conditions of vertical zoning is the increase seed productivity, which secures effective substitution of perished individuals. We have observed the maximal number of seed in inflorescences in sample Gornaya Saniba – 44% (115.8% to standard) (Fig. 16.6). High number of seed in inflorescences was found also in another wild sample Dargavski – 43% (113.1% to standard).

Introduced cultivar SibNIIK-10 belongs to cultivars created in ecological-geographic and climatic conditions of Siberian region. Tested for economic and biological traits in conditions of foothill zone of North Ossetia – Alania, it had low height and low seed number in inflorescences, though according to data of cultivars' originators from Siberian Scientific Research Institute for Fodder in its native region the cultivar grew to the height 83 cm.

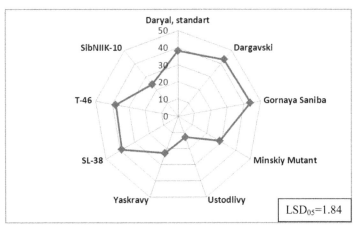

FIGURE 16.6   Seed number in inflorescences of red clover samples, %.

Introduced from Belarus cultivars Yaskravy, Minskiy Mutant, Ustodlivy had nearly the same productivity, which didn't exceed the level of standard. Plants of cultivar Yaskravy had small height and seed number in inflorescences, while in cultivar Ustodlivy the latter index didn't exceed 13%. Apparently, introduced cultivars SibNIIK-10, Yaskravy, Minskiy Mutant and Ustodlivy for their economic and biological traits don't suit as initial samples for creation of hayfield-pasture cultivars in conditions of North Ossetia – Alania, because they lack sufficient adaptation reserves allowing them to adjust to ecological-geographic and climatic conditions of particular locality.

Our studies showed that standard cultivar Daryal had maximal height. But this cultivar was less productive than SL-38 and T-46 and had less number of seeds in inflorescences than native populations.

Productivity of green mass of wild clover samples Gornaya Saniba and Dargavski is lower in comparison with introduced cultivars, but they have higher percent of seed number in inflorescences, which is an adaptive reaction to sharp variation of meteo-climatic factors in the highlands.

Introduced cultivar SL-38 displayed itself as the most productive sample. Among the cultivars of Belarussian selection it has the largest seed number in inflorescences.

16.4   CONCLUSIONS

- On the stage of stooling the green mass of most clover samples has a good quality. On the stages of budding and beginning of flowering (which are especially important for evaluation of economic and biological traits) content of dry matter, protein, ash, calcium and sugar has an insignificant positive dynamics in all tested samples, while content of cellulose has increased significantly. Dynamics of percent content of fat, potassium and calcium varied among samples on different stages of plant development both in positive and in negative direction.

- The following samples are of greatest genetic value for further selection on economic and biological traits: recommended for North Ossetia cultivar Daryal, introduced from Belarus cultivar SL-38, local wild samples Dargavski and Gornaya Saniba.
- Cultivar SL-38 has high adaptive capacity. Coincidence of reaction norm of genotype of this cultivar with main ecological and climatic parameters of introduction zone, which is the criterion of successful adaptation, allows its successful introduction in conditions of North Ossetia – Alania and its use in further selection for creation of hayfield-pasture cultivars along with recommended cultivar Daryal and wild clovers Gornaya Saniba and Dargavski (which have maximal adaptation to highland conditions).
- Use of mentioned samples in further selection work permits to solve successfully the problem of combination of high yield, seed productivity, nutrient value and environmental tolerance.

KEYWORDS

- **Budding**
- **Ecological-geographical natural habitats**
- **Hayfield-pasture**
- **Inflorescence**
- **Introduction**
- **Red clover**
- **Seed number**

REFERENCES

1. Bekuzarova S. A. (2006). Selection of clover/Bekuzarova, S. A. *Vladikavkaz. Publishing House Gorsky State Agrarian University.* 176 p. (In Rus.).
2. Klever v Rossii. (2002). Ed. by Shpakova, A. S., Novoselova, A. S., Kutuzova, A. A. et al. Voronezh, E. A. *Bolhovitinov Publishing House.* 297 p. (In Rus.).
3. Csekat of J., & Ribok H. (1989). Chrome in of Blakstonia of koniezyny of biatei/of *Poznanskie Copartnership of Sciences, 67,* 23–29. (In Pol.).
4. Bekuzarova, S. A., & Chebotaeva. S. A. (1988). Economic biological and biochemical description of standards of world collection of clover grassland Backlogs of increase of

production and use of forage/Bekuzarova, S. A. *Ordzhonikidze. – Publishing House IR.,* 54–58. (In Rus.).

5. Pleshkov, B. P. (1976). Workshop on plant biochemistry/B.P. Pleshkov. Moscow. Publishing House Kolos *("Ear")*. 250 p. (In Rus).

6. Kul'turnaja flora: V. XIII. Long-term leguminous herbares/Ed. by N.A. Muhinoj, A.K. Stankevich. – Moscow. – Publishing house Kolos *("Ear")*. 1993, 335 p. (In Rus.).

7. Baskakova L. E., Baskakov, L. E., Smorodova-Bianchi, G. B., & Yagunova, G. (1975). Chemical composition clover in conditions of Leningrad region *Bulletin of All-Union institute of plant growing of N.I. Vavilov. 55*, 33–40. (In Rus.).

8. Korjakina, V. F., & Smetannikova, A. I. (1970). Physiology of clove and alfalfa Korjakina, V. F. Physiology of clover Physiology of agriculture plants. Moscow. *Publishing house of the Moscow state university. 4*, 279 p. (In Rus.).

CHAPTER 17

A LECTURE NOTE ON RESTORATION OF BIODIVERSITY IN MOUNTAIN PLANT COMMUNITIES BY MEANS OF SELECTION

S. A. BEKUZAROVA

CONTENTS

ABSTRACT

Native species of mountain phytocenoses were studied. Methods of selection of valuable genotypes for formation of hayfield-pasture cultivars were elaborated. Methods of evaluation of selection samples for high competitive capacity were worked out. Rules of selection of plants with the maximal number of generative organs taking into consideration the altitude above sea level are established.

17.1   INTRODUCTION

The most important chain of the adaptive approach in selection is the elaboration of principles and methods of phytocenotic selection, that is, creation of competitive varieties capable to adapt in mixed crops in meadow diversity of mountain hayfields and pastures.

For sustainable development of mountain phytocenoses' biodiversity it's necessary to set up systems of environmentally differentiated varieties of fodder crops with tolerance to extreme and destabilizing environmental conditions. Environment-evolutionary principles are becoming dominant in selection strategy of fodder crops in recent years. They are based on the theory of adaptive system of plant growing and provide for creation of geographically and environmentally differentiated varieties [1].

As it is known [2, 3], wild plants from natural ecosystems have such valuable traits as longevity, frost hardiness, drought resistance and high concentration of nutrient elements.

The existing varieties of meadow-pasture grasses, as practice shows, are unable to form agrocenosis in specific mountain conditions, because they don't have enough productive longevity. Recommended varieties of red clover (*Trifolium-pratense*) aren't efficient in undersow due to low competitive capacity with native species of legumes. They have low survival rate of shoots, and even those plants that have survived don't live for long and soon drop out of the grass stand, which leads on practice to the excessive labor and capital costs [4, 5].

For mountain regions with complex environmental conditions cultivars with resistance to stress are needed. Such cultivars are lacking nowadays, because in Russia selection for potential productivity is traditionally preferred. Combination of high productivity and environmental resistance is a hard task. Decrease of adaptation level of modern cultivars is the result of limitation of their genetic base due to the use of a small number of genotypes as well as long and intensive selection in the constantly repeating environmental conditions.

For the long period of selection studies we have defined that creation of most productive cultivars in mountain environmental zone by well-known effective methods is unavailable for the cultivars of hayfield-pasture type, because created populations have the main drawback–the low adaptive capacity in conditions of vertical zonal differentiation of mountain slopes. Besides, studied selection samples in one-species crops had the minimal competitive capacity. All cultivars in selection are tested in one-species crops. We suggest using mixed crops in evaluation of competitive capacity.

For creation of competitive cultivars for mountain phytocenoses selection samples were evaluated in the mountains (600, 900, 1200 and 2000 m above sea level), sowing selected plants in the mixture with grasses and motley grasses of the wild flora.

Among gramineous plants we have chosen timothy-grass (*Phleumpratense*), because it belongs to loose-bush plants, which have the node of bushing out on the small depth (1–5 cm). The loose-bush gramineous plants (*Phleumpratense, Festucapratensis, Dactylisglomerata, Arrhenatherumelatius*) have over-ground shoots coming out from one node of bushing under the acute angle to the main shoot, forming as the result a loose bush. Every year new shoots grow in this bush, and each of them has its own node of bushing out. In their turn new shoots grow from those nodes, and the bush increases in its volume, but it remains loose, because new shoots after coming out of the ground stand not far from each other. Loose-bush gramineous plants have denser root rosette than rhizome species.

The choice of gramineous loose-bush component is based on its capacity to form the dense sod and to replace legumes. This biological peculiarity of the loose-bush gramineous plant provides for a possibility to evaluate the selection sample of clover in the tough conditions of a phytocenosis.

Gramineous and legume grasses have different demand for warm, light and nutrients. Legumes take from soil larger amount of calcium, manganese and chlorine, while the gramineous plants exceed legumes in the uptake of phosphorus and silicon. Calcium and chlorine are contained mainly in lower layers of soil with limestone streaks coming from the mother rock, and legume grasses with long roots can better take them from these layers. Due to nodule bacteria living on legume roots the gramineous plants are better supplied with nitrogen. At the expense of their dying out roots they supply with nitrogen nutrition the components of gramineous grass stands.

The process of bushing out in gramineous plants usually begins 1–1.5 months after emergence of shoots over ground. Formation of shoots happens at the expense of photosynthesis in the green parts of the plant and not at the expense of reserve substances. In natural pasture plant communities the loose-bush gramineous plants using their sod-forming capacity can replace the legume component (especially red clover, bird's foot trefoil, sainfoin). So for the evaluation on their competitive ability legumes should be sown with such gramineous plants. This biological trait of gramineous plants provides for a possibility to evaluate the legume component in tough conditions.

Evaluation of legumes in grass mixtures is fulfilled both on the quantity of shoots per plant in the first year of life and on the quantity of survived plants per square measure during all years of trial. If in the pure crop the number of shoots reaches 12–15 dependent upon the sample, in the mixture with gramineous plants the number of shoots per plant doesn't exceed 3–5. The number of flowering shoots per stem reduces from 7–10 in the pure crop to 2–3 in the mixture.

The optimal proportion of legume component in natural conditions of an ecosystem (i.e., under ideal ratio of grasses on pastures) should be 40–50% [2].

If the promising sample stood the test in comparison with the standard in pure crop and individual standing on complex of traits, but received low marks for competitiveness (9-grade international system), it would be classified unfit for formation of a meadow-pasture cultivar. However, it can be used as basic material for varieties of field fodder production.

Evaluation of samples in phytocenotic selection on competitiveness includes registration of number of legume shoots, height of plants, number of generative organs and number of seeds in inflorescences. According to the procedure adopted by CMEA countries [7] competition capacity in grass stands is determined by 9-point system:

1–very bad, when clover plants are depressed, develop poorly and fall out from grass stand the next year;

3–bad, when after weak growth plants have few shoots on 2nd and 3rd year;

5–medium, when more than half of plants develop well and flower in the year of sowing;

7–good, when 70% of legume plants survive on 2nd and 3rd year with good bushiness (4–5 shoots) and flowering inflorescences;

9–very good with survival of all sown legume plants.

But legumes, including clover, in natural ecosystems usually compete with motley grasses, which are dominant in the grass stand (above 50%). Due to this competition many legume species drop out of the grass stand.

Newly created cultivars of meadow-pasture legume grasses undersown in mountain phytocenoses have low adaptive capacity. So the binary mixture with the gramineous component isn't efficient enough for the selection evaluation of samples.

The aim of this study is the selection of the most adaptive plants for the complex of traits, among which the main is the competitiveness in the grass mixture including gramineous component (timothy-grass) and motley grass representative (*Poteriumpolygamum*).

17.2   MATERIALS AND METHODOLOGY

To achieve the mentioned aim clover and timothy-grass were sown in ratio 1:2 (one part of clover and two parts of timothy). *Poteriumpolygamum* was added in the amount of 15–20% of clover-timothy mixture. Competitive samples of legumes were selected on the 2nd year. Those samples were considered to be competitive, which survived. On their base a new meadow-pasture cultivar was formed.

Poteriumpolygamum belongs to family *Rosaceae*. This plant has high fodder qualities, a taproot, high winter hardiness, longevity, resistance to cold and drought. For such qualities it's included into the mixture of motley grasses. In the year of sowing *Poteriumpolygamum* develops a vigorous root system and a rosette of leaves. It makes possible to determine the most productive and competitive plants in the early period of development of selection samples in collection nurseries. In the mixture with the legume-gramineous component *Poteriumpolygamum* secures the high yield of forage mass.

We have included in the experiments evaluation of 18 samples of meadow clover in the collection nursery. Zoned variety of red clover from North Caucasian Region–Daryal–was used as standard. The square of each plot equated 5 square m. The studied samples were compared with the zoned variety known for its high longevity and quite stable productivity from year to year. On each plot we have sown the seed mixture of clover (7.5 g), timothy-grass (4 g) and *Poteriumpolygamum* (2.3 g, which consists 20% of the mixture of clover and timothy).

17.3   RESULTS AND DISCUSSION

The results of the experiments are shown in the Table.17.1.

TABLE 17.1   Evaluation of Red Clover Samples on Competitiveness in Mixture With Timothy-Grass and *Poteriumpolygamum*

Name of samples	Number of stems per 1 m² dependent on years of life		Survival of plants of the second year of life in comparison with survival on the first year of life, %
	First year	Second year	
Daryal–standard	12.5	4.2	33.6
Alan	15.2	4.8	31.6
Nart	13.8	4.0	28.9
Wild populations			
Iraf	12.6	5.6	44.4
Gizel	10.1	5.2	51.5
Dzinaga	11.4	6.4	56.1
GornayaSaniba	12.2	7.2	59.0
Dargavs	10.4	5.8	55.8
Synthetic populations			
Syn 305–03	10.8	6.8	62.9
Syn300–09	11.0	6.0	54.5
Syn314–08	11.8	5.8	49.1

TABLE 17.1 *(Continued)*

Name of samples	Number of stems per 1 m² dependent on years of life		Survival of plants of the second year of life in comparison with survival on the first year of life, %
	First year	Second year	
Syn 316–08	10.9	5.9	54.4
Syn 319–08	12.8	6.4	50.0
Syn 320–08	12.4	6.8	54.8
Syn 321–08	11.6	6.8	58.6
Syn 322–08	12.1	7.2	59.5
TOS-31	11.4	6.5	57.0
SGP-189	12.6	6.0	47.6

Cultivars Daryal (standard), Alan and Nart decrease the number of stems on the second year of life, which means their poor competitiveness with gramineous and motley grass components. Their adaptation in the mixture is 29–33%. Wild forms from highly elevated mountain districts of North Ossetia (villages Dzinaga, GornayaSaniba, Dargavs) have greater competitive capacity over 50%.

Synthetic populations formed from native species of mountain ecosystems have the maximal competitiveness measured by the number of survived plants.

In the selection of plants for creation of complex hybrid population we took into consideration not only competitiveness, but also a group of economically valuable traits: yield of overground mass, high seed productivity, resistance to diseases, fodder merits, longevity and winter hardiness. Correlations are calculated between all those traits, and it makes possible to select best genotypes in a short time.

Wild samples Dargavs, Iraf, Dzinaga, GornayaSaniba, which were a part of synthetic populations, distinguished themselves by good fodder merits. The length of their stems in the phase of stooling was 5–7 cm less than by zoned cultivars. But in the flowering period in mixture crops the wild forms reached the level of well-known cultivars. They also had advantage on the number of leaves per stem (2–6% above all other studied samples). The rate of leaf cover of stems (58–69%) was the largest among the samples, which originated from the highlands most elevated above sea level. The protein content in plant samples increased under the same rule of vertical zoning of natural habitats.

Biochemical analysis of wild introduced samples showed that the red clover populations had high content of protein and low content of cellulose in the phase of stooling (27% and 14.5%, respectively). In the phase of flowering the content of protein in the absolutely dry substance reached 19.7–23.2%, while the content of cellulose was 17.2–20.1%. In the flowering phase the content of these substances slightly decreased, but it was relatively high in comparison with zoned varieties. Plants of wild-growing forms contained 0.6–0.8% of phosphorus, 2–4% of sugar, 8–10% of ashes, which was slightly above of the qualitative characteristics of selection samples grown on the elevation of 600 m above sea level (village Mikhailovskoye).

It's important in the selection process of the red clover to create basic material with increased resistance to diseases, especially to root rot, anthracnose, ascochytose, leaf spot of clover and mildew. With the aim of receiving such cultivar samples were evaluated in natural conditions of mountains and foothills, on the infectious background, in mixed and pure crops.

Evaluation of samples showed the advantage of wild forms and complex hybrid populations, which were formed on the base of plants introduced from mountain regions. Synthetic populations Syn 305–03, Syn 321–08 and Syn 322–08 received high estimates for disease resistance (according to the method of All-Russian Institute for Plant Protection).

Incidence of the most widespread diseases in the region (anthracnose, ascochytose, leaf spot of clover) didn't exceed 1.5–1.8 points, while other samples were affected on the level 3.5–4 points.

In selection of plants in grass mixtures on the second year of life the seed productivity was taken into consideration as one of the main traits for cultivars of hayfield-pasture type.

Our studies for many years (1970–2012) showed that seed productivity varied due to climatic factors. Trials of clover in different agro-ecological zones led us to the conclusion that the optimal period of seed formation is the sum of positive temperatures (above 10°) per vegetation 1207–1648° with the quantity of precipitation 445–639 mm, and in the flowering period if seed yield reaches 1.5–1.8 q/ha, the hydro-thermic coefficient (HTC) should be 1.52–3.12. Seed yield decreases to 0.5–1.2 q/ha along with HTC growth.

Unlike steppe and foothill regions the climate of mountains has its peculiarities. With the ascent to altitudes 1300–2200 m above sea level the short-wave ultra-violet radiation becomes more intensive, plant vitality raises and stimulating influence of ultra-violet rays suppresses partly the negative temperature effect. Under sharp overfall of day and night temperatures active flowering, fruit and seed set are going. So for ripening of seeds in mountain conditions higher HTC and consequently lesser sum of effective temperatures are needed. It was determined on the results of fulfilled studies that with rise of mountain elevation wild plants (unlike cultivars) had greater percent of set seeds in comparison with foothills. Evidently this rule can be explained by the fact that low night temperatures in the mountains prevent transformation of stored in the day sugar into starch and other substances. It's known that sugar impedes freezing, and so nectar retains high quality, which is very important for such insect-pollinated plant as clover.

Comparison of native populations, cultivars and formed complex hybrid populations on different elevation made possible to show the environmental influence on seed set. We took into consideration temperature regime of air and soil, sum of precipitation, humidity and acidity of upper layer of soil (where roots are situated). It was found that on the same elevation, but on different soils seed set was different. For example, on elevation 900 m above sea level seed set was 27.5% on acid soil (pH 4.47) and 46.8% on nearly neutral soil (pH 6.45). On elevation 2000 m above sea level on two plots with pH 6.44 and 6.15 seed set of clover was 49.5 and 47.0%, respectively. It was also determined that the number of weak seeds changed in dependence upon soil acidity with the maximal proportion (above 50%) at pH 4.47. Seed set in inflorescences is higher on 9.5–27.1%, where acidity is 6.0 or above it.

It's found [9] that in selection for seed productivity it's possible to receive positive results, using phenotypic selection on following traits: coloration of flowers, number of generative shoots, size of flowering heads and inflorescences. Seed production depends heavily upon content of starch in the root crown in the period of flowering (correlation coefficient $r = 0.63$), content of sugar in nectar ($r = 0.78$) and presence of pollinators ($r = 0.95$).

17.4   CONCLUSIONS

Complex evaluation of selection samples in various conditions growing in mountains and foothills in natural plant communities, in pure and mixed crops secures creation of valuable basic material for formation of a meadow-pasture variety with such traits as high competitiveness, qualitative characteristics and maximal seed productivity.

Determined regularities of clover plants development with taking into consideration the vertical zoning make possible rational phenotypic selection and creation on this base of new cultivars for restoration of biodiversity of mountain hayfields and pastures.

KEYWORDS

- **Adaptation**
- **Clover**
- **Competitiveness**
- **Cultivars**
- **Genotypes**
- **Phytocenoses**
- **Seed Productivity**
- **Selection**
- **Selection samples**

REFERENCES

1. Shamsutdinov, Z. I. & Kozlov, N. I. (1996). Importance of genetic collection in intensification of fodder crop selection. *Selection and seed-growing., (3–4),* 9–12 (in Russ.).
2. Tyuldyukov, V. A. (1988). Theory and practice of grass farming/V.A. Tyuldyukov. Moscow. *Rosagropromizdat,* 286 p. (in Russ.).
3. (1997). Fodder production in Russia Collection of works of All-Russian Research Institute for *Fodder. Moscow*, 428 p. (in Russ.).
4. Foster, C. A. (1971). A study of the theoretical expectation of F_1 hybridity resulting from bulk interpopulation hybridization in herbage grasses. *Agr. Sci, 76(2)*, 293–300.
5. Taylor, N. 1968. Polycrossprogeny tenting of clover (*Trifoliumpratense* L.) *Crop. Sci. 8(4)*, 451–454.
6. Bekuzarova, S. A. & Dzugayeva, L. A. (1999). A method to define adaptive qualities of red clover's selection samples. Patent № 2201076. Published on 20.01.1999 (in Russ.).
7. Leokene, L. V., et al. (1983). Wide unified classificatory of CMEA and international classificatory of CMEA. Leningrad. Scientific-technical council of CMEA member-states on collections of wild and cultivated species of plants. *All-Russian Institute for Plant Growing,* 41 p (in Russ.).
8. Shamsutdinov, Z. S., et al. (1993). Methodic directions on selection and primary seed-growing of perennial grasses. Moscow. *Publishing house of Russian Agricultural Academy,* 112 p (in Russ.).
9. Bekuzarova, S. A. (2006). Selection of red clover/S.A. Bekuzarova. *Vladikavkaz*, 176 p. (in Russ.).

CHAPTER 18

BIOLOGY OF DEVELOPMENT OF PHYTOPATHOGENIC FUNGI OF FUSARIUM LINK AND RESISTANCE OF CEREALS TO IT IN CLIMATIC CONDITIONS OF TYUMEN REGION

N. A. BOME, A. JA BOME, and N. N. KOLOKOLOVA

CONTENTS

ABSTRACT

The differences were discovered between the cultivars of spring wheat for resistance to phytopathogenic fungi of the genus *Fusarium* in laboratory seed germination and seedling morphometric parameters. The effect of temperature (20 °C, 10 °C, and 5 °C) was studied on the rate of development of *Fusarium nivale Ces.* (beginning of active growth, sporulation, and diameter of the colony). Infection load in a field experiment decreased the selection and valuable features.

18.1   INTRODUCTION

Crop growing conditions in different areas of the Tyumen region formed unevenly. The climate is influenced by cold arctic air masses of the Arctic Ocean, the Asian continent, as well as dry winds blowing from Kazakhstan and Central Asia. The climate is typically continental, and all the climatic factors vary greatly over the years, both in tension and in development time, creating a cultivar of combinations. There are elements of the climate reminding the western region (dry summer periods), circumpolar areas (very short and cold growing season) and the deserts of the south (dry, oppressive weather from spring to fall) [1].

The agricultural areas of the Tyumen region are characterized by harsh cold winters, relatively short summers, short springs and autumns, a small frost-free period, and sharp changes in temperature during the year and even during the day.

One of the causes of the yield decrease in agricultural crops, including cereals, is the growth of infection of the most dangerous diseases. Plants suffer both from pathogens belonging tosoil pathological complex (root rot, *Fusarium* wilt, etc.) and from air-spread infections (rust, Septoria, smut disease, powdery mildew, etc.) [2].

Phytopathogenic fungi of the genus *Fusarium* belong to the most dangerous among more than 350 species of toxigenic fungi known in agriculture [3]. It is shown that the contamination of seeds of spring wheat can occur both in the hidden and explicit form, and to a large extent it is determined by the varietal characteristics [4].

18.2   MATERIAL AND METHODOLOGY

According to the results of our research 7 genera of pathogenic fungi were singled out in the microflora of grains of spring wheat, barley, and rye cultivars of different ecogeographical origin and different years of harvest. Of these genera, *Alternaria*, *Helminthosporium*, *Trichothecium*, *Tilletia*, *Fusarium* are representatives of field microflora, while *Mucor* and *Penicillium* belong to mold species. The fungal spores of the genus *Alternaria* dominated on most grains of all cultivars.

Pathogens from the genera *Helminthosporium* and *Fusarium* of the most harmful type, causing root rot and spot, were detected.

Taking into account the fact that pathogens of the genus *Fusarium* are common enough in the cereals (both spring and winter forms) in Tyumen region, and can cause significant yield losses, we have conducted laboratory and field studies on the biology of this genus. The experiments included method of phytopathological analysis of seeds with the calculation of the disease index [5–9].

18.3   RESULTS AND DISCUSSION

In our experiment conducted in the laboratory on four cultivars of soft spring wheat, dependence was observed of seeds' ability to normal germination from their contamination by pathogens. Cultivar Tyumenskaya 80 had the lowest laboratory germination of seeds among cultivars- 88.2% at the maximal level of infection. In less contaminated cultivars (Saratovskaya 57, Comet, Mir 11) indicators of seeds germination ability were higher: Mir 11–98.5%, Comet – 99.0%, Saratovskaya 57–99.3%.

Fungi of *Fusarium* genus are the most common pathogens among soil infections. They cause disease of roots and root collar, which leads to the death of productive stems, and the empty spike of infested plants.

According to the index of the disease of affected seedling cultivars Comet, Tyumenskaya 80, Mir 11 were classified as middle susceptible (RB = 28.15–30.45% 21.59–28.43% 20.40–29.50%, respectively), and cultivar Saratovskaya 57 – as low susceptible (RB = 12.48–18.48%) (Table 18.1).

TABLE 18.1   Evaluation of Spring Wheat Samples on Infectious Background for Resistance to *Fusarium sp.*

Cultivar	Phytopatholog- ical analysis of seeds		Benzimidazole method		The rots of root		The score
	P_δ, %	Score	P_δ, %	Score	P_δ, %	Score	
Tyumenskaya 80	21.59	2	24.00	4	25.55	2	8
Comet	28.15	1	35.43	1	21.87	3	5
Saratovskaya 57	18.48	4	34.35	2	21.11	4	10
Mir 11	20.40	3	24.27	3	28.89	1	7

Note: **>40%** – susceptible, 20–40% – middle susceptible, 10–20% – low susceptible.

A stronger root growth and inhibition of vegetative parts were observed in the study of morphometric parameters of the background of the infected seedlings (Table 18.2).

TABLE 18.2   Quantitative Traits Indicators of Spring Wheat Samples on Infectious Background of *Fusarium sp.*

Cultivar	Option	Length of sprout $X\pm m_x$, cm	Length of roots $X\pm m_x$, cm
Comet	Control	23.52±0.61	14.09±0.43
	Experiment	21.75±0.81	20.39±0.81*
Saratovskaya 57	Control	22.09±0.95	18.53±0.33
	Experiment	20.83±0.71	19.30±0.96
Mir 11	Control	24.25±0.75	19.28±0.71
	Experiment	19.63±0.75ˣ	20.72±0.62
Tyumenskaya 80	Control	26.02±0.60	18.96±0.57
	Experiment	19.81±0.67ˣ	24.96±0.42ˣ

Note: '*' is the differences were statistically significant at $P < 0.05$.

Cases of stimulation of the growth processes of infected plants are described [10]. Often this phenomenon is temporary and connected to the physiological characteristics of the pathogen. Intensive growth of roots and lagging behind of overground parts of the plants can probably be explained by the fact that the introduction of the pathogen into the roots of the plants leads to blockage of vascular system, disrupts the transport of water and dissolved substances, reduces the rate of photosynthesis, and therefore, produces a delay of plant development.

Comprehensive assessment on the grounds that characterizes the intensity of seed germination, seedling variability of quantitative traits and primary root system, have allowed to identify cultivars of spring wheat Tyumenskaya 80, Mir 11, Saratovskaya 57, as the most resistant to infection.

Productivity of winter crops forms is dependent upon a number of biotic (pathogens) and abiotic (temperature, rainfall, etc.) factors. Pathogenic fungi that cause disease play negative role in plant growth and development. In particular the snow mold, which is caused by *Microdochium nivale* (Fr.) Samuels and I.C. Hallett (*Fusarium nivale* Ces. ex Berl. and Voglino), is dangerous. It is widely specialized facultative parasite, always present in the soil.

One of the factors that determine the development of the fungus is the temperature. *Fusarium nivale* Ces. begins to develop at 5 °C, the optimal growth is observed at 11–17 °C [11, 12]. In our laboratory studies performed with U.B. Trofimova [13] the effect of temperature on the rate of development of the fungus was studied. By cultivating the fungus in the oven at 20 °C, 10 °C and 5 °C on potato glucose agar in Petri dishes in the three-fold repetition we determined the diameter of the colony and especially sporulation.

The lowest rate of growth of the fungus was recorded at 5 °C. Beginning of the growth in this variation was observed on the 8th day after sowing. Fungal colonies reached the diameter of Petri dishes on 42th day, with sporulation recorded only on 56th day (Fig. 18.1).

At a temperature of 10 °Con 4th day of the experiment diameter of the colony was equal to 12.5 mm, and after 28 days Petri dish was completely occupied by the fungus. In this variant sporulation happened much earlier – on the 16th day. The fastest growth of *Fusarium nivale* Ces. colony was observed at 20°

C. The active beginning of growth was already evident on 2nd day, sporulation was observed on the 6th, while on 8th day of the experiment the colony's diameter reached 90 mm.

Beginning of the growth	2 days	4 days	8 days
	20°C	10°C	5°C
Start sporulation	↑ 6 days ↓	↑ 16 days ↓	↑ 56 days ↓
The diameter of the colony	87.7±2.03 mm	33.0±2.00 mm	90.0±0.00 mm

FIGURE 18.1   Effect of temperature on the development of fungus *Fusarium nivale* Ces.

Development of snow mold is determined by weather conditions of the spring period and isn't observed every year, so any conclusion on plants' resistance can only be made in the years of strong manifestation of the disease.

One of the conditions to obtain reliable results in the determination of resistance is the creation of an artificial background ensuring optimal infection load. This background on the experimental site was created by application into soil of an aqueous suspension of spores and mycelia of pure 14-day culture of *Fusarium nivale* Ces. Infectious load was 10^6 conidia/mL of inoculum (500 mL/m² of soil). Infection was carried out in autumn in the phase of bushing out before snow cover. Estimate of snow mold infection of plants was carried out 10 days after snow melting in the early resumption of the growing season according to methodical guidelines of Kobylyansky [14], on a scale worked out by Andreev and Molchanova [15].

A study of infection in vivo and hard infectious background revealed that harmfulness of snow mold manifested in the reduction of such morphometric characteristics of winter rye as plant height, leaf area, and productivity traits. Decline of more than 50% was noted in leaf area per 1 m², number of grains per plant, grain weight per spike and plant. There was a strong development of disease on the infectious background, which resulted in lower yields compared to the control samples on average by 38.1%.

18.4 CONCLUSIONS

In the growing season, characterized by a long warm autumn, conditions favorably evolved for active growth of the pathogen. Effect of pathogen was aggravated by soil and air drought in spring and summer. In the experimental variant with infectious load inhibition of growth processes was observed, which manifested in significant reduction in breeding-valuable features to 26.21–67.70%.

To the group of resistant cultivars of winter rye belonged Chulpan, Ilmen Iset and Supermalysh 2, wave to middling susceptible– Voshod 1, susceptible– 8s-191 Rossianka × Getera, Desnyanka × Imerig, Tetra and Siberia.

KEYWORDS

- **Cultivar**
- *Fusarium*
- **Phytopathogenic fungi**
- **Snow Mold**
- **Spring Wheat**
- **Stability**
- **Winter Rye**

REFERENCES

1. Ivanov, P. K. (1971). Spring wheat/Moscow. Publishing house Kolos (*"Ear"*), 328 p. (In Russian).
2. Kosogorova, E. A. (2002). Protection of field and vegetable crops from diseases/Tyumen. Publishing house of the Tyumen State University, 244 p. (In Russian).
3. Kudayarova, R. R. (2007). Mitotoksiny. Problems and prospects of the development of innovation in agricultural production. All-Russian Scientific and Practical Conference of the XVII specialized exhibition Agro Complex. Ufa. Bashkir State Agrarian University. Part 2, 79. (In Russian).
4. Khairulin, R. M. & Kutluberdina, D. R. (2008). The prevalence of fungi of the genus *Fusarium* in grain of spring wheat in the southern forest of the Republic Bashkortostan. *Bulletin of the Orenburg State University, 12*, 32–36. (In Russian).
5. Naumova, N. A. (1970). Analysis of seeds to fungal and bacterial infection. Leningrad. Publishing house Kolos (*"Ear"*), 32 p, (In Russian).

6. (1981). Evaluation of crops for resistance to diseases in Siberia. Guidelines. – *Novosibirsk*, 48 p, (In Russian).

7. Mikhailina, N. I. (1983). Comparative evaluation of methods for determining the severity of root rot of spring wheat *Agricultural Biology, 4,* 95. (In Russian).

8. (1977). Guidance on the study of the stability of the grass to the agents of diseases of the conditions for nonchernozem zone of the Russian Soviet Federative Socialist Republic / Leningrad. – *All-union institute of plant protection*, 60 p. (In Russian)

9. Zrazhevskaya, T. G. (1979). Determination of the resistance of wheat to common root rot *Mycology and phytopathology, 13(3),* 58.

10. Rodigin, M. N. General phytopathology Moscow. – Publishing house High School. (1978). 365 p. (In Russian)

11. Rubin, A. (1970). Crop physiology; IV. Leguminous plants. Perennial grasses. Cereals (rye, barley, oats, millet, and buckwheat). Moscow. Publishing house Moscow State University, 654p. (In Russian).

12. Yakovlev, N. (1992). Phytopathology. Programmed instruction/Moscow. Publishing house Kolos ("*Ear*"), 384 p. (In Russian).

13. Trofimova, U. B. & Bome, N. A. (2006). Parameters of snow mold damage and resistance of winter rye to illness. *Journal of Plant Protection. St. Petersburg – Pushkin, 1,* 33–36. (In Russian).

14. Kobylyansky, V. D. (1981). Guidelines for the study of the world collection of rye. (Eds.) Leningrad. All-*Union Institute of Plant Growing of Vavilov N I.*, 20 p. (In Russian).

15. Andreev, V. & Molchanov, O. (1987). Snow mold of winter grains (Methods of study and control measures). Moscow. *Research Institute of Technic-Economic Researches*, 46 p. (In Russian).

CHAPTER 19

THE BASIC CHARACTERISTICS OF CONTINUOUSLY CULTIVATED SOD-PODZOL SOIL OF NORTHERN TERRITORIES (IN THE CASE OF TOBOLSK DISTRICT, TYUMEN REGION)

I. A. DUDAREVA (CHERKASHINA), and N. A. BOME

CONTENTS

ABSTRACT

In this chapter, the main characteristics of a well-cultivated soddy-finely podzolic soils of the Tobolskraion, Tyumen region. Here presented also the results of study of morphological characteristics of physical and chemical indicators and elemental composition of the soil type.

19.1 INTRODUCTION

Nowadays comprehensive and detailed study of the main characteristics, which are designated properties of certain sod-podzolsoil for growing of cultivated plants in adverse agroclimatic conditions of northern territories are of relevance [1–6].

Tobolsk district is situated in the northern part of Tyumen region, in subboreal forest zone and occupies 17.222 km². Summer time climate is formed under the influence of cyclones mowing from the west. However, intrusion of arctic air causes cooling and frosts in the beginning and in the end of summer period. Anticyclones of the Central Asia enhance climate continentality in wintertime, which leads to relative severity of the period [7]. Territories of the region are characterized by severe cold winter and short frost-free period. Not only annual but daily sharp temperature fluctuations are observed, especially at spring. Climate instability is related to unhampered intrusion of arctic air masses from the north and dry air masses from Kazakhstan.

By its hydrological-climatic conditions Tobolsk district belongs to highly humid zone and zone of insufficient heat supply [8].

Climatic conditions have essential influence on soil formation process and determine geographical and physic-chemical uniqueness of soil cover. This region differentiates by the great cultivar of soils. Main soil types are: floodplain, podzols, sod-podzols, gray forest soils, alluvial-meadow soils, boggy soils, black soils, sodium soils, ash gray soils, loessial loam.

Big spatial and time contrast of edaphic-climatic characteristics complexities growing conditions of cultivated plants [2]. That's why it is essential to know morphological, physical and chemical peculiarities of soil cover and its ultimate composition for realization of ecological and biological potential of crop cultivar [6, 9, 10].

The works goal is study of morphological features, physical and chemical characteristics and elemental ultimate composition of soil from experimental plot with the account of meteorological factors.

The research was conducted in 2009–2011 years on the experimental plot in Malaya Zorkaltseva village, Tobolsk district of Tyumen region, according to agroclimatological zoning situated in subboreal region.

Data on air temperature in Tobolsk district for 1900–2008 years were obtained on "Joint hydrometeorological station of Tobolsk district." Analysis of multiannual data showed that average monthly air temperatures in the period from November to March has negative values, and from April to October–positive, mean yearly air temperature was 0.6 °C.

Yearly variation of temperatures is characterized by minimum in January–February (−17.2°C) and maximum in July (+18.6°C). Mean temperature of cold period from November to March is −13.2°C. The coldest months were January and February with minimal temperatures −48.5°C (1964) and −47.7°C (1967). Duration of period with mean daily temperature above 0 °C is 190 days, above +5°C and +10 °C–157 and 116 days, respectively. According to multiannual data minimal July temperature is +39.6°C (1901). Spring begins after 10 April with the passage of daily temperature through 0 °C, but in the end of May and in the beginning of June snowfalls are possible. However, at times warm and even dry weather (+15–23°C) sets because of moving of dry warm air masses from Kazakhstan. Time of passage of mean air temperature through +10°C is accepted to denote the beginning of summer (21 May–10 June).

Characterizing thermal regime of soils form Tobolsk district it is possible to remark that they undergo rather protracted and deep freezing in winter; slow thawing and warming of soils is common in spring.

By the results of profiling it was established that soil of the plot is well-cultivated residual carbonate, sod-podzol on ancient alluvium deposits. The surface is billowy, without erosional features, profile character is simple unbroken. Signs of textural and structural profile dissimilarity are layers of heavy grain texture up to B-horizon and light grain texture in C-horizon. Gleization signs are absent, calcareousness features are weak. Profile strength is 110 cm. Parent rock material–sand alluvial deposits of the first terrace above the Irtysh river floodplain. Soil profile composition is the next: Ap (0–38 cm), E (38–48 cm), EB (48–76 cm), Bh, f, al (76–93 cm), C (93–110 cm).

Productivity of plants cultivated on the soil with high cryogenic load is determined by the level of warm and water supply during their growth and development. The object of research in our field experiment was soft summer wheat.

The 2009–2011 years vegetation seasons were considerably differed both between them and with mean multiannual values. Observed fluctuations of mean daily air temperatures were from +10.8°C (May) to +17.2 °C (July) in 2009, from +8.9°C (September) to +17.5 °C (July)–in 2010 and from +10.8°C (May) to +18.0 °C (June)–in 2011. Minimal daily temperatures were observed in May 2009 and 2011 and in September 2010; maximal temperatures–in July 2009 and 2010 and in June 2011. It is possible to characterize the years of the research as warm with sharp fluctuations of daily and monthly temperatures. The overage above multiannual data made: in 2009–2.8 °C, in 2010–0.5 °C, in 2011–3.2 °C. Sum of active temperatures was above the normal (1500–1700°C) at the average for the whole period on 219 °C and made 1977.4 °C in 2009, 1855.4 °C in 2010 and 1925.3 °C in 2011.

Precipitation total during plant vegetation was close to norm in 2009 and in 2011 and made 311.3 mm and 358.2 mm in each year, respectively. 2010 year vegetation period belongs to critical by precipitation (221.9 mm), which is lower than norm for 73.1 mm. Months characterized by shortage of moisture were determined: July and September 2010 (precipitation totals were 19.9 and 30.0 mm, respectively); may, august and September 2011 (precipitation totals were 8.5, 46.6 and 39.5 mm, respectively). In some periods minimal precipitation values were noted: 2009 (July–91.4 mm, August–89.6 mm), 2010 (July–61.4 mm, August–78.7 mm), 2011 (June–162.5 mm, July–101.1 mm).

Laboratory analyzes were made on the basis of accredited laboratory "Ecotoxycology" (POCC RU. 0001.516420) of Tobolsk complex scientific station, Ural division of the Russian academy of sciences.

Sample collection for the research was made by soil sampling tube according to Russian state standard (GOST)2816–89, Ruling Documents (RD) 52.18.156–99, GOST 17.4.3.01.

Soil moisture is important characteristic for passing of ontogeny stages by organism from the moment of seed germination. Moreover, this characteristic has immediate impact on soil chemical composition because of having influence on transition of chemical elements from immobile forms to mobile.

Soil moisture regime for certain period was determined according to GOST 28268–89 during vegetation (May, June, July, August, September). Moisture content in the soil (39.4%) was sufficient and favorable for summer wheat seed germination and sprout forming in June 2011. Soil moisture was low in 2009 and 2010 (11.7 and 13.6%, respectively), which was reflected by field germination rate and biological resistance of plants. The important period in water consumption is thought to be booting and ear formation stage, that is, period of reproductive organs formation, which comes to be in July in our research. Maximally hard conditions for wheat were observed in July 2010, which was characterized by precipitation deficit on the background of increased air temperatures. Soil moisture during this period was no higher than 6.8%. Low soil moisture (9.0%) was observed in August 2011, when milk stage of grains took course and plants consumes 20–30% of all moisture during vegetation period. Warmth and water regime influenced on soil chemical properties, the degree of mobility of different elements and plant ability to consume them through its root system.

Soil acidity is stated by the negative logarithm of the hydrogen ion concentration pH. This characteristic determines availability of chemical elements for plant organism. It is worth to mention that amelioration of sod-podzolic soils leads to changes in qualitative composition of organic matter, decrease of fulvicacids composition and increase of lime humates. At the same time as a result of amelioration the base exchange capacity increases and composition of exchangeable cations changes: increasing of consumed Ca^{++} and Mg^{++} and decreasing of exchange H^{+} and Al^{+++}.

Reaction of soil solution markedly changes because of saturation of soil by Ca^{++} and Mg^{++} cations: pronounced acidity, which is characteristic to virgin soils, gradually replaced by subacidic and sometimes neutral and weakly alkaline reaction.

Biological activity of soil microbal flora–nitrate bacteria and azotobacter, which does not occur in virgin soils and weakly cultivated soils or occurs in very fractional amount, intensifies because of it [11].

19.2 MATERIAL AND METHODOLOGY

Soil of the experimental plot belongs to weakly alkaline type and has medium pH 7.70. It is known that nutrients and chemical elements for wheat plants will be available under pH range 6.0–7.5. If pH level is lower, key nutrients will be either less available or become toxic for plants. Therefore, soil pH of experimental plot can be referred as satisfying to requirements of the culture.

Dry residue (solid residue) is a characteristic of soil salinity, it is determined by the ratio of anions and cations in the soil solution. In normal conditions it can't exceed 0.30%, in soil samples it is equal to 0.35%. Salinity is determined by salt content in soil solution. Salts are need to be formed mostly sodium, calcium and magnesium cations with chloric and sulfuric anions. Potash cations, bicarbonate, carbonate and nitrate anions can make insignificant part. Thus results by dry residue in soil, which were obtained in this experiment, allow drawing conclusion that anion and cation amount is optimal and they are the main compounds of the soil in current agroclimatic conditions.

19.3 RESULTS AND DISCUSSION

The amount of anions in soil is considerably less than the amount of cations. Anion-cation balance is shifted toward cations (Fig. 19.1).

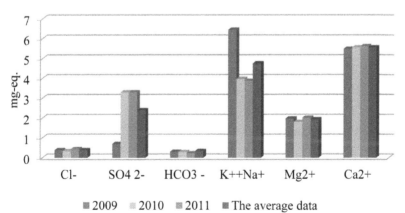

FIGURE 19.1 The content of anions and cations in the soil of experimental site (Cl^-; SO_4^{2-}; HCO_3^-; $K^+ + Na^+$; Mg^{2+}; Ca^{2+}).

On the basis of the obtained data on biogenic substances, it may be concluded that the nitrogen in the soil is presented in three forms: ammoniacal (NH_4^+), nitrate (NO_2^-), nitrite (NO_3^-), there is a sufficient amount of it in the soil (Fig. 19.2).

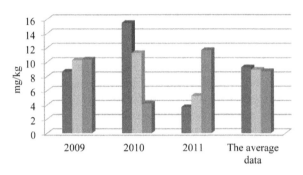

NH4+   NO2-   NO3-

FIGURE 19.2   Nutrients in the soil of experimental site (NH_4^+; NO_2^-; NO_3^-).

Phosphorus is available in mobile forms $H_2PO_4^- \times HPO_4^-$, it is contained in the soil in large amount. Its accumulation is conspicuous over time, the maximum was observed in 2011, the minimum in 2009, respectively (Fig. 19.3).

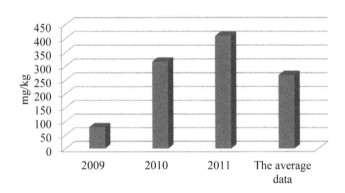

H2PO4- и HPO4-

FIGURE 19.3   Mobile forms of phosphorus in the soil of experimental site ($H_2PO_4^{-\times} HPO_4^-$).

The humus content (I. V. Tyurin method in V. N. Simakov modification), in the soil is not high and it varied slightly: year 2009–1.86%, year 2010–1.45% year 2011–1.76%.

The total content of the chemical elements (As, Ca, Cd, Co, Cr, Cu, Fe, Mg, Mn, Mo, Ni, Pb, Sb, Sr, Zn) was estimated in soil samples, as well as their stationary and mobile forms, by atomic emission methods of inductively coupled plasma on the spectrometer OPTIMA-7000 DV (Perkin Elmer).

According to the research years average data on the total composition, it was found that the topsoil contained, mg/kg: As–16.95 ± 0.18; Ca–456.00 ± 18.66; Cd–6.83 ± 0.00; Co–16.85 ± 0.00; Cr–16.43 ± 0.00; Cu–0.99 ± 0.19; Fe–301.73 ± 21.62; Mg–288.93 ± 29.66; Mn–96.65 ± 1.97; Mo–16.23 ± 0.00; Ni–17.04 ± 0.64; Pb–25.71 ± 1.64; Sr–1.89 ± 0.12; Zn–21.72 ± 1.28.

The reference points, such as clarks for chemical elements, are necessary for detection of natural soil-geochemical changes and correct assessment of the soil composition of elements, as well as hygienic standards: maximum and approximate permissible concentration (MPC) and (APC). The study used the conventional world soil clarke, proposed by D. P. Malyuga [12].

As a result of determination of chemical substance concentrations (C_c) in soil, it was found that chemical elements form two groups in background sample comparing to conventional world Clarke [6]. The excess over the MPC is not revealed.

The elements of the 1st group–Co, Mo, Pb, Cd are characterized by an increased content in the soil relative to clarke, which indicates the accumulation of these substances, but not higher than the MAC level. The coefficient of the chemical substance concentration (Cs) is from 1.61 units − Co to 3.42 units − Cd. The increased content of these elements may be due to geochemical characteristics of the parent rock materials.

More numerous 2nd group of elements −Mn, Cr, Sr, As, Ni, Cu, Zn, Mg, Ca, Fe, shows the deficit relative to clarke, its critically low content in the soil, suggesting that there is an elements subtraction.

Sample preparation for the determination of chemical elements stationary forms, associated with the various soil components was performed to obtain the acid extracts in the microwave decomposition system under pressure Speedwave MWS-2 (made in Germany BERGHOF Products + Instruments GmbH), using the individually selected mode. This process applies extra pure HNO_3 and HCL acids, which additionally underwent the distillation in the purification system BSB-939-IR.

According to the content, in the form of stationary compounds, bounded with soil components, elements can be conventionally divided into two groups: calcium, iron, magnesium, manganese occur in a large amount in the stationary forms (from 64.98 mg/kg–Mn, to 284.90 mg/kg–Ca) and form a small cluster. Other chemical elements are poor in the soil (0.39 mg/kg–Mo, to 6.74 mg/kg–Pb) in stationary forms and form a large group of trace elements.

Mobile element forms were determined by the chemical fractionation. In the present study, after the comparative overview of the most commonly used chemical elements fractions and extraction agents, the Sposito method appeared to be effective for their recovery. However, some extraction agents of the method were replaced by those with similar properties and device-oriented. Furthermore the oxide fraction, extracted by the Pampura method, was added to the factions set, offered by Sposito. This fractionation allowed detection of the mobile element compounds in soil samples of 2009–2011 years, to determine their chemical fractions and ratio, to conduct a statistical analysis of samples.

Three fractions (of the total amount of mobile forms) were found to be dominant for As, Ca, Cd, Co, Mo, Ni, Pb: exchange fraction–26.7% (Ni)–29.1% (Co), organic–26.6% (As)–34.3% (Ca) and soluble–18.2% (Ca)–21.8% (Co). Elements are bounded with both the various soil components (mineral components, hydroxides and oxides, colloids) and the organic substance, in almost equal proportions, form stable complexes with them and play the principal role in plant nutrition. The water-soluble forms are most mobile, potentially more mobile, since they are transported by surface and ground waters, are easily involved in biogeochemical migration, and are available to plants. In addition, the exchange and the water-soluble fractions are the reserve of plants nutrition.

The most of the Mg and Sr content in the soil also falls at the organic (28.8% and 28.7%) and the exchange fraction (22.1% and 26.9%); 17.1% of magnesium of the mobile forms total content is contained in residual compounds. This is a strategic reserve, bounded with carbonates and bicarbonates Fe, Mn, Al. The manganese proportion in the organic fraction is 60.6%, and only 16.7%–in the exchange fraction. The iron is in the organic–39.1%, oxide–28.6%, residual fraction–26.2%, it is presented in the soil (mainly) in the form of stable complexes with organic substance. Its occurrence is considerably less in the form of cations and sesquioxides hydrates in colloid-soluble form. The copper is bounded with carbonates and bicarbonates, it is found only in the carbonate fraction (100%). The proportion of zinc oxide fraction is 91.3%, the residual fraction–8.7%. This element forms stable surface complexes and is released while the destruction of Fe and Mn hydroxides.

19.4   CONCLUSIONS

The studied morphological parameters, physical and chemical indices and composition of elements are the key indicators of the certain type soil characteristics and they determine fertility and optimality of crops growing conditions in the areas with a high cryogenic load. They allow the estimation of the soil productive balance, the processes occurring in a complex dynamic system of life, as well as the probability (extent) of pollution under the certain human impact.

The results of the soil characteristics basic indices study can be used in resolving the issues of management and sustainable utilization of soil resources, taking into account the regional features of climatic factors.

KEYWORDS

- **Bulk composition**
- **Concentration of a chemical**
- **Forms of chemical elements**
- **Fractionation**
- **Meteorological factors**
- **Soil**

REFERENCES

1. Golov, G. V. (2001). Soil and ecology agrophytocenosis Zeya-Bureya plain/Vladivostok. Dalnauca (*Far-Eastern science*), 162p.
2. Bome, N. A., Bome, A. J. & Belozerova, A. A. (2007). Sustainability of cultural plants to unfavorable factors of environment/Tyumen. Publishing house of the Tyumen state University, 192 p.
3. Kiriushin, V. I. (2007). Evaluation of soil quality and fertility of soils to form systems of agriculture and agro. *Soil Science, 7,* 873–880.
4. Perelomov, L. V, Pinsky, J. L. (2003). The forms of Mn, Pb and Zn in the gray forest soils Upland. *Soil Science, 6,* 682–691.
5. Bolshakov, V. A. (2002). Trace elements and heavy metals in soils. *Soil Science, 7,* 844–849.
6. Ozyorskiy, A. Y. (2008). Fundamentals of environmental Geochemistry: textbook. Manual/Krasnoyarsk. IPK SFU (Regional institute of in-plant training ofSiberian federal university), 316p.
7. Agroclimatic resources of the Tyumen region. Leningrad. Gidrometeoizdat (Publishing house of the State committee of the USSR on hydrometeorology and control of natural environment), 1972, 150p.
8. Lihenko, I. E. (2004). Selection of spring soft wheat for the conditions of the Northern Urals, that is, Lihenko: Dissertation of doctor of agricultural Sciences. *Omsk,* 14–19.
9. Elnikov, I. I., Biryukova, O. A. & Kryschenko, V. S. (2009). The multidiagnostic value for the prediction of winter wheat yield and optimal content of available phosphorus in calcareous chernozem. *Agrochemicals, 11,* 7–15.
10. Darmaeva, N. N., Haydapova, D. D., Badmaev, N. B. & Nimaeva, O. D. (2009). Agrochemical and physicomechanical properties of frozen soils, which determine their potential sustainability in agricultural use. *Agrochemicals, 11,* 16–21.
11. Long-cultivated sod-podzolic and soils, their classification and description. http://big-archive.ru/geography/pedology/46.php
12. Malyuga, D. P. (1963). Biogeochemical method explorations for mineral deposits. Leningrad. Academy of Sciences of the USSR, 264p.

CHAPTER 20

A LECTURE NOTE ON THE EFFICIENCY OF THE USE OF CROP PRODUCE FOR ALTERNATIVE FUEL PRODUCTION

DORONIN ANDRIY VOLODYMYROVYCH

CONTENTS

ABSTRACT

The perspectives of the use of sugar beets and their processed products, wheat and corn for the alternative fuel production were discussed in the paper. The suggestions how to ensure the competitiveness of bioethanol production using processed sugar beets in Ukraine were discussed; the processing of molasses for bio-ethanol appeared to be the most cost-effective. The cost calculations of the bio-ethanol production from processed sugar beet depending on the price of beets, type of bio-raw materials and processing technology were presented. The predictive calculation of bioethanol production based on sugar beet productivity in Ukraine was made.

20.1   INTRODUCTION

The influence of internal and external factors on the competitiveness of domestic agricultural enterprises requires the manufacturers to intensely form competitive advantages of products and to ensure its competitive position in the market. However, in view of an urgent problem of providing our country with price-affordable energy carriers, it becomes appropriate to speed up the alternative fuel production.

Sugar and starch crops are used world-wide for the bio-ethanol production, including the output of sugar beet processing. Sugar beet industry has always played an important role in Ukraine's economy and social development of society. Alongside with sugar production, and as the country needs energy carriers very much, it is quite relevant to use sugar beets and the output of sugar beet processing for the production of bio-ethanol as an alternative fuel.

A considerable contribution to the development of the competitiveness of sugar beet companies, including the explanation of the ways of its enhancement, was made by Varchenko [1], Zaiets [2], Kaletnik [3], Kodenska [4], Royik [5], Sabluk [6] and Shpychak [7]. The essence of innovative technologies of the alcohol industry is studied by Shyian, Sosnytskyi, Oliynichuk [8] and others.

The goal of our research is to develop practical recommendations, which help ensure the competitive manufacturing of bioethanol using the output of sugar beet processing.

The developed world and European Union countries make great efforts to replace traditional kinds of fuel with bio-fuels. Thus, the EU directive RED (Renewable Energy Directive) 2009/28/EU establishes the use of 10% of renewable energy in transport and 20% of renewable energy in the structure of gross energy consumption to 2020 [9] as mandatory parameters.

The Law of Ukraine "On Alternative Fuels" [10] envisages that from 2013 it will be recommended to add at least (not less than) 5% of bio-ethanol into gasoline; this rule will become mandatory in 2014–2015. From 2016 a required content of bioethanol in motor gasoline, which is produced and/or sold on the territory of Ukraine, will not be less than 7 percent.

Ukraine is obliged to consider the European standards concerning the use of biofuels in the context of its entry to the European Energy Community. Therefore, the country has the obligation to bring the biological component in motor fuel up to 10% by the year of 2020.

20.2   MATERIALS AND METHODOLOGY

The data received from the State Statistics Department of Ukraine and the author's own calculations were used in the process of writing an article. Prices and production cost were translated at the official exchange rate of Hryvnia to the U.S. dollar, established by the National Bank of Ukraine for the appropriate period. The prices were given excluding value added tax (VAT – 20%), budget subsidies and surcharges. The coefficients of ratio between the cost of production of sugar beet processing and prices of sweet roots were calculated based on the fact that the price of sugar beets is 1.0. The methods of system analysis and logic generalization were used to study the experience of bioethanol production from crop produce; the comparative analysis was used in the process of analyzing the statistical information; the economic-mathematical modeling was used to develop a polynomial model which describes the level of sugar beet productivity in Ukraine; the monographic method helped substantiate the necessity of diversification of the products of sugar beet industry, induction and deduction facilitated the summarizing of the results of the research; an abstract logic method was applied to make conclusions and proposals.

20.3   RESULTS AND DISCUSSIONS

Bio-ethanol production is possible at the ethyl and sugar factories reequipped for this production. The manufacture in several areas is possible at sugar processing factories, namely: the workshop producing ethanol is mounted – the plant produces sugar by the traditional technology, and it manufactures bio-ethanol using the products of sweet root processing, or only the bio-ethanol production from sugar beet (crude juice) is planned. Also the bioethanol production is possible when starch crops such as wheat and corn are used; the products of their processing are used for human nutrition.

Over the last 12 years sugar beet, wheat and corn production has undergone the significant changes for the better in Ukraine (Table 20.1) [11].

TABLE 20.1 The Indices of Sugar Beet, Wheat, Corn Production in UKRAINE in 2000–2012

Index	Year					2012 in % to	
	2000	2005	2010	2011	2012	2000	2011
Sugar beets							
Harvested area, thousand ha	747.0	623.3	492.0	515.8	448.9	60.1	87.0
Production, thousand t	13198.8	15467.8	13749.2	18740.5	18438.9	139.7	98.4
Yield, thousand ha	17.67	24.82	27.95	36.33	41.08	232.5	113.1
Wheat							
Harvested area, thousand ha	5161.6	6571.0	6284.1	6657.3	5629.7	109.1	84.6
Production, thousand t	10197.0	18699.2	16851.3	22323.6	15762.6	154.6	70.6
Yield, thousand t	1.98	2.85	2.68	3.35	2.80	141.4	83.6
Corn							
Harvested area, thousand ha	1278.8	1659.5	2647.6	3543.7	4371.9	341.9	123.4
Production, thousand t	3848.1	7166.6	11953.0	22837.8	20961.3	544.7	91.8
Yield, thousand t	3.01	4.32	4.51	6.44	4.79	159.1	74.4

Following the technology elements of sugar beet cultivation, the main ones are plant nutrition and pest (disease, weed) management, together with soil and climatic conditions ensured the increase of crop productivity by 2.3 times – from 17.67 t/ha in 2000 to 41.08 t/ha in 2012. The area, the sugar beets were harvested from in the respective period, decreased by 39.9% (from 747.0 thousand hectares in 2000 to 448.9 thousand ha in 2012). The gross production of sugar beets increased by 39.7% (from 13198.8 thousand tons in 2000 to 18,438.9 thousand tons in 2012). The increase of sugar beet gross production during this period was due to the increase of their productivity. Thus, the tendency of the area optimization under sugar beet crops along with the yield increase through the intensive use of land resources is observed in Ukraine.

As for wheat production, the yield of the crop increased by 1.4 times – from 1.98 t/ha in 2000 to 2.8 t/ha in 2012. The gross production of wheat increased by 54.6% (from 10197.0 thousand tons in 2000 to 15,762.6 thousand tons in 2012). The increase of wheat gross production is caused by the increase of productivity and that of the harvested area – by 9.1% (from 5161.6 thousand hectares in 2000 to 5629.7 thousand hectares in 2012). Accordingly, the corn yield for grain increased by 1.6 times – from 3.01 t/ha in 2000 to 4.79 t/ha in 2012 The gross production of corn increased by 5.4 times (from 3848.1 thousand tons in 2000 to 20,961.3 thousand tons in 2012). The increase of the corn gross production to a large extent is due to the increase of the harvested area by 3.4 times (from 1278.8 thousand hectares in 2000 to 4371.9 thousand hectares in 2012).

The market condition of agricultural crops remains unattractive for sugar beet growers (Table 20.2) [12–16]. The competition with other crops (wheat, corn) pushes farmers to reduce the sugar beet fields. Thus, the price of wheat increased from $166.5/t in 2011 to $194.6/t in 2012, or 16.9%, respectively, for corn – from $170.0/ton to $190.3/ton, or 12.0%, while the prices of sugar beet decreased from $65.2/t in 2011 to $53.8/t in 2012, or 17.5%. Within a year the level of wheat production profitability declined from 17.6 to 11.8%, that of corn – from 38.6 to 19.8%, and sugar beet – from 36.5 to 15.7%, respectively.

TABLE 20.2   The Economic Efficiency of Sugar Beet, Wheat, Corn Production in Ukraine in 2008–2012 (Agricultural Companies)

Index	Year					2012 p. in % to	
	2008	**2009**	**2010**	**2011**	**2012**	**2008**	**2011**
Sugar beets							
Total cost of 1 ton, US $	39.7	39.2	52.6	47.7	46.5	117.1	97.5
Average price of 1 ton, US $	42.5	53.7	61.4	65.2	53.8	126.6	82.5
Level of profitability, %	7.1	37.0	16.7	36.5	15.7	-	-
Wheat							
Total cost of 1 ton, US $	122.1	96.6	125.5	141.6	174.1	142.6	123.0
Average price of 1 ton, US $	143.6	102.2	137.5	166.5	194.6	135.5	116.9
Level of profitability, %	17.6	5.7	9.6	17.6	11.8	-	-
Corn							
Total cost of 1 ton, US $	125.9	92.0	120.5	122.6	158.9	126.2	129.6
Average price of 1 ton, US $	139.2	111.7	156.5	170.0	190.3	136.7	112.0
Level of profitability, %	10.6	21.5	29.9	38.6	19.8	-	-

In the context of the world urgent problem–the lack of foodstuff–the international community may most likely prohibit the bio-ethanol production using corn and wheat. The by-products of sugar beet processing are not used directly for food, which is a relevant confirmation of the expediency to use it for the bioethanol production. In addition, there will be no need of reducing the sugar beet areas but rather of their expanding which in turn will create additional jobs in the sugar beet industry.

Foreign experience of sugar beet production and analysis of the current state of the domestic sugar beet production show that an important factor of improving the competitiveness of production in the investigated area is rational distribution of sugar beet fields. The resolving of this issue will improve the yield and sugar content of roots.

On the basis of the soil potential (fertility of the main soil types), the peculiar features of the climate conditions which are determined by the interaction of such factors as incoming solar radiation, atmospheric circulation, moisture supply, the researchers of the Institute of bioenergy crops and sugar beets of the NAAS of Ukraine defined a beetroot zone–the most favorable zone (as to its soil and climatic conditions) for sugar beet cultivation [5].

The most favorable growing area for this important agriculture crop (the zone of sufficient moisture, the rainfall/precipitation is over 550 mm per year), which allows to produce the sugar beet yields within 55–60 t/ha, are the western regions of Ukraine – Volyn, Ivano-Frankivsk, Lviv, Rivne, Ternopil and Khmelnytskyi. The less favorable zone (unstable moisture, the rainfall is 450–480 mm per year) includes Vinnytsia, Zhytomyr, Kyiv, Poltava, Sumy, Cherkasy and Chernihiv regions, there you can get guaranteed yield of beets – 50–55 t/ha [5].

Still less favorable area of beet cultivation zone (the zone of low moisture, the rainfall is 430–480 mm per year) is in Kirovohrad and Chernivtsi regions; there you can get the yields of sugar beets at 45–50 t/ha.

The rest of the regions, including those where sugar beets are cultivated, are not favorable for sugar beet production because of their soil and climatic conditions.

In Ukraine the developed sugar beet production is a universal basis for the production of bioethanol (Table 20.3).

TABLE 20.3   The Calculation of the Output of Bioethanol from Various Types of Raw Materials by the Different Yields

Raw	The output of bioethanol from 1t of production, t	The output of bioethanol in calculating per 1 ha depending on the yields of culture, t	
		Yield	Output of bioethanol
Sugar beets (crude juice)	0.074–0.079	40.0	2.96–3.16
		50.0	3.70–3.95
		60.0	4.44–4.74
Molasses (processing of sugar beet into sugar)	0.222–0.237	1.56	0.35–0.37
		1.95	0.43–0.46
		2.34	0.52–0.55

TABLE 20.3 *(Continued)*

Raw	The output of bioethanol from 1t of production, t	The output of bioethanol in calculating per 1 ha depending on the yields of culture, t	
		Yield	Output of bioethanol
		3.0	0.71–0.93
Wheat	0.237–0.311	4.0	0.95–1.24
		5.0	1.19–1.56
		4.0	1.28–1.38
Corn	0.321–0.346	5.0	1.61–1.73
		6.0	1,93–2.08

The greatest output of bioethanol per unit area at the appropriate level of yield can be obtained from the sugar beets. However, in the processing of sugar beet into sugar we get the molasses, and depending on its quality the output of bioethanol from 1t can be 0.222–0.237 t.

Considering the world experience of using sugar beets for bioethanol production as an alternative fuel, it would be appropriate to implement it at the sugar processing factories of Ukraine. The need to diversify a subcomplex of sugar beet production is determined not only by the country's high dependence on energy resource import, but also by the necessity to have additional facilities to process the excess production, taking into account the cyclical and risk nature of sugar beet production.

The calculation of the cost of bioethanol production from different bioraw materials shows that the most competitive bioethanol production is from molasses (Table 20.4).

TABLE 20.4   The Competitiveness of Bioethanol Production Depending on Bioraw in Ukraine, 2012

Type of bioraw	The average price of 1 ton bioraw, USD. USA	The need of bio-raw for the production of 1 ton of bioethanol, t	Prime cost of bioethanol, dollars. USA	
			1 t	1 l
Sugar beets	53.8	12.65–13.49	1447.5	1.14
Molasses	81.5	4.22–4.50	934.7	0.74
Wheat	194.6	3.21–4.22	1313.1	1.04
Corn	190.3	2.89–3.11	1184.0	0.94

Note. The official exchange rate of the National Bank of Ukraine in 2012: $100 = UAH 799.1.

Based on the polynomial model which describes the level of sugar beet productivity in Ukraine within the period of 1913–2012, the prediction of sugar beet yields (Fig. 20.1), which envisages its increase by 14%, was made.

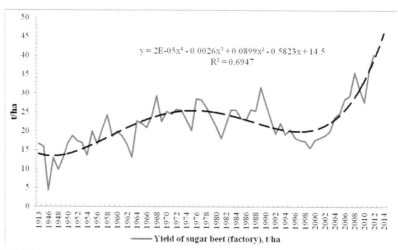

$$y = 2E\text{-}05x^4 - 0.0026x^3 + 0.0899x^2 - 0.5823x + 14.5$$
$$R^2 = 0.6947$$

Yield of sugar beet (factory), t/ha

FIGURE 20.1   The dynamics of sugar beet yields in Ukraine by the years.

The intensive consumption growth and condition of bioethanol world market enable to quickly build up the capacity of its production in Ukraine.

The European Union countries are interested in bioethanol import from our country. Every year the bioethanol market is growing by 3% in Europe, this is a significant potential for Ukraine to increase export [3].

Thus, there are the prerequisites for biofuel application in Ukraine, in particular, by means of the diversification of the output production of sugar beet industry.

The competitiveness of bioethanol production using the output of sugar beet processing depends on several factors: the prices of sugar beets, the quality of the processed products of sweet roots and the technology of its manufacture.

In Ukraine the price of sweet roots affects the competitiveness of bioethanol production from the sugar beet processing products significantly.

Thus, when the price of roots was $42.3/t, the cost of bioethanol production was $1316.1/1 t ($1.04 per 1 l). By raising the price of sugar beet by 51.5% – to $64.1/ton, the cost of bioethanol production was $1566.1/1t ($1.24/1l) or increased by 19% (Table 20.5).

TABLE 20.5 The Calculation of the Cost of Bioethanol Production from Sugar Beet (Crude Juice) Depending on their Prices in Ukraine, 2012

Index	Sugar beets price, USD/t		
	Minimum	**Medium**	**Maximum**
	42.3	**53.8**	**64.1**
Raw material expenses, $/t	724.8	856.1	974.8
Processing of raw material, $/t	591.3	591.3	591.3
Production cost of bioethanol, $/t	1316.1	1447.5	1566.1
$/l	1.04	1.14	1.24

The official exchange rate of the National Bank of Ukraine in 2012: $100 = UAH 799.1.

The similar dependence is observed when bioethanol is produced from the molasses in Ukraine (Table 20.6).

TABLE 20.6 The Calculation of the Cost of Bioethanol Production from Molasses Depending on its Price in Ukraine, 2012

Index	Molasses price, USD/t		
	Minimum	**Medium**	**Maximum**
	71.1	**81.5**	**90.9**
Raw material expenses, $/t	299.6	343.4	383.0
Processing of raw materials, $/t	591.3	591.3	591.3
Production cost of bioethanol, $/t	890.9	934.7	974.3
$/L	0.70	0.74	0.76

Note: The official exchange rate of the National Bank of Ukraine in 2012: $100 = UAH 799.1.

When the price of molasses increases by 27.8% – from $71.1 to 90.9/ton, the cost of bioethanol production will increase by 9.4% – from $890.9 to 974.3 per 1 ton, or $0.7 to 0.76 per 1 L.

These researches have shown that provided the molasses is not a commodity and in turn it has no price, and sugar refineries process it into bioethanol, then its cost of production will decrease significantly.

In addition, the combined production of sugar for the consumer market and bioethanol from the molasses enhances the competitiveness of the sugar beet industry enterprises considerably. Thus, when the price of sweet roots is $53.8/t, the factory can produce sugar by the price of $698.5 /t and bioethanol from the molasses – $934.7/t; however, if it produces only sugar or bioethanol from sugar beet then the gross income will significantly reduce. The similar correlation is observed when the price of sugar beet is different (Table 20.7).

TABLE 20.7 The Cost of Sugar, Molasses, Bioethanol Production Depending on the Price of Sugar Beet in Ukraine, 2012

Produce	Sugar beet price, USD/t			Coefficients of ratios
	Minimum	Medium	Maximum	
Sugar beets	42.3	53.8	64.1	1.0
The cost of sugar, molasses and bioethanol production, USD/t				
Sugar	608.7	698.5	778.6	12.1–14.4
Molasses	71.1	81.5	90.9	1.4–1.7
Bioethanol from molasses	890.9	934.7	974.3	15.2–21.1
The cost of bioethanol production from sugar beet (crude juice), USD/t				
Bioethanol from beets	1316.1	1447.5	1566.1	24.4–31.1

The official exchange rate of the National Bank of Ukraine in 2012: $100 = UAH 799.1.

The experience of the foreign countries and price situation of sugar beet in Ukraine indicate that to develop the technologies which use cheaper and economically better raw material for bioethanol production is an important task for sugar beet production, as the use of sugar beet and the products of their processing for the manufacture of other commodities, besides sugar, makes them more competitive in comparison with other crops.

The sugar beet production should be concentrated at large well-developed farms that are able to implement intensive technologies with the use of modern agricultural machinery, fertilizers, integrated crop and soil protection, to improve sugar beet yield and quality, to reduce their costs as important factors of sugar quality improvement according to the requirements of the consumer sugar, which exist in EU countries, and ensuring the competitiveness of sugar in the domestic and foreign markets.

The quality and the type of raw material have the effect on the increase of the competitiveness of bioethanol production.

The raw material for ethanol production, besides molasses and sugar beets, can be the by-products of sugar beets with high sugar content, in particular the green blackstrap molasses and syrup. Processing of 1 t of green blackstrap molasses- 394 L or 311 kg and 1 tone of syrup – 375 l (297 kg) gives the largest output of bioethanol. The output of bioethanol from molasses is 1 t – 300 l (237 kg), 1 t of sugar beet – 100 L (79 kg).

The reduction of the bioethanol cost and the increase of its competitiveness depend greatly on the production technology. Bioethanol technology consists of two phases: the production of raw ethanol and its further dehydration. Azeotropic distillation, adsorption on molecular sieves and evaporation through the membrane are used for ethanol dehydration of [8].

The cost of bioethanol production from sugar beet (crude juice), green molasses, syrup and molasses, depending on the technology used, was calculated by us (Table 20.8).

TABLE 20.8 The Competitiveness of Bioethanol Production from the Products of Sugar Beet Processing with the use of Different Technologies, 2012

Bioraw	Production cost of bioethanol using different technologies, $					
	Azeotropic distillation		Adsorption on molecular sieves		Evaporation through the membrane	
	1 t	1 l	1 t	1 l	1 t	1 l
Sugar beets	1447.5	1.14	1409.9	1.11	1378.2	1.09
Green blackstrap molasses	1194.8	0.94	1157.3	0.91	1125.6	0.89
Syrup	1646.4	1.30	1608.8	1.26	1577.1	1.24
Molasses	934.7	0.74	897.2	0.71	865.5	0.69

The official exchange rate of the National Bank of Ukraine in 2012: $100 = UAH 799.1.

Bioethanol with the lowest cost is obtained by the evaporation through the membrane independently of the type of raw material for processing. Thus, when the bioethanol production from the sugar beet (crude juice) by the azeotropic distillation was used, the production cost of 1 ton of bioethanol was USD 1447.5 ($1.14 per 1 L), and when the evaporation through the membrane was used – $1378.2 ($1.09 per 1 L) or decreased by 4.8%.

The analogical dependence occurred when the bioethanol production from green blackstrap molasses, syrup and molasses took place. The lowest production cost of bioethanol from molasses was obtained while using all three processing technologies. Thus, the production of bioethanol from the molasses by azeotropic distillation the production cost of 1 ton of bioethanol was $934.7 ($0.74 per 1 L), and it decreased by 7.4% – to $865.5/t ($0.69 per 1 l) when the evaporation through the membrane was used.

The calculations of the bioethanol production cost from sugar beet by-products show that the competitiveness ensures the bioethanol production from molasses using the evaporation through the membrane.

It should be noted that the dehydration of the ethanol by the azeotropic rectification requires considerable operational and energy expenses. Ethanol dehydration technologies using the adsorption on molecular sieves and evaporation through the membrane are less power consuming ones. However, the ethanol dehydration by the evaporation through the membrane requires significant capital investment and the smooth/uninterrupted operation of the factory.

According to the data of Ukraine's State Statistics, the average annual consumption of gasoline in the country varies within 4.7 million tons within years, including agriculture – 0.2 million tons (Table 20.9) [17–21].

TABLE 20.9   Fuel Consumption and Calculation of Bioethanol Needs in Ukraine.

Year	Consumed gasoline		The needs of bioethanol for mixtures with gasoline			
			6%		10%	
	Total, th. t	Including agriculture, thousand t	Total, th. T	Including the agriculture, thousand t	Total, th. t	Including the agriculture, thousand t
2007	4821.8	299.0	289.3	17.9	482.2	29.9
2008	5061.1	288.5	303.7	17.3	506.1	28.9
2009	4696.1	216.0	281.8	13.0	469.6	21.6
2010	4632.7	231.6	278.0	13.9	463.3	23.2
2011	4401.3	224.5	264.1	13.5	440.1	22.5

With the small deviations, these amounts of gasoline are forecasted for future. There are no significant disruptions in providing agricultural enterprises with this type of fuel. However, the price of gasoline is growing by 7% almost every year. Specific annual increase of fuel prices occurs in April and September which coincides with the main field processes in agriculture – sowing and harvesting time. Due to low solvency of agricultural enterprises it may have a negative effect on proper timing of production processes. And the price of oil increases world-wide every year. Therefore, the solution consists in production and use of alternative fuel types, first of all for agriculture.

The growth of sugar beet production by 2015 and 2020 will facilitate the implementation of the Law of Ukraine "On Alternative Types of Fuels," in the part of bioethanol addition into gasoline, as well as reduce the dependence of agricultural producers on the market prices of fuel and its import (Table 20.10).

TABLE 20.10 The Estimated Calculation of Bioethanol Production Based on the Productivity of Sugar Beet in Ukraine

Index	Forecast			
	2015		2020	
	Optimistic	Pessimistic	Optimistic	Pessimistic
Gasoline consumption, total, th. t	4723		4723	
including agriculture, th. t	252		252	
share of bioethanol in gasoline mixtures, %	6		10	
Required bioethanol amount, th. t	283		472	
including agriculture, th. t	15.1		25.2	
Sugar beet production, th. t	26,000	18,200	35,000	25,000
Processing of sugar beets into sugar, th. t	15,090		14,775	
bioethanol, th. t	10,000	2473	19,000	9350
Molasses output of, th. t	604		473	
Bioethanol production from molasses, th. t	143		112	
Bioethanol production from beets, th. t	790	195	1501	739
Bioethanol production, total, th. t	933	338	1613	851

By the foretold forecast by the year of 2020, the significant growth of agricultural production in Ukraine is creating favorable conditions for the renewable energy development, in particular for the increase of biofuel production [22]. Also, the steady fuel price increase promotes fuel production and use in Ukraine.

In order to ensure the predicted volumes of bioethanol production in Ukraine, it is necessary to stimulate the producers to do the renewal of fixed assets for both sugar beet and sugar and bioethanol production. The best option is the construction of the combined workshops at sugar and ethanol factories which will produce bioethanol from sugar beet by-products in the season of sugar beet harvesting, and in the off-season – from the waste of grain of headed crops or corn.

The solution of these issues requires the development of a national comprehensive program for sugar industry enhancement in Ukraine which will take into account world tendencies as to the use of sugar beet produce for bioethanol manufacture as an alternative fuel.

20.4   CONCLUSIONS

Sugar beet high yields are most likely produced through the application of intensive technologies with the use of all agro-biological cultivation practices together with their high quality performance in optimal terms. In Ukraine sugar beet production has to be concentrated in the most favorable areas for sugar beet cultivation, where soil and climatic conditions guarantee high indicators of root productivity and quality.

The bioethanol production and use, in particular from the by-products of sugar beet processing, will allow to establish the production of environmentally clean alternative fuel, to create new jobs, to increase profits of the enterprises, to reduce the dependence of Ukraine on imported fuel and to develop new markets for the products of sugar beet manufacture. Producing bioethanol from molasses can meet the demand for this product at a lower price and improve the competitiveness of the enterprises of sugar beet industry as well. The solution of these issues will increase the competitiveness of the produce manufactured by businesses of AIC (agro-industrial complex) of Ukraine both in domestic and foreign markets.

20.5   ACKNOWLEDGMENTS

The author expresses his gratitude for the help in preparing materials to National Association of Sugar Producers of Ukraine "Ukrtsukor" and personally to Mykola Yarchuk, Petro Borysiuk and Mykola Kalinichenko. The special thanks to Vitaliy Sosnitskyi for the consultation and help on the issues of innovative technologies of ethanol industry. The sincere words of gratitude are for his scientific advisor, Professor Maria Kodenska.

KEYWORDS

- **Bioethanol**
- **Competitiveness**
- **Molasses**
- **Production cost**
- **Sugar**
- **Sugar beet production**
- **Sugar beets**

REFERENCES

1. Varchenko, O. M. (2009). The world and national experiences of the sugar market regulation: Monograph, Bila Tserkva, 334 p. (in Ukr).
2. Zayets, O. S. (1998). The sugar market: issues, trends and practices, Kyiv: *Naukova Dumka*, 365 p. (in Ukr).
3. Kaletnik, G. M. (2009). The perspectives of bioethanol production in Ukraine *Agricultural equipment and machinery, (2)*, 50–55. (in Ukr).
4. Kodenska M. Y. (2011). The justification of the need of the investment projects development in sector of bioethanol production development based on sugar beet production *AhroInKom, (4–6)* [Electronic resource], Mode of access: http://www.nbuv.gov.ua/portal/Chem_Biol Agroin/2011_4–6/KODENSKA.pdf., free, From the screen. (in Ukr).
5. Royik, M. V. (2001). Beets, *Kyiv: XXI vik – RIA TRUD-KYIV*, 320 p. (in Ukr).
6. Sabluk, P.T., Kodenska, M. Y., Vlasov V. I., et al. (2007). Sugar beet production in Ukraine: problems of revival, perspectives of development Monograph, *Kyiv: IAE NNC*, 390 p. (in Ukr).
7. Spychak, A. M. (2009). The Economic problems of biofuel production and food security of Ukraine *Economy APK, 8*, 11–19. (in Ukr).
8. Shiyan, P. L., Sosnitsky, V. V. & Oliynichuk, S. T. (2009). The Innovative technologies of alcohol industry. Theory and Practice: Monograph. *Kyiv. Askaniya*, 424 p. (in Ukr).
9. DIRECTIVE 2009/28/EC OF THE EUROPEAN PARLIAMENT AND OF THE COUNCIL of 23 April 2009 on the promotion of the use of energy from renewable sources and amending and subsequently repealing Directives 2001/77/EC and 2003/30/EC [Electronic resource], Mode of access: http://eur-lex.europa.eu/LexUriServ/LexUriServ.do?uri=OJ:L:2009:140:0016:0062:EN: PDF, free, From the screen.
10. The Law of Ukraine *"About alternative types of fuel,"* from January 14th, 2000 *(1391-XIV)*: as of the 21.07.2012 The Verkhovna Rada of Ukraine, Off. Publ.–*Kyiv, Parl. publishing house,* 2000, *12*, 94, (The library of official publications). (in Ukr).
11. Vlasenko, N. S. (2013). The plant growing of Ukraine: Statistical Digest for 2012. Kyiv: State Statistics Committee of Ukraine, 110 p. (in Ukr).
12. The Basic economic indexes of agricultural production in agricultural enterprises: The Statistical Bulletin for 2008, Kyiv: State Statistics Committee of Ukraine, (2009). 76 p. (in Ukr).
13. The Basic economic indexes of agricultural production in agricultural enterprises: The Statistical Bulletin for 2009, *Kyiv: State Statistics Committee of Ukraine*, (2010). 80 p. (in Ukr).
14. The Basic economic indexes of agricultural production in agricultural enterprises: The Statistical Bulletin for 2010. *Kyiv, State Statistics Committee of Ukraine*, (2011). 88 p. (in Ukr).
15. The Basic economic indexes of agricultural production in agricultural enterprises: The Statistical Bulletin for 2011, *Kyiv: State Statistics Committee of Ukraine*, (2012). 88 p. (in Ukr).
16. The Basic economic indexes of agricultural production in agricultural enterprises: The Statistical Bulletin for 2012, *Kyiv: State Statistics Committee of Ukraine*, (2013). 88 p. (in Ukr).

17. Lupenko, Y. O., Mesel-Veselyaka, V. J. (2012). The Strategic directions of agriculture development for the period till 2020 in Ukraine, 2nd publ., revis. and enlar. *Kyiv, NSC IAE,* 218 p. (in Ukr).
18. The Statistical Yearbook of Ukraine for the 2007/Ed. O.G, Osaulenko, *Kyiv: State Statistics Committee of Ukraine,* (2008). 571 p. (in Ukr).
19. The Statistical Yearbook of Ukraine for the 2008/Ed. O.G, Osaulenko, *Kyiv: State Statistics Committee of Ukraine,* (2009). 566 p. (in Ukr).
20. The Statistical Yearbook of Ukraine for the 2009/Ed. O.G, Osaulenko, *Kyiv: State Statistics Committee of Ukraine,* (2010). 566 p. (in Ukr).
21. The Statistical Yearbook of Ukraine for the 2010/Ed. O.G, Osaulenko, *Kyiv: State Statistics Committee of Ukraine,* (2011). 559 p. (in Ukr).
22. The Statistical Yearbook of Ukraine for the 2011/Ed. O.G, Osaulenko, *Kyiv: State Statistics Committee of Ukraine.* (2012). 558 p. (in Ukr).

CHAPTER 21

ORIENTATION CONTROLLED IMMOBILIZATION STRATEGY FOR B-GALACTOSIDASE ON ALGINATE BEADS

M. S. MOHY ELDIN, M. R. EL-AASSAR, and E. A. HASSAN

CONTENTS

ABSTRACT

In recent years, enzyme immobilization has gained importance for design of artificial organs, drug delivery systems, and several biosensors. Polysaccharide based natural biopolymers used in enzyme or cell immobilization represent a major class of biomaterials which includes agarose, alginate, dextran, and chitosan. Especially, Alginates are commercially available as water-soluble sodium alginates and they have been used for more than 65 years in the food and pharmaceutical industries as thickening, emulsifying and film forming agent. Entrapment within insoluble calcium alginate gel is recognized as a rapid, nontoxic, inexpensive and versatile method for immobilization of enzymes as well as cells. In this research, the formulation conditions of the alginate beads entrapment immobilized with the enzyme have been optimized and effect of some selected conditions on the kinetic parameter, Km, have been presented. β-galactosidase enzymes entrapped into alginate beads are used in the study of the effect of both substrate diffusion limitation and the mis-orientation of the enzyme on its activity, the orientation of an immobilized protein is important for its function. Physicochemical characteristics and kinetic parameters; Protection of the activity site using galactose as protecting agent has been presented as a solution for the mis-orientation problem. This technique has been successful in reduction, and orientation-controlled immobilization of enzyme. Other technique has been presented to reduce the effect of substrate diffusion limitation through covalent immobilization of the enzyme onto the surface of alginate beads after activation of its OH-groups.

21.1   INTRODUCTION

Recently an increasing trend has been observed in the use of immobilized enzymes as catalysts in several industrial chemical processes. Immobilization is important to maintain constant environmental conditions order to protect the enzyme against changes in pH, temperature, or ionic strength; this is generally reflected in enhanced stability [1]. Moreover, immobilized enzymes can be more easily separated from substrates and reaction products and used repeatedly. Many different procedures have been developed for enzyme immobilization; these include adsorption to insoluble materials, entrapment in polymeric gels, encapsulation in membranes, cross-linking with a bifunctional reagent, or covalent linking to an insoluble carrier [2]. Among these, entrapment in calcium alginate gel is one of the simplest methods of immobilization. The success of the calcium alginate gel entrapment technique is due mainly to the gentle environment it provides for the entrapped material. However, there are some limitation such as low stability and high porosity of the gel [1]. These characteristics could lead to leakage of large molecules like proteins, thus generally limiting its use to whole cells or cell organelles [3].

Enzymes are biological catalysts with very good prospects for application in chemical industries due to their high activity under mild conditions, high selectivity and specificity [4–7]. However, enzymes do not fulfill all of the requirements of an industrial biocatalyst or biosensor [8, 9]. They have been selected throughout natural evolution to perform their physiological functions under stress conditions and quite strict regulation. However, in industry, these biocatalysts should be heterogeneous, reasonably stable under conditions that may be quite far from their physiological environment [10] and retain their good activity and selectivity when acting with substrates that are in some instances quite different from their physiological ones. β-galactosidase, commonly known as lactase, catalyzes the hydrolysis of β-galactosidic linkages such as those between galactose and glucose moieties in the lactose molecule. While the enzyme has many analytical uses, being a favorite label in various affinity recognition techniques such as ELISAs, or enzyme-linked immunosorbent assays, its main use is the large scale processing of dairy products, whey, and whey permeates.

Immobilization of the enzymes to solid surface induces structural changes, which may affect the entire molecule. The study of conformational behavior of enzymes on solid surface is necessary for better understanding of the immobilization mechanism. However, the immobilization of enzymes on alginate beads is generally rapid, and depends on hydrophobic and electrostatic interactions as well as on external conditions such as pH, temperature, ionic strength and nature of buffer [11, 12]. Enzymes denaturation may occur under the influence of hydrophobic interactions, physicochemical properties of the alginate beads or due to the intrinsic properties of the enzyme.

One of the main concerns regarding immobilized proteins has been the reduction in the biological activity due to immobilization. The loss in activity could be due to the immobilization procedure employed, changes in the protein conformation after immobilization, structural modification of the protein during immobilization, or changes in the protein microenvironment resulting from the interaction between the support and the protein. In order to retain a maximum level of biological activity for the immobilized protein, the origins of these effects need to be understood, especially as they relate to the structural organization of the protein on the immobilization surface. The immobilization of proteins on surfaces can be accomplished both by physical and chemical methods. Physical methods of immobilization include the attachment of protein to surfaces by various inter actions such as electrostatic, hydrophobic/hydrophilic and van der Waals forces. Though the method is simple and cost-effective, it suffers from protein leaching from the immobilization support. Physical adsorption generally leads to dramatic changes in the protein microenvironment, and typically involves multipoint protein adsorption between a single protein molecule and a number of binding sites on the immobilization surface. In addition, even if the surface has a uniform distribution of binding sites, physical adsorption could lead to heterogeneously populated immobilized proteins. This has been ascribed to unfavorable lateral interactions among bound protein molecules [13]. The effect of ionic strength on protein adsorption has also been studied, and a striking dependence on the concentration of electrolyte was noticed. For example, in the case of adsorption of apotransferrin on a silicon titanium dioxide surface, it was found that an increase in ionic strength resulted in a decrease in adsorption. The increase in ionic strength decreases the negative surface potential and increases the surface pH. This leads to an increase in the net protein charge, creating an increasingly repulsive energy barrier, which results in reduced effective diffusivity of molecules to the surface [14]. Several research groups have attempted to develop a model that can qualitatively and quantitatively predict physical adsorption, especially as it relates to ion-exchange chromatography. Models have been developed to predict average interaction energies and preferred protein orientations for adsorption. Adsorption studies on egg white lysozyme and o-lactalbumin adsorbed on anion-exchanger polymeric surfaces have shown that it is possible to determine the residues involved in the interaction with the support [15, 16]. It has been also possible to calculate the effective net charge of the proteins. Entrapment and microencapsulation are other popular physical methods of immobilization and have been discussed elsewhere [17]. The immobilization of enzymes through metal chelation has been reviewed recently [18].

Attachment of proteins by chemical means involves the formation of strong covalent or coordination bonds between the protein and the immobilization support. The chemical attachment involves more drastic (non mild) conditions for the immobilization reaction than the attachment through adsorption. This can lead to a significant loss in enzyme activity or in binding ability (in the case of immobilization of binding proteins and antibodies). In addition, the covalent and coordinate bonds formed between the protein and the support can lead to a change in the structural configuration of the immobilization protein. Such a change in the enzymatic structure may lead to reduced activity, unavailability of the active site of an enzyme for the substrates, altered reaction pathways or a shift in optimum pH [19, 20]. In the case of immobilized antibodies and binding proteins this structural change can lead to reduced binding ability.

Oriented immobilization is one such approach [21]. Adsorption, bioaffinity immobilization and entrapment generally give high retention of activity as compared to covalent coupling method [22–26]. Use of macroporous matrices also helps by reducing mass transfer constraints. Calcium alginate beads, used here, in that respect, are an attractive choice. Calcium alginate beads are not used for enzyme immobilization as enzymes slowly diffuse out [31]. In this case, binding of β-galactosidase and galactos to alginate before formation of calcium alginate beads ensured the entrapment of these individual enzymes and undoubtedly, also contributed to enhanced thermostabilization. Such a role for alginate in fact, has already been shown for another alginate binding enzyme.

In this work, β-galactosidase enzyme entrapped into alginate beads are used in the study of the effect of both substrate diffusion limitation and the mis-orientation of the enzyme on its activity, Physicochemical characteristics and kinetic parameters. Protection of the activity site using galactose as protecting agent has been presented as a solution for the mis- orientation problem. This technique has been successful in reduction, but not in elimination of the effect of mis-orientation on the activity as well as the kinetic parameters; Km and Vm. Other technique has been presented to reduce the effect of substrate diffusion limitation through covalent immobilization of the enzyme onto the surface of alginate beads after activation of its OH-groups. The impact of different factors controlling the activation process of the alginate hydroxyl groups using p-benzoquinone (PBQ) in addition to the immobilization conditions on the activity of immobilized enzyme have been studied. The immobilized enzyme has been characterized from the bio-chemical point of view as compared with the free enzyme.

21.2 MATERIALS AND METHODS

21.2.1 MATERIALS

Sodium alginate (low viscosity 200 cP), β-Galactosidase (from Aspergillus oryzae) p-Benzoquinone (purity 99+ %): Sigma-Aldrich chemicals Ltd. (Germany); calcium chloride (anhydrous Fine GRG 90%): Fisher Scientific (Fairlawn, NJ, USA); ethyl alcohol absolute, Lactose (pure Lab. Chemicals, MW 360.31), Galactose: El-Nasr Pharmaceutical Chemicals Co. (Egypt); glucose kit (Enzymatic colorimetric method): Diamond Diagnostics Co. for Modern Laboratory Chemicals (Egypt); Tris-Hydrochloride (Ultra Pure Grade 99.5%, MW 157.64): amresco (Germany); sodium chloride (Purity 99.5%): BDH Laboratory Supplies Pool (England).

21.2.2 METHODS

21.2.2.1 PREPARATION OF CATALYTIC CA-ALGINATE GEL BEADS BY ENTRAPMENT TECHNIQUE

The Ca- alginate gel beads prepared by dissolving certain amount of sodium alginate (low viscosity) 2% w/v in distilled water with continuous heating the alginate solution until it becomes completely clear solution, mixing the alginate solution with enzyme [β-galactosidase enzyme as a model of immobilized enzyme (0.005 g)]. The sodium alginate containing β-galactosidase is dropped (drop wise) by 10 cm^3 plastic syringe in (50 ml) calcium chloride solution (3% w/v) as a safety cross-linker to form beads to give a known measurable diameter of beads. Different aging times of beads in calcium chloride solution are considered followed by washing the beads by (50 mL) buffer solution, and determination of the activity of immobilized enzyme. In case of oriented immobilization, galactose with different concentrations was mixed first with the enzyme-buffer solution before mixing with the alginate.

21.2.2.2   ALGINATE BEADS SURFACE MODIFICATION

The Ca-alginate gel beads prepared by dissolving sodium alginate (low viscosity) in distilled water with continuous heating the solution until become completely clear to acquire finally 4% (w/v) concentration. The alginate solution was mixed with p-benzoquinone (PBQ) solution (0.02 M) and kept for four hours at room temperature to have final concentration 2% (w/v) alginate and 0.01 M (PBQ). The mixture was added dropwise, using by 10 cm^3 plastic syringe to calcium chloride solution (3% w/v) and left to harden for 30 min at room temperature to reach 2 mm diameter beads. The beads were washed using buffer-ethanol solution (20% ethanol) and distilled water, to remove the excess (PBQ), before transferring to the enzyme solution (0.005 g of β-galactosidase in 20 ml of Tris-HCl buffer solution of pH=4.8) and stirring for 1 h at room temperature then the mixture was kept at 4 °C for 16 h to complete the immobilization process. The mechanism of the activation process and enzyme immobilization is presented in Scheme 21.1.

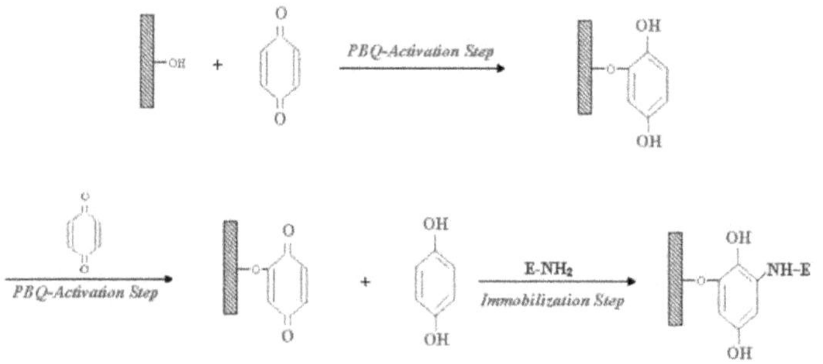

SCHEME 21.1   Mechanism of Activation and Immobilization Process.

21.2.2.3   DETERMINATION OF IMMOBILIZED ENZYME ACTIVITY

The catalytic beads were mixed with 0.1 M Lactose-Tris-HCl buffer solution of (pH 4.8 with stirring, 250 rpm, at room temperature for 30 min. Samples were taken every 5 min to assess the glucose production using glucose kit. Beads activity is given by the angular coefficient of the linear plot of the glucose production as a function of time.

21.3   RESULTED AND METHODS

21.3.1   ENTRAPMENT IMMOBILIZATION

The impact of different factors affecting the process of enzyme entrapment and its reflection on the activity of the immobilized enzyme, its physicochemical characters and its kinetic parameters have been studied and the obtained results are given below.

21.3.1.1   EFFECT OF ALGINATE CONCENTRATION

It's clear from Fig. 21.1 that increasing the concentration of alginate has a linear positive effect on the activity. Increasing the activity with alginate concentration can be explained in the light of increasing the amount of entrapped enzyme as a result of the formation of as more densely cross-linked gel structure [28]. This explanation has been confirmed by the data obtained in case of immobilized and/or entrapped amount of enzyme.

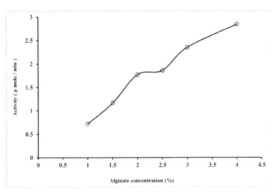

FIGURE 21.1   Effect of alginate concentrations on beads catalytic activity.

21.3.1.2   EFFECT OF CROSS-LINKING TEMPERATURE

Opposite behavior of the activity has been obtained with increasing the temperature of cross-linking process in $CaCl_2$ solution (Fig. 21.2). Such results could be explained based on increasing cross-linking degree along with the temperature of calcium chloride solution, which leads finally to decrease the amount of diffusive lactose substrate to the entrapped enzyme and hence reduce the activity of the immobilized enzyme.

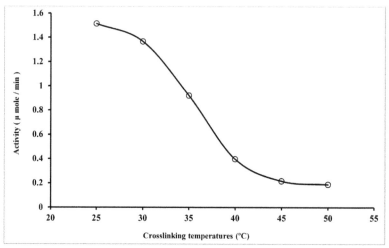

FIGURE 21.2   Effect of crosslinking temperature on beads catalytic activity.

21.3.1.3   EFFECT OF CROSS-LINKING TIME

Figure 21.3 shows the effect of increasing the cross-linking time in $CaCl_2$ solution on the activity of entrapped enzyme. It's clear that a reduction of about 20% of the activity has been detected with increasing the time from 30 to 60 min. Further increase has no noticeable effect on the activity.

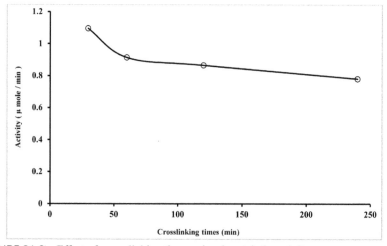

FIGURE 21.3   Effect of cross-linking time on beads catalytic activity.

21.3.1.4 EFFECT OF CACL₂ SOLUTION CONCENTRATION

The effect as given in Fig. 21.4 of $CaCl_2$ concentration on the activity of entrapped enzyme shows that increase in $CaCl_2$ concentration decreases the activity linearly. Such results could be interpreted in the light of increasing the gel cross-linking density and hence reducing the amount of diffusive lactose.

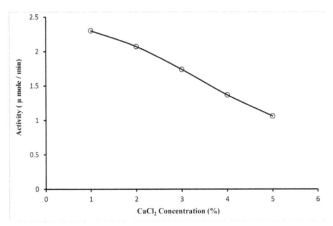

FIGURE 21.4   Effect of $CaCl_2$ concentration (%) on beads catalytic activity.

21.3.1.5 EFFECT OF ENZYME CONCENTRATION

The dependence of the catalytic activity of the beads on enzyme concentration is illustrated in Fig. 21.5. It is clear that increasing the enzyme concentration increases the activity linearly within the studied range. Increasing the amount of entrapped enzyme is the logic explanation of such results.

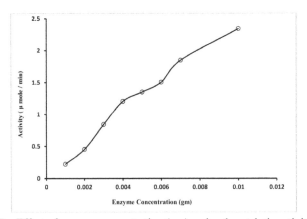

FIGURE 21.5   Effect of enzyme concentration (gm) on beads catalytic activity.

21.3.1.6 EFFECT OF SUBSTRATE'S TEMPERATURE

Shift of the optimum temperature towards lower side has been detected upon entrapping of the enzyme (Fig. 21.6). A suitable temperature found to be 45 °C for the entrapped enzyme compared with 55 °C for the free form. Such behavior may be explained in the light of acidic environment due to the presence of free unbinding carboxylic groups. This explanation is confirmed by the higher rate of enzyme denaturation at higher temperatures, 50–70 °C, in comparison with the free form.

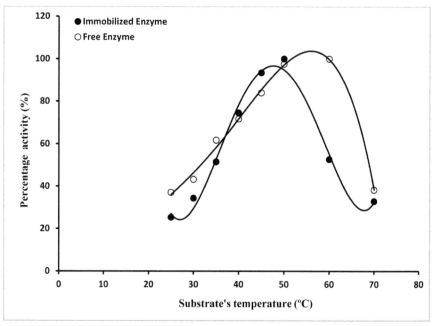

FIGURE 21.6   Effect of substrate's temperature (°C) on beads catalytic activity.

21.3.1.7 EFFECT OF SUBSTRATE'S PH

Similar behavior of the immobilized enzyme to the free one has been observed in respond to the changes in the substrate's pH (Fig. 21.7). The optimum pH has not shifted, but the immobilized enzyme shows higher activity within acidic region, pH 2.5–3.0. This resistance to pH could be explained due to the presence of negatively charged free carboxylic groups un-binding by calcium ions. This result is in accordance with those obtained by other authors [29].

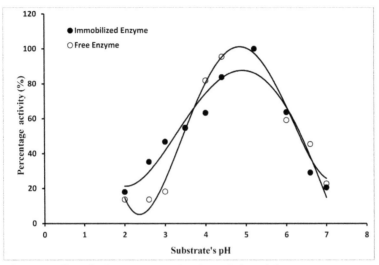

FIGURE 21.7   Effect of substrate's pH on beads catalytic activity.

21.3.1.8   KINETIC STUDIES

Figure 21.13 presents the effect of the immobilization process on the kinetic parameters especially *Km* which can be considered as the reflection of presence or absence of substrate diffusion limitation.

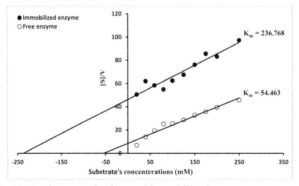

FIGURE 21.8   Han's plot curve for free and immobilized enzyme.

A Han's plot curve indicates how much the immobilization process can be affected as medicated from the kinetic parameters of immobilized enzyme. That the *Km* value has become five times higher than that of the free one is a refection of powerful diffusion limitation of the substrate.

In addition, mis-orientation of the active sites as a result of the immobilization process could be another factor affecting the accessibility of the active sites to the substrate. To study this effect, galactose has been added during the immobilization process in different concentrations to protect the active site (Fig. 21.9).

The *Km* values have been reduced with increasing the galactose concentration reaching minimum value at 30 μ mole (Fig. 21.16). Beyond this concentration, the *Km* value starts to increase again.

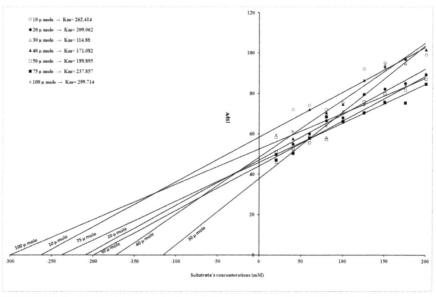

FIGURE 21.9   Han's plot curve for immobilized enzyme with addition of galactose.

The protection of the active sites has a positive refection on the activity of immobilized enzyme (Fig. 21.10). The reduction in the *Km* value and activity enhancement proved the role of mis-orientation but still the fact that diffusion limitation of the substrate has the main role.

To solve the problem of diffusion limitation, the enzyme was used to be covalently immobilized on the surface of the alginate beads after activation with *p*-benzoquinone. The impact of different factors controlling the activation process of the alginate hydroxyl groups using *p*-benzoquinone (PBQ) in addition to the immobilization conditions on the activity of immobilized enzyme have been studied. The immobilized enzyme has been characterized from the bio-chemical point of view as compared with the free enzyme.

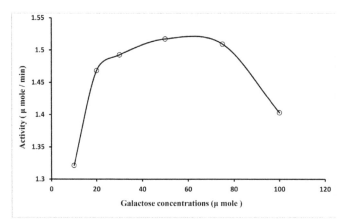

FIGURE 21.10   Effect of galactose concentration on beads catalytic activity.

21.3.2   ACTIVITY OF THE BEADS [30]

21.3.2.1   THE EFFECT OF ALGINATE CONCENTRATION

From Fig. 21.11 it's clear that increasing the concentration of alginate has negative effect on the activity. A reasonable explanation could be obtained by following the amount of immobilized enzyme, which decreases in the same manner. This could be explained a s a result of increasing the crosslinking density and a direct reduction of the available pores surface area for enzyme immobilization. Indeed, the retention of activity has not affected so much since the decrease rate of both activity and amount of immobilized enzyme is almost the same.

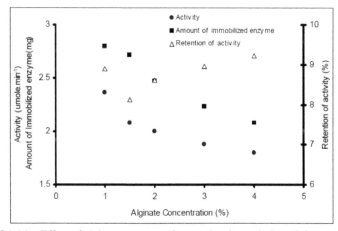

FIGURE 21.11   Effect of alginate concentrations on beads catalytic activity.

21.3.2.2   EFFECT OF CACL$_2$ CONCENTRATION

As shown in Fig. 21.12, the activities of the beads have not been affected so much by changing the concentration of CaCl$_2$ solution. Unexpectedly, the amount of immobilized enzyme has been increased gradually with concentration of CaCl$_2$ solution. This behavior could be explained according to the impact of concentration of CaCl$_2$ solution on the number of the formed pores inside the beads which offering additional surface area for enzyme immobilization. Since the activity was found almost constant and the amount of immobilized enzyme increased under the same conditions, so it is logic to have a reduction of the retention of activity.

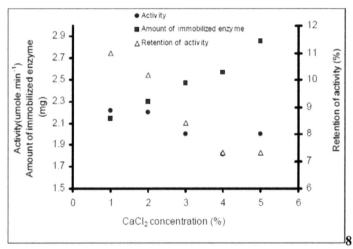

FIGURE 21.12   Effect of CaCl$_2$ concentrations on beads catalytic activity.

21.3.2.3   EFFECT OF AGING TIME

It has been found that changing the aging time from 30 to 240 min has a clear effect on the catalytic activity of the beads. This may be due to the crosslinking effect of CaCl$_2$, which affected directly on the pores size producing pores smaller than the enzyme size and hence reducing the available pores surface area for immobilizing enzyme. Since the amount of immobilized enzyme reduced with higher rate than the activity did, so the retention of activity affected positively with aging time increase (Fig. 21.13). It is clear from the results here that not only the pores surface area but also the pores size affects the amount of immobilized enzyme and hence the activity.

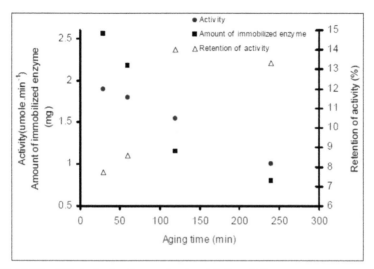

FIGURE 21.13   Effect of aging time on beads catalytic activity.

21.3.2.4   EFFECT OF AGING TEMPERATURE

The dependence of the catalytic activity of the beads on aging temperature is illustrated in Fig. 21.14. From the figure, it is clear that increasing the aging temperature resulted in increasing the activity in gradual way regardless the cross bonding decrease of the amount of immobilized enzyme. This takes us again to the combination between the pores size distribution and pores surface area. The best retention of activity has been obtained with the highest aging temperature.

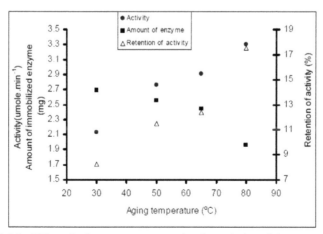

FIGURE 21.14   Effect of aging temperatures on beads catalytic activity.

21.3.2.5   KINETIC STUDIES

The kinetic parameters, K_m and V_m, of the immobilized enzyme under selected conditions, aging time and temperature, have been calculated from the Hanes Plot curves and tabulated in Tables 21.1 and 21.2, respectively. In addition, the Diffusion coefficient (De) of lactose substrate has been calculated [31] and related to the K_m and V_m values.

TABLE 21.1   Kinetic Parameters and Diffusion Coefficient of Immobilized Enzyme Prepared under Different Aging Time

Parameters	Aging time (minutes)			
	30	60	120	240
Km (mmol)	61.00	64.66	133.86	221.80
Vm (umol.min^{-1})	2.94	2.46	4.00	3.36
De (cm^2.s^{-1})	4.9×10^{-11}	4.6×10^{-11}	9.73×10^{-11}	0.184×10^{-11}

TABLE 21.2   Kinetic Parameters and Diffusion Coefficient of Immobilized Enzyme Prepared under Different Aging Temperature

Parameters	Aging temperature (°C)			
	30	50	65	80
Km (mmol)	61.32	73.44	79.17	265.70
Vm (umol.min^{-1})	2.99	3.24	2.27	5.62
De (cm^2.s^{-1})	5.49×10^{-11}	13.98×10^{-11}	7.3×10^{-11}	10.8×10^{-11}

From the obtained results it's clear that increasing the crosslinking degree of the beads through increasing either the aging time or aging temperature in CaCl$_2$ solution has affected directly the Km values which tended to increase. This could be explained based on the diffusion limitation of the lactose into the pores of the beads. This conclusion could be supported by taking into account the obtained data of Diffusion coefficient, which in general is higher than the value of immobilized enzyme on beads free of diffusion limitation of the substrate; 1.66×10^{-8} cm^2.s^{-1}. On the other hand V_m values are in general less than that of the free enzyme regardless that some of the obtained K_m values are equal to that of the free enzyme.

21.3.3   BIOCHEMICAL CHARACTERIZATION OF THE CATALYTIC ALGINATE BEADS

21.3.3.1   EFFECT OF SUBSTRATE'S TEMPERATURE

Similar temperature profile to the free enzyme has been obtained for our immobilized form (Fig. 21.15). An optimum temperature has been obtained at 55 °C. This similarity indicates that the microenvironment of the immobilized enzyme is identical to that of the free one, so absence of internal pore diffusion; illustrates the main drawback of the immobilization process, and the success of surface immobilization to overcome this problem. The confirmation of this conclusion has come from determination of the activation energy for both the free and immobilized form [29], which was found very close in value; 6.05 Kcal/mol for the free and 6.12 Kcal/mol for the immobilized forms, respectively.

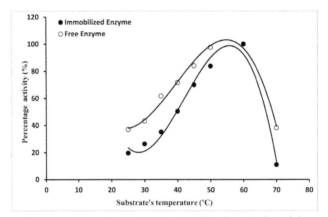

FIGURE 21.15   Effect of substrate's temperature on beads catalytic activity.

21.3.3.2   EFFECT OF SUBSTRATE'S PH

Different behaviors of the immobilized enzyme compared with those of the free one have been observed in respond to the change in the catalytic reaction's pH (Fig. 21.16). Both slight and broaden of the optimum pH for the immobilized form has been obtained since the "optimum activity range," activity from 90–100%, was found from pH value of 4.0 to 5.5. This shift to the acidic range could be explained by the presence of negative charges of unbinding carboxylic groups [29]. This behavior is advantage for use this immobilized form in the degradation of lactose from whey waste, which is normally has a low pH. It's enough to mention here that at (pH 3.0) the immobilized form beads retained 65%, of its maximum activity compared with 20% of the free enzyme.

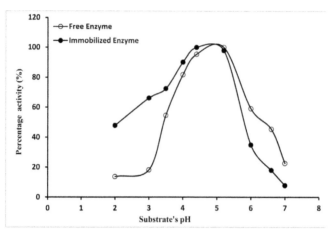

FIGURE 21.16   Effect of substrate pH on beads catalytic activity.

21.3.3.3   *OPERATIONAL STABILITY*

The operational stability of the catalytic beads was investigated (Fig. 21.17). The beads were repeatedly used for 21 cycles, one hour each, without washing the beads between the cycles. The activity was found almost constant for the first five cycles. After that, gradual decrease of the activity has been noticed. The beads have lost 40% of their original activity after 21 h of net working time. No enzyme was detected in substrate solution, so we cannot claim the enzyme leakage as the cause for activity decline. Incomplete removal of the products from the pores of the beads, which can cause masking of the enzyme active sites could be a proper explanation. Washing the beads with suitable buffer solution between the cycles and using higher stirring rate may help in solving this problem.

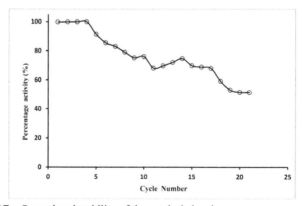

FIGURE 21.17   Operational stability of the catalytic beads.

21.4 CONCLUSION

From the obtained results we can conclude that the activation of alginate's OH groups by using *p*-benzoquinone offers a new matrix for covalent immobilization of β-galactosidase enzyme. Both the temperature and the pH of the activation process were found to be determining factors in obtaining high catalytic activity beads. The pH of the enzyme solution is found to have a pronounced effect on the catalytic activity of the beads. Three hours of immobilization time was found optimum using 5 mg of enzyme-buffer solution (pH range 4.0–4.5). The wide "optimum pH range" of the activity of the immobilized β-galactosidase makes it's to be recommended to be used in the degradation of lactose in whey waste to reduce its BOD and the production of de-lactose milk to serve the peoples suffering from lactose tolerance. Finally, the immobilization of β-galactosidase on the surface of *p*-benzoquinone activated alginate beads has succeeded in avoiding the diffusion limitation of lactose.

KEYWORDS

- **Alginate's OH groups**
- **Biopolymers**
- **Diffusion limitation of lactose**
- **Drug delivery systems**
- **High catalytic activity beads**
- **Immobilization of β-galactosidase**

REFERENCES

1. Cheetham, P. S. J. (1985). Principles of industrial enzymology: Basis of utilization of soluble and immobilized enzymes in industrial processes, Handbook of Enzyme Biotechnology (Wiseman, A., ed.) Ellis Horwood Limited, *Chichester,* 74–86.
2. Klibanov, A. (1983). Immobilized enzymes and cells as practical catalysts. *Science, 219,* 722–727.
3. Smidsrod, O. & Skjak-Braek, G. (1990). Alginate as immobilization matrix for cells. *Trends Biotech., 8,* 71–78.
4. Pollard, D. J., & Woodley, J. M. (2007). Biocatalysis for pharmaceutical intermediates: the future is now. *Trends Biotechnol, 25,* 66–73.
5. Woodley, J. M. (2008). New opportunities for biocatalysis: making pharmaceutical processes greener. *Trends Biotechnol, 26,* 321–327.
6. Schmid, A., Dordick, J. S, Hauer, B., Kiener, A., Wubbolts, M., & Witholt, B. (2001). Industrial biocatalysis today and tomorrow. *Nature, 409,* 258–268.
7. Straathof, A. J. J., Panke, S., & Schmid, A. (2002). The production of fine chemicals by biotransformations. *Curr Opin Biotechnol. 13,* 548–556.

8. Schoemaker, H. E., Mink, D., & WubboLts, M. G. (2003). Dispelling the myths – biocatalysis in industrial synthesis. *Science;299*, 1694–1697.
9. Meyer, H-P. (2006). Chemocatalysis and biocatalysis (biotransformation): some thoughts of a chemist and of a biotechnologist. *Org Proc Res Dev. 10*, 572–580.
10. Durand, J., Teuma, E., & Gómez, M. (2007). Ionic liquids as a medium for enantioselective catalysis. *C R Chim, 10*, 152–177.
11. Trivedi, A. H., Spiess, A. C., Daussmann, T., & Buchs, J. (2006). Effect of additives on gas-phase catalysis with immobilized Thermoanaerobacter species alcohol dehydrogenase ADH T), *Appl. Microbiol. Biotechnol. 71*, 407–414.
12. Solis, S., Paniagua, J., Martinez, J. C., & Asmoza, M. (2006). Immobilization of papain on mesoporous silica: pH effect, *J. Sol-Gel Sci. Technol. 37*, 125–127.
13. Johnson, R. D. Wang, A. G., & Arnold, E. H. (1996). *J. Phys. Chem. 100*, 5134.
14. Ramsden, J. J. Prenosil, J. E. (1994). *J. Phys. Chem., 98*, 5376.
15. Noinville, V., Calire, V-M., & Bernard, S. J. (1995*). Phys. Chem., 99*, 1516.
16. Sadana, A. (1993). *Bioseparation, 3*, 297.
17. Marek, P. K. J. & Coughlan, M. E. (1990). in: Protein Immobilization (R. E Taylor, ed.), MarcelDekker, New York, 13–71.
18. Kennedy, J. E., & Cabral, J. M. S. (1995). Artif Cells, *Blood Substitutes Immobilization Biotechnol., 23*, 231.
19. Mosbach, K., & Mattiasson, B. (1976). *Methods Enzymol., 44*, 453.
20. Trevan, M. D. (1980). Immobilized Enzymes: An Introduction and Applications in Biotechnology, Wiley, Chichester.
21. Wilchek, M., & Miron, T. (2003). Oriented versus random protein immobilization. *J. Biochem Biophys Methods, 55*, 67–70.
22. Gupta, M. N., & Mattiasson, B. (1992). Unique applications of immobilized proteins in bioanalytical systems. In: Suelter, C. H., Kricka, L., editors. Bioanalytical applications of enzymes, New York: John Wiley & Sons Inc., *36*, 1–34.
23. Mohy Eldin, M. S., Serour, E.S, Nasr, M., & Teama, H. (2011). *J Appl Biochem Biotechnol, 164*, 10.
24. Mohy Eldin, M. S., Serour, E., Nasr, M., & Teama, H. (2011). *J Appl Biochem Biotechnol, 164*, 45.
25. Mohy Eldin, M. S. (2005). *Deutsch lebensmittel-Rundschau, 101*, 193.
26. Mohy Eldin, M. S., Hassan, E. A., & Elaassar, M. R. (2005). *Deutsch lebensmittel-Rundschau, 101*, 255.
27. Smirsod, O., & Skjak-Braek, G. (1990). Alginate as immobilization matrix for cells. *Trends Biotechnol, 8*, 71–8.
28. Maysinger, D., Jalsenjak, I., Cuello, A. C. (1992). *Neurosci. Lett. 140*, 71–74.
29. Bergamasco, R., Bassetti, F. J. Moraes, F. De & Zanin, G. M. (2000). *Braz. J. Chem. Eng. 17*, 4.
30. Mohy Eldin, M. S., Hassan, E. A., & Elaassar, M. R. (2004). β-Galactosidase Covalent Immobilization on the Surface of P-Benzoquinone-Activated Alginate Beads as a Strategy For Overcoming Diffusion Limitation. III. Effect of Beads Formulation Conditions on the Kinetic Parameters, The 1st International Conference of Chemical Industries Research Division. 6–8 December.
31. White, C. A., & Kennedy, J. F. (1980). *Enzyme Microb. Technol. 2*, 82–90.

CHAPTER 22

REGENERATION POTENTIAL OF PEAR CULTIVARS AND SPECIES (*PYRUS* L.) OF THE COLLECTION OF THE NATIONAL DENDROLOGICAL PARK "SOFIYIVKA" OF NAS OF UKRAINE

ANATOLIY IV. OPALKO, NATALIYA M. KUCHER,
and OLGA A. OPALKO

CONTENTS

ABSTRACT

Specific features of nonmorphogenetic posttrauma regeneration, which facilitates the healing of possible plant damage, were studied on several representatives of the genus *Pyrus* L. The comparison of rates and intensity of wound healing to the dates of artificial notchings makes it possible to conditionally classify vegetative period of the cultivars and species studied based on their regeneration potentials into such stages as—regeneration rise, relative decrease, a second rise wave, and rather fast damping. The tendency of higher dependence of regeneration potential on temperature fluctuations rather than on precipitation amount and hydrothermal coefficient was established. The suggestion was made that a regeneration ability indicator confirms indirectly the level of ecological adaptation of the genotypes under study, and the periods of the highest regeneration activity can be favorable for vegetative propagation, including propagation by cutting and grafting, in vitro (micropropagation through plant tissue culture) and other technological processes accompanied by plant damage.

22.1   INTRODUCTION

*Pyrus*L.is a member of the subtribe Pyrinae, tribe Pyreae, subfamily Spiraeoideae (formerly Maloideae), family Rosaceae. Among widely distributed in Euro Asia 40 species of the genus *Pyrus*, *Pyruscommunis* L. and *P. pyrifolia* (Burm.) Nak. are the main edible pear species; however, *P. communis* are of the greatest economic value [1]. Mostcultivars, grown in temperate zones, were created from this species. Many researchers identify intraspecific taxon— *P. communis* sub sp. *domestica* (Medik.) Domin [2],—which enumerates over five thousand pear cultivars [3], and only a small percentage of them are cultivated commercially [4]. Pear cultivars grown in Japan, China and other countries of Eastern Asia belong mostly to *P. pyrifolia* [1, 4].

About seven percent of pears are grown in commercial orchards of Ukraine, and 15%—in home orchards. According to this indicator pears are second after apples, the latter occupies 75 and 40% of the areas allocated for fruit and berry crops, correspondingly. Last decades witness a positive tendency in pear production in compliance with a national program of horticulture development in Ukraine approved in 2008; it is planned to increase pear orchards to 20.8th ha for the period up to the year of 2025 [5]. At present Ukraine is among top 20 world pear producers; its annual harvest is 102–177th tons (metric tons) and it is the best index for former Soviet Union countries. Last decades show an increasing tendency of worldwide pear production from 8–897 in 2001 to 15–945th tons in 2011. This tendency is true for Ukraine except for quite unstable pear production in various years [6]. With this in view, the requirements of quantity and quality are higher, including unification of planting pears; it also explained the necessity to do research concerning the improvement of advanced vegetative propagation of both cultivars and clonal (vegetative propagated) pear rootstocks and their close wild relatives, which can be used in a pear-breeding program.

Pear is an allogamous plant and, as a result, economic and other features are not preserved when seed propagation is used. Thus, only vegetative propagation, including clonal pear rootstocks, can provide unified planting pears. All methods of vegetative propagation are associated with plant damage; they are based on the plants' ability to regenerate, which in turn defined the subject/direction of our research.

Various localized damages occur on the plants (especially perennial plants) during their life period. Stems of tender plants (and other organs) can be easily damaged by strong wind, frost, pouring rain, insects, different animals due to poor handling etc. So in the process of evolution they develop adaptive mechanism of protection against damage, that is, the ability to recover. In some cases natural damage (also a severe one) can be a common and necessary stage of the development of an individual (plants and animals), which results in breakage, and death of some parts [7]. The ability to repair can be considered as one of the adaptive modifications, it is proportionally equal to intensity and duration of the effect of artificial and natural damaging factor. Adaptive modification takes place when intensity and duration of a damaging factor is within the limit defined by previous evolution history of a species. Non-adaptive changes, when a plant does not show adequate response and may die, can occur, provided intensity and duration of a damaging factor exceeds a standard [8].

The problem of regeneration has always been of great interest for scientists-biologists, crop growers and livestock breeders, professionals and amateurs, in particular horticulturists [9, 10]. All signs of posttrauma regeneration can be classified into two large groups—morphogenetic regeneration, when lost parts and organs reproduce, and also a new organism from one part of a primary body can develop, including the one from a separate cell in vitro; and nonmorphogenetic posttrauma regeneration which results in healing of all possible wounds [8]. The cases of so-called compensatory regeneration were described, when, after removing a part of a stem or a root, a plant reproduces a lost organ (morphogenetic regeneration), or when all the leaves but one were removed, and this only leaf grows in size considerably and provides leafless plants with photosynthesis products [11].

The results of posttrauma self-repair in plants can be explained by cambial activity [8], which, depending on phylogenetic peculiarities, is seen in basic specific and varietal distinctions concerning the ability to regenerate; both potential productivity and ecological adaptability of plants depend on it [11]. Among other factors associated with the rate and the whole course of regenerative processes, the following should be mentioned: ontogenetic peculiarities of an individual, its physiological condition, and endogenous and exogenous factors of chemical, physical and biological nature as well. These are various chemical compounds, wound provocative, ionizing radiation, temperature and moisture of the air and soil, photo-period, phyto-sanitary condition, ontogenesis phase and alike [8].

In the mid-thirties of the previous century N.P. Krenke applied quantitative methods in studying age variation of somatic characteristics and formogenesis and regeneration factors, specific aspects of joining grafting components and reasons for the formation of chimeras in plants. In his works [7], which are still actual, the importance of regeneration ability for the success of vegetative plant propagation is emphasized; this became a good foundation for further theoretical and applied research [8, 10 and 11]. Krenke classified all stress factors as natural and artificial, normal and abnormal (e.g., grafting of a cutting in an upside-down position), with integrity breakage of some parts or their separation from a plant, including further analysis of regeneration potential of a separated part or those parts which remained in an initial organism. The factors of grafting plant parts (structural, physiological, combined-factorial) were described, when translocation of some elements of a plant itself or introduction of unnatural elements to a plant took place, as well as other changes of plant details/parts affected by natural factors, and also bending, twisting, centrifugation and other artificial effects [7]. Thermal, chemical and radiation stresses can be of natural origin; they can result from by-effects of technical-genetic activity of man or conscious special impact, etc. [8].

Plants, as well as their parts, change both quantitatively (mass, size, etc.) and qualitatively in the process of individual development. Plants move from the first (juvenile) phase, which lasts from seed germination to fruiting of a seedling, to an adult one, and then they get old. Such ontogenetic variation can be typical for both an individual seedling and a clone itself. Long-lived clones of grape and fruit trees keep their inheritance in most cases; however, their current properties do not always coincide with the features described by distant pomologists. Some typical physiological, anatomical and morphological signs and properties of the earlier life period of a plant can be less seen; instead others can appear [8, 12]. Young seedlings differ from adult trees in size, leaf form and edge serration. Young seedlings of various breeds have thorns, which are not seen on adult trees grown from these seedlings. There are also distinctions, which concern morphology and a deviation angle of laterals from a stem, productive morphogenesis, etc. In addition to the above-mentioned and many other differences, young seedlings are mostly characterized by better regenerative-morphogenetic potentials than adult and older trees [12].

Completion of a juvenile phase of ontogenesis is usually associated with the development of floral buds, but a lower part of a young tree crown may remain in a juvenile stage whereas flowers are formed in its upper part and a young tree enters an adult phase getting an ability of fruiting [12].

The ideas of interrelationship between individual and historic development of living organisms appeared back in XIX century, when, according to Zhuchenko [13], basic provisions of bio-genetic "*law of Friedrich Müller (1864) and Ernst Haeckel (1866)*" [13] were proclaimed. According to this law ontogenesis recapitulates major phylogenesis stages of a group, to which an individual organism belongs [13]. It is possible to regenerate roots from green cuttings of various woody plants such as coniferous, oak; most of fruit tress and others belong to hard-rooting species, if grafting of 1–3 year-old seedlings is done. Most of the researchers believe that advantages in regeneration ability of green cuttings taken from young seedlings can be explained by the factor of juvenility; therefore, a seedling (in compliance with a bio-genetic law) did not totally lose the properties of ancestral forms to regenerate roots from stem parts [8, 9, 14, 15]. There is evidence that the plants, which were described in old books as those capable of adventive root formation, became hard-rooted ones after many years [8, 9]. At present physiological stress called trauma/ damage is considered to be an inductor of adaptive response of an organism, which facilitates regeneration [8, 16].

22.2 MATERIALS AND METHODOLOGY

Post-trauma regeneration processes in perennial woody plants and their dependence on meteorological factors were estimated according to the ability of some representatives (species and intraspecific taxons) of *Pyrus* L. to nonmorphogenetic post-trauma regeneration. Pear cultivars Bere Desiatova, Umans'kaiuvileina, Kniahynia Ol'ha and Sofiia, few rootstocks, and also cultivars of basic species *Pyruscommunis* L. and several other wild relatives of pears were studied.

Pear cultivars and species studied were grown in the collections of the National Dendrological Park "Sofiyivka" of NAS of Ukraine, situated in the Central-Dnieper elevated region of Podilsko-Prydniprovsk area of the Forest-Steppe Zone of Ukraine. The area is characterized by temperate-continental climate with unstable humidification and considerable temperature fluctuations. Average many-year amount of precipitation per year is 633.0 mm, its amount being 300–310 mm at +10°C, which corresponds to the precipitation amount in dry southern areas of Ukraine. Average many-year air temperature is +7.4°C.

To evaluate regeneration ability the recommendations of I.A. Bondorina [17] were used, that is, notchings (10–12 mm long, 15.5 mm wide) (Fig. 22.1) with a special cutter (Fig. 22.2) were made on one-year-old shoots of the previous year of the plants studied every decade from March till October. It's important to notch into the cambium.

FIGURE 22.1   The wound made with manual cutter (notching 10–12 mm long and 15.5 mm wide).

FIGURE 22.2   The manual notch cutter.

The wound, where the notching was made, was covered with transparent scotch-tape to avoid infection and withering (Fig. 22.3a). The equation to calculate regeneration coefficient was adapted to a 9-point scale of the evaluation of regeneration efficiency [8, 10, 14]. The overgrowth of the wound was observed every decade with help of a magnifying glass. To compare results used of a 9-point scale. The intensity of callus genesis was estimated at 1 point, if callus formation did not occur or its surface was less than 5% of the wound (Fig. 22.3b). Objects with callus areas equal to 85.5–100% were estimated at 9 points (Fig. 22.3c).

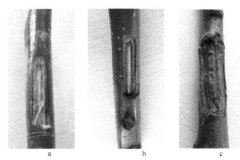

FIGURE 22.3   Estimation of the wound overgrowing: (a) wound covered with transparent scotch-tape to avoid infection and withering; (b) surface of callus occupies less than 5% of the wound (1 point); (c) surface of callus occupies 85.5–100% of the wound (9 points).

Regeneration coefficient was calculated in units of regeneration coefficient (urc) according to Opalko's equation [8, 14]:

$$R = \frac{S^2}{n_1 + n_2},$$

where R – regeneration coefficient, urc; S – intensity of callus genesis, points; n_1 –number of days after notching was made to the appearance of the first signs of callus;

n_2 – number of days after notching was made to the completion or termination of callus development.

The statistical processing of the experimental data was carried out by the methods of R.A. Fisher [18]. To calculate precipitation amount and air temperature sums, the data of Uman meteorological station was used. The hydrothermal coefficient of Selyaninov [19] was calculated with the formula:

$$HTC = \frac{\sum Q}{0.1 \sum T},$$

where HTC – hydrothermal coefficient; $\sum P$ – precipitation amount; $\sum T$ – sum of temperatures higher than +10°C for some period of time.

22.3   RESULTS AND DISCUSSION

The comparison of rates and intensity of wound healing to the dates of artificial notchings makes it possible to classify vegetative period of the cultivars and species studied based on their regeneration potentials into such stages as—regeneration rise, relative decrease, a second rise wave, and rather fast damping. Besides, substantial interspecific and intervarietal distinctions in general regeneration ability were observed, indices of regeneration coefficient fluctuating in dependence on notching terms/dates. Most likely seasonal variations of regeneration coefficient indices were due to the fluctuation of weather/meteorological conditions, namely, precipitation amount and air temperature.

In 2007–2012, average annual precipitation amount was higher in 2010 alone (by 104.1 mm). Plant moisture support in 2011 was close to average many-year indices, in 2008 and 2009 precipitation deficit was recorded (115.6 and 109.5 mm), whereas in 2007 and 2012 the deficit amounted to 217.0 and 160.1 mm, correspondingly. All deviations/variations of average daily temperatures from average many-year air temperature in the years studied, except for the year of 2011, tended to the increase of many-year data, the temporal variations of temperature being much smaller in a relative sense than those of precipitation.

It was found out that the most favorable conditions to show regeneration potential for pear cultivars Bere Desiatova, Umans'kaiuvileina, Kniahynia Ol'ha and Sofiia were in the year of 2010. Generalized average seasonal indices of regeneration coefficient in 2010 were 5.42 urc with coefficient variation over 70%. In 2007–2009, average regeneration potential was lower by 2.32–2.38 urc. In 2012 regeneration exceeded the indices of the years of 2007–2009, but it was lower (0.66 urc) than the results of 2010. During the years of investigation the cultivars studied had higher regeneration coefficients (over 3 urc) from May till the second decade of August including, which coincided with the period of active plant vegetation. And the index of over 6 urc was recorded when notchings were made from the second decade of June till the end of the first decade of July with the maximal index 6.61 urc at the beginning of the second decade of June.

The speed at which callus developed depended on the term the notchings were made. When notchings were made in the third decade of March in all years of investigation, the index of regeneration coefficient of pear cultivars did not exceed two units. The most favorable period appeared to be the end of March in 2007 (1.96 urc), whereas this indicator was only 1.32 urc in 2010. In the first decade of April in all years of investigation the indices of regeneration coefficient increased by 0.37–0.98 urc, and in the first half of the second decade by 0.10–0.58 urc. Slight decrease of average regeneration coefficient (0.48–0.51 urc) was recorded on the shoots damaged in the third decade of April in 2007 and 2008, whereas in 2009 and 2010 in the same period the indices of average-taxon regeneration coefficient continued to increase, and they exceeded those of the previous notching term by 0.20–0.47 urc (Table 22.1).

TABLE 22.1 Indices of Pear Average-Tax on Regeneration Coefficient Depending on the Notching Terms in 2007–2010

Notching dates	Regeneration coefficient, urc×				
	2007	2008	2009	2010	$\bar{X}$
21–23.03	1.96	1.55	1.55	1.32	1.60
01–03.04	2.45	2.53	2.49	1.69	2.29
08–14.04	3.03	3.02	2.59	2.20	2.71
19–22.04	2.52	2.54	2.79	2.67	2.63
02–04.05	2.46	3.16	2.96	3.86	3.11
12–15.05	4.67	4.63	4.49	4.32	4.53
21–23.05	5.60	5.39	5.60	4.89	5.37
01–05.06	5.33	5.03	5.36	6.67	5.60
09–14.06	4.17	4.26	4.64	7.10	5.04

TABLE 22.1   *(Continued)*

Notching dates	Regeneration coefficient, urc×				
	2007	**2008**	**2009**	**2010**	**X̄**
21–22.06	5.58	5.73	5.69	9.42	6.61
01–04.07	4.27	4.64	4.35	12.01	6.32
08–10.07	3.96	4.36	3.84	12.79	6.24
15–20.07	4.86	4.38	4.35	8.26	5.46
25–29.07	3.93	4.35	4.25	8.22	5.19
06–09.08	2.16	2.72	2.89	10.29	4.52
16–19.08	2.04	1.96	2.69	8.72	3.85
25–31.08	1.91	1.60	1.67	2.95	2.03
07–10.09	1.24	1.16	1.17	3.70	1.82
17–22.09	1.02	0.94	1.06	1.29	1.08
26–29.09	0.63	0.69	0.58	0.68	0.65
08–14.10	0.09	0.10	0.16	0.85	0.30
X̄	3.04	3.08	3.10	5.42	3.66
S^2	1.67	1.68	1.67	3.81	
$V_{\%}$	54.87	54.67	53.81	70.34	

× X̄ —mean; S^2—variance; $V_{\%}$— coefficient of variation

In the first decade of May, 2007 average regeneration coefficient, as compared with the notchings made in the third decade of April, decreased by 0.06 urc, and in 2008–2010—increased by 0.17–1.19 urc, the smallest difference being in 2009 and the largest one—in 2010. During the second-third decades of May the indices of regeneration coefficient continued to intensively increase in all years of investigation. In 2007, when notchings were made at the beginning of the third decade of May, the maximal/highest index of regeneration coefficient was recorded—5.60 urc, and in 2008 and 2009—5.39 and 5.60 urc, correspondingly. When notchings were made in the first and second decades of June, the indices of regeneration coefficient decreased slightly in 2007–2009, and at the beginning of the third decade of June the peak of the second regeneration coefficient rise was recorded. In the first decade of July the indices of regeneration coefficient decreased gradually by 1.37–1.85 urc.

In 2010, regeneration coefficient increased bit-by-bit, having reached its maximum 12.79 urc at the end of the first decade of July. When notchings were made in the second decade of July in 2007 and 2009, the regeneration coefficient rise was observed, its indices being 4.86 and 4.35 urc, correspondingly. In 2008, a similar regeneration coefficient rise did not take place in the same period. In 2010, in the second-third decades of July the indices of regeneration coefficient decreased almost by 8 urc, and in the first decade of August they increased by 10.29 urc. When notchings were made in subsequent terms, regeneration coefficients gradually decreased in all years of investigation and they were zero-proximity at the end of September or at the beginning of October.

The tendency of higher dependence of regeneration coefficient on temperature fluctuation than on precipitation amount was identified in earlier researches/experiments with the representatives of *Corylus* L. [10] and *Malus* Mill. [14]. This tendency was confirmed in the trials with the above-mentioned pear cultivars in 2009 and 2010. Correlation coefficients of the indices of regeneration coefficient, received in 2009 with precipitation amount (P, мм), air temperature (T, °C) and hydrothermal potential (HTC), prove higher contingency of regeneration coefficient with HTC and slight connection with precipitation amount when notchings are made (Table 22.2).

TABLE 22.2   Correlation Coefficient Between Indices of Pear Regeneration Ability and Precipitation Amount, Temperature and HTC When Notchings Are Made

Cultivar	2009			2010		
	P, мм	T, °C	HTC	P, мм	T, °C	HTC
Bere Desiatova	0.14	0.51	0.49	0.35	0.82	0.50
Umans'kaiuvileina	0.26	0.51	0.73	0.46	0.73	0.38
Kniahynia Ol'ha	0.30	0.44	0.56	0.36	0.87	0.22
Sofiia	0.19	0.39	0.53	0.30	0.79	0.29

$p > 0.95$.

Taking into consideration the fact that in warm months when precipitation supply/support is very important, and the sums of effective temperatures, decreased by one order, are nearly equal with evaporation amount, we can assume that HTC denomination in fact characterizes evaporation, and the relationship of precipitation amount and evaporation is the indicator of moisture support of the area in a vegetative season [17].

HTC value for regeneration increases in the years of precipitation deficit. Owing to the fact that from May till the end of August of 2009 average monthly air temperature was 0.4–4.3°C lower, and in turn evaporation was lower than in the same period of abundant precipitation in the summer of 2010, contingency of regeneration ability with HTC was higher in 2009. Precipitation amount and air temperature in 2010 exceeded the indicators in the year of 2009, which affected considerable increase of correlation coefficients between indices of regeneration ability of pear cultivars and average monthly air temperature.

In 2012, the experiment went on and another pear species *Pyrussalicifolia* Pall was added/involved. Meteorological conditions of pear vegetative season in 2012 were characterized by increased average monthly air temperature by 1.8–4.4°C, as compared with average many-year indices and monthly precipitation deficit (except for June), and accordingly by very low HTC. Under these conditions contingency of regeneration ability of light-loving xerophyte and mesoterm *P. salicifolia* with HTC was much higher than that of long-naturalized *P. Communis* (Table 22.3).

TABLE 22.3   Correlation Coefficients Between the Indices of Regeneration Ability of Pear Species and Precipitation Amount, Temperature and HTC When Notchings are Made

Species	P, мм	T, °C	HTC
Pyruscommunis L.	0.14	0.75	0.10
Pyrussalicifolia Pall.	0.13	0.66	0.54

p>0.95.

Correlation coefficients between the indices of regeneration ability of both pear species and precipitation amount were almost equally insignificant, but contingency of regeneration ability with average air temperature was rather high, *P. Communis* predominating. It is worth mentioning that during the whole season regeneration ability of *P. Communis* was higher that that of *P. salicifolia.* Maximal index of regeneration coefficient of *P. Communis* amounted to 10.12 urc in the variant when notching was made on August 20, whereas the best indicator of *P. Salicifolia* hardly reached 2.45 urc (the notching was made on July 11). All this gives every ground to assume that the indicator of regeneration ability can prove the level of ecological adaptation of the genotypes studied.

22.4   CONCLUSION

The intensity of nonmorphogenetic posttrauma regeneration in *Pyrus* species studied was not the same and changed in the years of investigation: it depended more on temperature fluctuation than on precipitation amount and hydrothermal coefficient. The assumption can be made that periods of the highest regenerative activity can be favorable for cutting, grafting, microcloning, etc., as well as for various technological processes, associated with trauma/damage.

22.5 ACKNOWLEDGMENT

This material is partly based on the work supported by the National dendrological park "Sofiyivka" of NAS of Ukraine (№ 0106U009045) together with Uman National University of Horticulture (№ 0101U004495) in compliance with their thematic plans of there search work. We thank corresponding member of NAS of Ukraine Ivan Kosenko, PhDs. Michael Nebykov and Aleksandr Serzhuk as well as graduate student Vitaliy Adamenko for their help with scientific-technical programs, for consultations and discussion.

KEYWORDS

- **Adaptive modifications**
- **Clonal pear rootstocks**
- **Hydrothermal coefficient**
- ***In vitro***
- **Juvenility**
- **Notching**
- **Ontogenesis**
- **Physiological stress**
- **Regeneration**
- **Vegetative propagation**

REFERENCES

1. Bell, R. L. & Itai, A. (2011). *Pyrus*. Wild crop relatives: genomic and breeding resources, temperate fruits [Ed. Chittaranjan Kole]. Berlin; Heidelberg: *Springer,* Ch. 8, 147–178.
2. Catalogue of Life: 2010. Annual checklist. Catalogue by Royal Botanical Gardens Kew. URL: http://www.catalogoflife.org/annual-checklist/2010/details/species/id/7179820(Accessed 12 June 2013).
3. Yamamoto, T. & Chevreau, E. (2009). Pear Genomics. Genetics and Genomics of Rosaceae, Plant Genetics and Genomics: *Crops and Models, 6,* [Eds. Kevin M. Folta, Susan E. Gardiner]. N.Y.: *Springer Science + Business Media, LLC.* Ch. 8, 163–188.
4. Monte-Corvo, L., Goulao, L. & Oliveira, C. (2001). ISSR analysis of cultivars of pear and suitability of molecular markers for clone discrimination. *Journal of the American Society for Horticultural Science, 126(5),* 517–522.
5. (2008). Horticulture development sectoral Program of Ukraine for the period until 2025. *Kyiv: Zhytelev,* 76 c. (In Ukrainian).
6. (2011) **Pear Production.** Food and Agriculture Organization of the United Nations FAO-STAT. URL: http://faostat.fao.org/site/567/DesktopDefault.aspx?PageID=567#ancor

7. Krenke, N.P. (1950). Regeneration of Plants. Moscow; Leningrad: *Academy of Sciences of the USSR Press,* 667 p. (In Russian)

8. Kosenko, I. S., Opalko, O. A. & Opalko, A. I. (2008). Posttravmatic regeneration processes at plants. Autochthonous and alien plants. The collection of proceedings of the National dendrological park "*Sofiyivka*" *of NAS of Ukraine, 3–4,* 10–15. (In Ukrainian)

9. Hartmann, H. T. & Kester, D. E. (1975). Plant propagation: principles and practices. Englewood Cliffs: *Prentice-Hall Inc.,* 622 p.

10. KosenkoI, S., Opalko, A. I. & Sergienko, N. V. (2011). Posttraumatic regeneration processes of the representatives of *Corylus* L. genus. Proceedings of the Russian Scientific conference with international participation "Botanical gardens in the developing world: theoretical and applied research" (July 5–7, 2011, The Tsytsin Main Moscow Botanical Garden of RAS)[Ed. Aleksandr Demidov]. Moscow: *KMK Scientific Press Ltd.,* 347–350. (In Russian).

11. Jusufov, A. G. (1978). On the evolution of the phenomena of regeneration of plants. The science of plant regeneration. *Makhachkala,* 53–55. (In Russian).

12. Visser, T. (1964). Juvenile phase and growth of apple and pear seedlings.*Euphytica, 13, (2),* 119–129.

13. Zhuchenko, A. A. (1988). Adaptive potential of cultivated plants (genetic and ecological bases). *Kishinev: Shtiinca.* 768 s. (In Russian)

14. Opalko, O. A. & Opalko, A. I. (2006). Regenerative capacity as criterion of the use of representatives of the genus *Malus* Mill. in landscape compositions. Proc. *Tbilisi bot. Garden, 96,* 187–189. (In Russian).

15. Sinnott, E. W. (2012). Plant morphogenesis. Whitefish: *Literary Licensing, LLC.* 560 p.

16. Opalko, O. & Balabak, O. (1999). Physiological stress—rhizogenic inductor activity of garden plants cuttings. Bulletin of the Lviv State Agrarian University: *Agriculture, 4,* 179–181. (In Ukrainian).

17. Bondorina, I. A. (2000). Principles for improved the properties of ornamental woody plants grafting techniques: Abstract of thesis of candidate of biological sciences: 03.00.05. The Tsytsin Main Moscow Botanical Garden of RAS. Moscow, 21 p. (In Russian).

18. Fisher, R. A. (2006). Statistical methods for research workers. New Delhi: Cosmo Publications, 354 p.

19. Shein, E. V. & Goncharov, V. M. (2006). Agrophysics. Moscow: Feniks, 400 p. (In Russian).

CHAPTER 23

THE PROSTOR AND FERM KM COMPLEX PROBIOTIC ADDITIVES: BIOTECHNOLOGICAL PREPARATIONS FOR ENHANCING THE QUALITY OF DOMESTIC FISH MIXED FEED

D. S. PAVLOV, N. A. USHAKOVA, V. G. PRAVDIN,
L. Z. KRAVTSOVA, C. A. LIMAN, and S. V. PONOMAREV

CONTENTS

ABSTRACT

The ProStor and Ferm-KM-1 complex probiotic preparations based on solid-state fermentation of beet pulp with a probiotic association (three strains of *Bacillus subtillis, Bacillus licheniformis*, a complex of lactic acid bacteria) have been developed. A *Cellulomonas* microorganism has been additionally introduced to the Ferm KM-1 probiotics composition. Some fish mixed feed formulations with use of the preparations have been developed. In experiments, the efficiency of new mixed fodders for the young of carp and sturgeon has been demonstrated.

The prerequisite of effective economic management in modern industrial fisheries is increasing their productivity that is directly linked to the use of complete and cost- effective feed. The most important task is to create and use in fish farming the feed, which:

- is maximally needed for the body to ensure its vital functions;
- has growth/development stimulating properties as well as prevention and antistress characteristics;
- ensures the environmental safety of feed produced.

In order to improve the quality of fish mixed feed, some enzyme, probiotic, prebiotic and probiotic/enzyme combination feed additives are used, as well as complex probiotic preparations enriched with phytocomponents.

Probiotic fodder preparations are regarded as a potential alternative to feed antibiotics, so the use of probiotics is considered an essential point of *obtaining* ecologically clean feed [1–5]. Probiotic preparations balanced with phytochemicals, show an enhanced biological activity due to the combination of the actual probiotic effect and the action of a phytobiotic.

Probiotics are live microbial supplements that have a beneficial effect on the body by improving the intestinal microbial balance, and stimulate the metabolism and immune processes. Probiotics are widely used in mixed feed for fish [6–9]. In themselves, probiotics do not provide a significant amount of nutrients for producing more products. But their biological potential improves fish health, enhances productivity levels, and better use of feed.

The determining factor of the probiotics efficiency is, in many ways, the technology of formulating these preparations. Modern biotechnology approaches to the development of probiotic preparations imply, firstly, the use of different types of microorganisms in certain combinations, and, secondly, their production in a form allowing their long-term storage at normal temperatures, and granulation.

The technology for production of the biologically active complex probiotic preparations ProStor and Ferm KM-1 is based on a partial solid phase fermentation of beet pulp with a probiotic association. The final product includes biomass of probiotic bacteria forming a biofilm on the surface of a phyto-carrier, products of their metabolism, phytosubstrate biotransformation products, prebiotics–pectins of beet, and phytocomponents. The bacterial composition of the preparations contains vegetative cells of three strains – *Bacillus subtillis, Bacillus licheniformis*, and a lactic acid bacteria complex. The ProStor preparation contains in the probiotic association a unique strain *Bacillus subtillis* – a producer of hydrolase class enzyme, which has antiinflammatory and antiviral effects, stimulates the immune reactions of the body. A cellulolytic *Cellulomonas* microorganism is additionally introduced to the Ferm KM-1 probiotics composition and capable of both synthesizing enzymes that break down cellulose, and producing lysine, the essential amino acid. Depending on the type of fish and their food, the effect of biological action of bacteria varies. Therefore, the preparation Ferm KM-1 increases the digestibility of all feed components, and, to the upmost degree, of fiber in case when the feed mix contains a lot of fiber, which is important, for example, for the carp. For the sturgeon on the protein diet, the preparation increases the digestibility and protein digestibility of feed.

Probiotic bacteria have an enhanced viability and resistance to adverse environmental conditions for they are in the form of a biofilm on a phyto-carrier (Fig. 23.1). The preparations contain enzymes: cellulase, amylase, complex of proteases, lipase, as well as organic acids, biologically active substances, vitamins, amino acids, immunoactive peptides – products of probiotics metabolism. The preparations comprise phyto-particles that are a cellulose microsorbent.

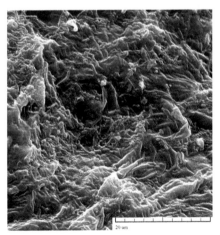

FIGURE 23.1   Microphotograph of the fermented sugar beet pulp with biofilm of probiotics.

The preparations are featured with combining probiotics and prebiotics (mannans and glucans on cell walls of yeast *Saccharomyces cerevisiae*), and phytobiotics of the medicinal plants – *Echinacea purpurea* and *holy thistle*. Echinacea has immunomodulatory properties. Echinacea preparations exhibit antibacterial, antiviral and antifungal properties. When intaking the Echinacea preparations at metabolic disorders, at the impact of different chemical compounds of toxic nature, contained in the feed (heavy metals, pesticides, insecticides, fungicides), a stimulation of the immune system has been observed.

Holy thistle is used for prevention of various liver affections. Preparations of holy thistle increase protective properties of liver to infection and poisoning, stimulate the formation and excretion of bile. The positive effect of the plant also affects the liver, and the whole digestive tract.

The special feature of the ProStor and Ferm-KM-1 products is the presence of yeast cell walls. They contain mannan oligosaccharides and beta-glucans, which effectively bond and absorb in the gastro-intestinal tract different pathogens. Beta-glucans have a stimulating effect and optimize the immune system.

The preparations increase the digestion and feed efficiency, growth rate, optimize the productive indices of fish, effective in the treatment and prevention of parasitic diseases.

The preparations that are hi-tech, bulk products of brown color, with slightly specific odor, which makes it easy to mix them with compound feed components. They tolerate forage production processes without loss of biological activity.

Warranty storage period of preparations – six months from the production date, subject to +30 °C temperature and relative humidity up to 75%.

The ProStor and Ferm KM-1 preparations are used in the feed for the young and adult fish (the carp and the sturgeon). The preparation is administered in the feed in the feed mills or farms, by mixing. They are applied daily to feed on recommended zootechnical dosage rates (for the carp 1.0–1.5 kg per ton of feed, for the sturgeon 1.5–2.0 kg per ton of feed). Side effects and complications at the use of preparations at the recommended doses have not been observed. There are no contraindications. Fish products after the use of preparations can be used without restrictions.

The efficacy of the ProStor preparation for fish is demonstrated in an experience with carp and sturgeon juveniles (Table 23.1). The preparation in an amount of 1.5 kg per ton of feed was introduced to the feed KM-2 M for the carp, and in the amount 2.0 kg per ton of feed OT-7 for the sturgeon. The under yearlings were kept in aquaria in groups of 15 animals.

TABLE 23.1 The Efficacy of the Prostor Preparation for Fish

Index	Carp		Bester (sterlet+beluga cross)	
	Experiment, 1.5 kg ProStor/t mixed feed KM-2M	Control, mixed feed KM-2M	Experiment, 2 kg ProStor/t mixed feed OT-7	Control, mixed feed OT-7
Absolute weight gain, g	8.9	6.25	14.2	6.6
% of control	142.4	100.0	215.1	100.0
Average 24-h weight grow rate, %	6.78	5.65	11.2	7.84
Food expenses, units	1.8	2.2	1.2	1.9
% of control	81.8	100.0	63.1	100.0
Survival rate, %	100	100	100	87

Fish breeding and biological indices of young carps and sturgeons as for absolute weight gain and average daily growth rate were higher than the ones of the control carp group, respectively, by 45% and 25%, and for control sturgeons, respectively, by 120% and 45%. The experimental sturgeon fry survival rate demonstrated was by 13% higher than the index of the control fish.

The cost of 1 kg of growth gain of the experiment carp was 23.4 rubles, which is 13% lower than in the controls (26.95 rubles). The cost of 1 kg of growth gain in the experiment *bester* equaled 21.2 rubles, which is 35% lower than in controls (33.0 rubles).

Feed cost indices (1.2 units) of pilot feed line with similar values of better-feed foreign companies.

In experiments on the cultivation in a closed water supply for young sturgeons on the Ferm-KM-1 diet at 0.1% in the production OT 7 feed of for young sturgeons, the *condition* factor, as well the absolute and average daily weight gain coefficient significantly increased (Table 23.2).

TABLE 23.2   Fish Breeding and Biological Indices of Growing for Two-Year Old Sturgeon Hybrids

Indices	Experiment versions	
	Control	**Experiment with Cellulomonas**
Weight, g: initial		
final	250.6±19.17	243.8±20.86
	292.5±22.4	3042±31.2
Fullton's condition factor, %	0.35 (100%)	0.39 (111%)
Absolute weight grow rate, g	41.9 (100%)	60.4 (144%)
24 h grow rate, g	1.32 (100%)	1.95 (148%)
24 average 24 h grow rate, %	0.50 (100%)	0.72 (144%)
Weight gain coefficient, unit	0.031 (100%)	0.045 (145%)
Food coefficient	1.2 (100%)	1.0 (83%)
Survival rate, %	100	100

The results of checking the efficiency of the incorporation of the ProStor and the Ferm KM probiotic preparation to the mixed feed for the sturgeon demonstrate higher industrial productivity rates for *Russian-Siberian* sturgeon hybrid. The obtained data as fishery/biological indices allow to recommend the use of the ProStor and Ferm KM-1at the large-scale mixed fodder production for they provide higher productivity figures, lowering the costs for feed and stable health conditions for the fish cultivated.

KEYWORDS

- Biofilm
- Feed
- Fish farming
- Phytobiotics
- Probiotics

REFERENCES

1. Bychkova, L. I., Yukhiмenкo, L. N., Khodaк, A. G., et al. (2008). Fish Farming, 2, 48 (in Russian).
2. Pokhilenкo, V. D. & Perelygin, V. V. (2008). News of Medicine and Pharmacy, 18(259), 56 (in Russian).
3. *Harbarth, S. & Samore, M. H.* (*2005*). Emerg Infect Dis., 11, 794
4. Pickering, A. D. (1993). Stress and Fish. Pickering, A. D. (ed.). London-N.Y. Acad. Press, 1,
5. Matsuzaki, T. I. (2000). Immunol Cell Biol., 78 (1), 67
6. Grozesku, Yu. N., Bakhareva, A. A. & Shulga, E. A. (2009). News Bulletin of Samara Scientific Center, RAS, 11, 1(2), 42 (in Russian).
7. Sariev, B. T. Tumenov, A. N. Bakaneva, Yu. M. & Bolonina, N. V. (2011). ASTU News Bulletin. Ser. Fish Farming. 2, 118 (in Russian).
8. Panasenкo, V. V. (2008). Fish Farming, 1, 74, (in Russian).
9. Ponoмaryov, S.V., Grozesku, Yu.N. & Bakhareva, A. A. (2006). Industrial fish farming. Moscow: Kolos. 320 p. (in Russian).

CHAPTER 24

VARIABILITY OF COMBINING ABILITIES OF MS (MALE STERILITY) LINES AND STERILITY BINDERS OF SUGAR BEETS AS TO SUGAR CONTENT

KORNEEVA MYROSLAVA OLEKSANDRIVNA,
and NENKA MAKSYM MYKOLAYOVYCH

CONTENTS

ABSTRACT

The expression of combining abilities of MS lines and O types as to sugar content, depending on the area of feeding (regular and extended) and the background of mineral fertilizers (normal and elevated) are shown in the article. Lines with stable expression of the characteristic of sugar content are marked. Genotypic structure of the variability of sugar content in different environments is investigated. Components of simple sterile hybrids are differentiated according to their genetic values.

24.1 INTRODUCTION

Sugar content is an important element of the productivity of sugar beet that is a selection goal in the development of hybrids based on CMS (cytoplasm male sterility). Many scientists pointed out that sugar content is characterized by a significant variation factor (from 15 to 21%), which was significantly lower compared to yield [1, 2]. Studying the variability of populations of different origins, it was found that the populations of the same variability were characterized by different absolute values of the sugar content, and vice versa [3]. Some scientists pointed at the appearance of transgress forms in the offspring with the frequency 0.7–1.4% [4].

Variability of sugar content depends either on the genotypic factors, or on the conditions of the environment and their interaction. The variability of this feature in populations depends mainly on the additive gene effects; in interline hybrids—on the additive and nonadditive effects [5–7]. However, the phenotypic expression of the sugar content is influenced a lot by other factors (environmental, agronomic, and others) that "mask" genetic parameters contributing to this feature, and create difficulties in the selection of genotypes.

"Cell method" (hexagonal method of organizing plants) with the intensity of selection of 15% was used to equalize differences caused by the environment in selection of some crops. Many authors indicated modifications in the areas of supply of sugar beet. So, Mazlumov [8] thought that the use of the extended area of supply could identify all capabilities of the genotypes secured by nature. He wrote that the extended area of supply influenced the variability of useful traits of beets more than special properties of soil or fertilizers. According to other researchers, it was shown that sugar content was higher in the offspring selected in the extended area of supply of the pioneers than in the selection on the normal area of supply. Moreover, the expansion of the phenotypic variance into components—genotypic and environmental, showed that the proportion of the genotypic variance to total phenotypic was higher in the extended area of supply [9].

24.2 MATERIALS AND METHODOLOGY

Maternal component represented by two types – simple sterile hybrids (SSH) derived from crosses of sterile (MS) lines with unrelated binder of sterility (BS) and

MS line—analogs of O type—was tested in the experiments in 2011–2012, conducted at the Institute of Bioenergy Crops and Sugar Beet in different environments. Backgrounds were as follows: normal background of fertilizing—the common area of supply (NBCA), normal background—the extended area of supply (NBEA), increased background of fertilizing—the common area (IBCA) and increased background—the extended area of supply (IBEA).

24.3 RESULTS AND DISCUSSIONS

24.3.1 SUGAR CONTENT OF SIMPLE STERILE HYBRIDS (SSH) IN THE ENVIRONMENTS NBCA AND NBEA

As the analysis of sugar content showed (Table 24.1), SSH was characterized by specific reaction to changes in the area of supply in the environments NBCA and NBEA.

TABLE 24.1 Sugar Content of SSH, Deviation from Average and Standard Environments of NBCA and NBEA, IBCSB of NAAS, 2011–2012

Simple sterile hybrids	Area of supply					
	NBCA			NBEA		
	Sugar content, %	Deviation from the average, %	Deviation from St, %	Sugar content, %	Deviation from the average, %	Deviation from St, %
MS 1/Ot 2	16.5	−0.5*	−4.3	17.8	0.8*	3.3
MS 1/Ot 3	16.7	−0.3	−3.3	16.0	−0.9*	−7.2
MS 1/Ot 4	16.6	−0.4*	−3.9	17.2	0.2	−0.2
MS 1/Ot 5	16.5	−0.4*	−4.1	18.1	1.1*	4.8
MS 2/Ot 1	17.3	−0.4*	0.5	17.2	0.2	−0.2
MS 2/Ot 3	16.9	−0.1	−2.2	15.9	−1.1*	−7.7
MS 2/Ot 4	17.1	0.1	−0.8	16.2	−0.8*	−6.2
MS 2/Ot 5	17.6	0.6*	1.4	17.5	0.6*	1.7
MS 3/Ot 1	17.3	0.4*	0.5	17.8	0.8*	3.3
MS 3/Ot 2	16.5	−0.5*	−4.5	16.3	−0.6*	−5.2
MS 3/Ot 4	17.3	0.4*	0.5	16.9	−0.1	−2.1
MS 3/Ot 5	17.8	0.8*	3.0	17.2	0.2	−0.2

TABLE 24.1   *(Continued)*

Simple sterile hybrids	Area of supply					
	NBCA			NBEA		
	Sugar content, %	Deviation from the average, %	Deviation from St, %	Sugar content, %	Deviation from the average, %	Deviation from St, %
MS 4/Ot 1	16.9	−0.1	−1.8	17.4	0.4*	0.8
MS 4/Ot 2	16.6	−0.4*	−3.9	16.5	−0.5*	−4.3
MS 4/Ot 3	16.6	−0.4*	−3.7	16.2	−0.8*	−5.8
MS 4/Ot 5	17.5	0.5*	1.3	17.2	0.2	−0.2
MS 5/Ot 1	17.1	0.2	−0.6	16.7	−0.3	−3.1
MS 5/Ot 2	16.4	0.6*	−5.1	17.7	0.7*	2.5
MS 5/Ot 3	17.1	0.2	−0.6	17.4	0.4*	1.0
MS 5/Ot 4	16.9	−0.1	−2.0	16.2	−0.7*	−5.8

Note: * – substantially 5% of the significance level.

As Table 24.1 shows, on NBCA significant positive deviation from the average medium value was observed in five SSH, on NBEA—in seven, that is, extended area (EA) of supply as the factor contributed to the appearance of increased sugar content. There is a specific reaction of genotypes: some hybrids reduced the feature value on the extended area (MS 1/Ot 3, MS 2/Ot 3, MS 2/Ot 4, MS 3/Ot 4, MS 3/Ot 5, MS 5/Ot 1), and other hybrids were not sensitive to such factor as the area of supply, and some of them—MS 5/Ot 2, MS 1/Ot 2, MS 1/Ot 4, MS 1/Ot 5, MS 4/Ot 1, MS 5/Ot 2—increased sugar content. It agrees with the observations of Mazlumov [8], who wrote that the sugar content on different areas of supply varied in a different way. There are plants in which the sugar content in the extended area of supply does not change, and increases. Selection of lines with high sugar content on this background improved the material on this feature significantly.

EA supply contributed to the manifestation of high sugar content in the hybrid MS 1/Ot 5, which showed the highest value of the index (18.1%). The similar tendency was observed in the combinations with high sugar content MS 3/Ot 1 (17.8%) and MS 5/Ot 2 (17.7%), which did not show themselves in the control variant NBCA. The best hybrids MS 1/Ot 2, MS 1/Ot 4, MS 3/Ot 1, MS 5/Ot 2 increased the group standard for 2.5…4.8% in the sugar content.

The range of variation in the sugar content on EA was higher than on CA. The amplitude values of this index varied from 15.9…18.1% (EA) and 16.4…17.6% (CA), with the difference of 2.2 and 1.2%, respectively.

Phenotypic variability of hybrids sugar content was divided on the genotypic and environmental components by means of dispersive analysis. It turned out that the effect of the genotype in the variant on NBEA was bigger than on NBCA, and was 86.6 vs. 69.2%, respectively. It indicated a good differentiating ability of such factors as the extended area of supply for the manifestation of the genotype. Genotypic variance was also divided into components (Figs. 24.1 and 24.2).

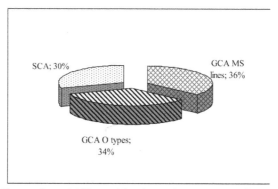

FIGURE 24.1   Genotypic variability and its share of sugar content feature of simple sterile hybrids, 2011–2012, environment NBCA.

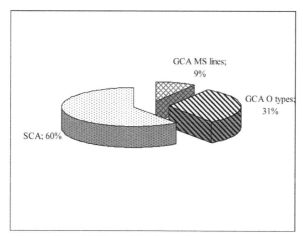

FIGURE 24.2   Genotypic variability and its share of sugar content feature of simple sterile hybrids, 2011–2012, environment NBEA.

There is a noteworthy fact that the additive effects of genes of both parental forms (SCA of MS lines + SCA O types) on the extended area of supply was lower than on the common one and made the total of 40 vs. 70%. While the share of nonadditive actions of gene was higher (60 vs. 30%). It shows that effects of components interaction became more significant on the extended area of supply (Figs. 24.1 and 24.2).

Significant influence of parental forms and their interactions in the environment of NBCA revealed the effects of the combining ability—general (GCA) and specific (SCA) (Table 24.2).

TABLE 24.2 The Effects of GCS and SCA of MS Lines and O Types, NBCA, 2011–2012

MS lines	Effects of GCS MS lines	Effects of SCA				
		Ot 1	Ot 2	Ot 3	Ot 4	Ot 5
MS 1	−0.38*	#	−0.16	0.46*	0	−0.31
MS 2	0.24*	0.05	#	0.04	−0.09	0
MS 3	0.27*	0.01	−0.39*	#	0.11	0.27
MS 4	−0.06	−0.05	0.04	−0.29	#	0.30
MS 5	−0.07	0.15	−0.15	0.25	−0.26	#

Note: * – substantially 5% of the significance level.

MS 2 and MS 3 lines were carriers of additive genes in the environment of NBCA. Significantly high effect of SCA was in the combination of MS 1/Ot 3 (+0.46*), but hybrid combination with their participation did not show competitive heterosis (deviation from St was—3.3%) because of the low effect of GCS of MS lines (−0.38*) (Table 24.1).

Combination of MS 2/Ot 5 (17.6%) had a significant difference of sugar content from the average population values (Table 24.1), which is due to the additive effect of the mother form—MS 2 (+0.24*) (Table 24.2).

Dispersive analysis of the data showed that effect of all components of the variation of genotypes was significant in the environment NBEA. This allowed determining the proportion (Fig. 24.2), as well as the effects of combining abilities of parental lines of SSH (Table 24.3).

TABLE 24.3 The Effects of GCA and SCA of the Sugar Content of MS Lines and O Types, Environment NBEA

MS lines	Effects of GCS MS lines	Effects of SCA				
		Ot 1	Ot 2	Ot 3	Ot 4	Ot 5
MS 1	0.30*	—	0.13	−0.78*	0.13	0.52*
MS 2	−0.27*	0.09	—	−0.31	−0.34	0.55*
MS 3	0.08	0.34	−0.23	—	0.01	−0.13
MS 4	−0.14	0.14	0.16	−0.40*	—	0.10
MS 5	0.03	−0.71*	1.15*	0.59*	−1.05*	—

Note: * – substantially 5% of the significance level.

Line MS 1 was the best among SCA in the environment of NBEA, which was well combined with MS Ot 5 (SCA=0.52 ×). Hybrid with their application had the highest sugar content—18.1% (Table 24.1). The components of hybrids MS 2/Ot 5, MS 5/Ot 2 and MS 5/Ot 3 had interaction effects that were significantly higher—0.55, 1.15 and 0.59, respectively, resulting in the increased level of sugar content in the hybrids, created with their participation—17.5, 17.7 and 17.4%, respectively.

So, nonadditive variance was dominated by EA compared with the common area (OA), and was 60 vs. 30% on the common background of fertilizing in the genotypic structure of variability of the feature of sugar content.

Extended area of supply is the factor in which the effect of the genotype is higher than normal (86.6% vs. 69.2%), which indicates the feasibility of selecting the best genotypes in this environment. Differentiating ability of the environment NBEA is higher than on NBCA (seven of the best hybrids vs. five were distinguished). The range of variation in the sugar content on EA was higher (2.2%) compared to the CA (1.2%). Variability of effects of GSA and SCA, as well as the specificity of the reaction of hybrids due to the change of the area of supply was distinguished. The best hybrids in the environment NBCA were hybrids MS 3/Ot 5 and MS 2/Ot 5, and in the environment NBEA—MS 1/Ot 5, MS 3/Ot 1 and MS 1/Ot 2.

24.3.2 SUGAR CONTENT OF SIMPLE STERILE HYBRIDS IN ENVIRONMENTS IBCA AND IBEA

Background of mineral supply influences the sugar content in a certain way, modifying its absolute value. The tested set of SSH was also tested on the elevated background of mineral supply in two variants—with standard (IBCA) and extended (IBEA) areas of supply. Sugar content of hybrids is given in Table 24.4.

TABLE 24.4 Sugar Content of SSH, Deviation from the Average and Standard in the Environments IBCA and IBEA, IBCSB of NAAS, 2011–2012

| Simple sterile hybrids | Area of supply | | | | | |
| | NBCA | | | NBEA | | |
	Sugar content, %	Deviation from the average, %	Deviation from St, %	Sugar content, %	Deviation from the average, %	Deviation from St, %
MS 1/Ot 2	17.7	−0.81*	3.1	18.2	1.16*	5.5
MS 1/Ot 3	17.7	0.74*	2.7	17.1	0.09	−0.7
MS 1/Ot 4	17.7	0.74*	2.7	16.8	−0.17	−2.3
MS 1/Ot 5	17.2	0.31	0.2	17.5	0.46*	1.4
MS 2/Ot 1	17.8	0.88*	3.5	17.4	0.43*	1.2
MS 2/Ot 3	15.5	−1.42*	−9.9	16.2	−0.84*	−6.1
MS 2/Ot 4	16.3	−0.62*	−5.2	16.2	−0.80*	−5.9
MS 2/Ot 5	17.1	0.14	−0.8	18.6	1.60*	8.0
MS 3/Ot 1	16.5	−0.42*	−4.1	17.8	0.83*	3.5
MS 3/Ot 2	16.6	−0.29	−3.3	17.9	0.86*	3.7
MS 3/Ot 4	16.6	−0.32	−3.5	15.3	−1.74*	−11.4
MS 3/Ot 5	17.6	0.64*	2.1	16.4	−0.60*	−4.8
MS 4/Ot 1	16.8	−0.09	−2.1	16.5	−0.47*	−4.0
MS 4/Ot 2	15.8	−1.09*	−7.9	16.6	−0.44*	−3.8
MS 4/Ot 3	16.2	−0.69*	−5.6	16.3	−0.74*	−5.6
MS 4/Ot 5	16.5	−0.42*	−4.1	16.6	−0.40*	−3.6
MS 5/Ot 1	17.3	0.81*	3.1	18.1	1.10*	5.1
MS 5/Ot 2	17.3	0.38	0.6	17.6	0.60*	2.2
MS 5/Ot 3	16.7	−0.22	−2.9	16.2	−0.74*	−5.6
MS 5/Ot 4	16.8	−0.12	−2.3	16.6	−0.44*	−3.8

Note: * – substantially 5% of the significance level.

Analysis of Table 24.4 showed that on the background of IBCA 5 the combinations were significantly higher than the average population value of sugar content, while on the background of IBEA—8 combinations. The rate of reaction of the studied genotypes on EA was specific: hybrids increased or lowered the sugar content, some of them showed stability. However, the range of variation characteristics was different in two environments. The difference between the highest (MS 1/Ot 2) and the lowest (MS 3/Ot 4) indexes of the sugar content was higher on the EA, it was 3.3% (abs. index). It was smaller in the environment IBCA—2.0% (hybrids MS 2/Ot 1 and MS 4/Ot 2). In relation to the standard the significant excess was 2.1...3.5% (IBCA) and 2.2...8.0% (IBEA). The high sugar content on EA was observed in hybrids MS 2/Ot 5 (18.6%), MS 5/Ot 1 (18.1%), MS 1/Ot 2 (18.2%). Combinations of MS 2/Ot 1 (17.8%) and MS 1/Ot 2, MS 1/Ot 3, MS 1/Ot 4—17.7% each were the best on IBCA.

Determination of the proportion in the total genotypic variability revealed that it is larger than on EA (92.4%) compared to CA (83.5%). This indicates a better differentiating ability of EA than CA.

Decomposition of genotypic variance (with the help of dispersive analysis) on the effects associated with different types of gene interactions showed that nonadditive effects of genes on EA had a larger proportion (50%) compared to CA (40%) (Figs. 24.3 and 24.4).

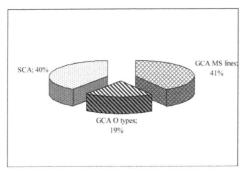

FIGURE 24.3 Genotypic variability and its share of feature of sugar content of simple sterile hybrids, 2011–2012, background IBCA.

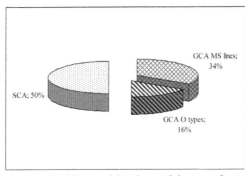

FIGURE 24.4 Genotypic variability and its share of feature of sugar content of simple sterile hybrids, 2011–2012, background IBEA.

Analysis of Figs. 24.3 and 24.4 showed that the effect of O types was approximately two times lower than that of MS lines: 19 vs. 41% (IBCA) and 16 vs. 34% (IBEA).

The significance of differences between SSH identified the effects of GCA and SCA components of hybridization and their expression depending on the area of supply (Tables 24.5 and 24.6). The variability of combining ability, depending on the environment was pointed out by many scholars. So, in some cases, SCA was more stable, in others—GCA [10, 11].

TABLE 24.5   The Effects of GCA and SCA on Sugar Content of MS Lines and O Types, IBCA

MS lines	Effects of GCS MS lines	Effects of SCA				
		Ot 1	Ot 2	Ot 3	Ot 4	Ot 5
MS 1	0.67*	-	−0.25	0.41*	0.30	−0.47*
MS 2	−0.24*	0.72*	#	−0.84*	−0.16	0.28
MS 3	−0.09	−0.74*	0.13	#	−0.01	0.62*
MS 4	−0.56*	0.07	−0.19	0.09	-	0.03
MS 5	0.22*	0.19	0.49*	−0.22	−0.46*	-

Note: * – substantially 5% of the significance level.

TABLE 24.6   Effects of GCA and SCA on Sugar Content of MS Lines and O Types, IBEA

MS lines	Effects of GCS MS lines	Effects of SCA				
		Ot 1	Ot 2	Ot 3	Ot 4	Ot 5
MS 1	0.67*	-	−0.25	0.41*	0.30	−0.47*
MS 2	−0.24*	0.72*	-	−0.84*	−0.16	0.28
MS 3	−0.09	−0.74*	0.13	-	−0.01	0.62*
MS 4	−0.56*	0.07	−0.19	0.09	-	0.03
MS 5	0.22*	0.19	0.49*	−0.22	−0.46*	-

Note: * – substantially 5% of the significance level.

As the analysis of Table 24.5 showed, the best of SCA were MS 1 (+0.67*) and MS 5 (+0.22*). These lines possess a set of additive genes that will always influence the display of sign in F_1. In specific combinations, in nonadditive effects of genes are expressed very well, associating with dominance and overdominance, pairs of MS 1 and Ot 3 (0.41*), MS 2, and Ot 1(0.72*), MS 3, and Ot 5 (0.62*), MS 5 and Ot 2 (0.49*) showed themselves better. MS 4 Line did not show itself in any combination, GCA was significantly lower (−0.56*), and all hybrids with its participation inferred standard for 2.1…7.9% (Table 24.4).

On EA on MS 1 line confirmed its high degree for GCA, and also showed itself on CA. GSA of this line was significantly higher (+0.4*), and MC 5 line had a positive but nonsignificant effect (0.14) (Table 24.6).

MS 4 line was characterized by a negative and significant effect on SCA both on CA and EA (Tables 24.5 and 24.6). Consequently, one can argue about the relative stability of effects of GSA, which can't be said about effects of SCA. Thus, the components of the hybrid MS 1/Ot 3 on CA had the effect of "plus" that changed to the effect of "negative" in EA, indicating that a significant expression of the effects of dominance and overdominance. Non-additive effects of genes of MC 4 line with O types on EA showed the contrast: with the binder of sterility Ot 1 were significantly negative (–0.58*), and with Ot 3—significantly positive (+0.60*), while on CA they had no significant effect on the display of sugar content in hybrid combinations with their participation.

MS 3 line also had contrast—with Ot 1 and Ot 2 effects of SCA were positive (0.37* and 0.96*), while with Ot 4 and Ot 5—negative (–0.75* and 0.58*). On CA MS 3 line from Ot 4 had a high positive effect of SCA. Consequently, the environment influenced the manifestation of nonadditive effects of genes, and this effect is specific for each genotype.

24.4 CONCLUSIONS

So, on the elevated background of fertilizing the extended area is the environment with a good differentiating ability, as 8 hybrids were revealed, significantly exceeding average population value (compared to 5 on IBCA). Genotypic variance for IBEA was also higher (92.4%) and the range of variability—wider (3.3%) compared to IBCA—83.5% and 2.0%, respectively.

The best hybrids that had a stable effect on the sugar content on both backgrounds were MS 1/Ot 2, MS 2/Ot 1, MS 5/Ot 1 and MS 2/Ot 5, as evidenced by the cumulative effect of GCA and SCA. SCA was more variable in relation to the area of supply than GCA. MC 1 and MC 5 lines were distinguished as carriers of additive genes and their stable expression.

KEYWORDS

- **Combining ability**
- **MS (male sterility) lines**
- **Simple sterile hybrids**
- **Sterility binders**
- **Sugar content**

REFERENCES

1. Savitskiy, V. F. (1940). Genetika saharnoy svekly/Savitskiy, V.F. *Kiev.*, 561–580.
2. Korneeva, M. A. (1987). Selektsionno-geneticheskoe izuchenie ishodnyih populyatsiy saharnoy sveklyi s tselyu sozdaniya kombinatsionno-tsennyih liniy—opyiliteley/Avtoref. dis. na soiskanie uch. stepeni kand. biol. nauk Spets. 03.00.15 – genetika. Institut molekulyarnoy biologii i genetiki. *Kiev.*, 20 p.
3. Balkov, I. Y. (1986). Geterozis saharnoy sveklyi po priznaku saharistosti/I.Y. Balkov, V.P. Petrenko, M.A. Korneeva Vesnik selskohozyaystvennoy nauki.–Moskva, *10*, 55–59.
4. Logvinov, V. A. (1985). Izmenchivost soderzhaniya sahara v korneplodah saharnoy sveklyi i otbor vyisokosaharistyih rasteniy/V.A. Logvinov, L.L. Chebotar Povyishenie effektivnosti proizvodstva saharnoy sveklyi na Severnom Kavkaze. *Krasnodar*, 48–53.
5. Roik, N. V. (2010). Kombinatsionnaya sposobnost opyiliteley saharnoy sveklyi razlichnoy geneticheskoy strukturyi po elementam produktivnosti. Korneeva Entsiklopediya roda Beta: Biologiya, genetika i selektsiya sveklyi. *Novosibirsk*, 525–541.
6. Korneeva, M. O. (2006). Uspadkuvannya tsukristosti topkrosnimi ChS gibridami tsukrovih buryakiv/M.O. Korneeva, P.I. Vakulenko Kiyiv: *Tsukrovi buryaki, 4*, 7–8.
7. Korneeva, M. O. (2007). Zastosuvannya aditivno-dominantnoyi modeli dlya otsinki liniy tsukrovih buryakiv/M.O. Korneeva, E.R. Ermantraut, M.V. Vlasyuk Zbirnik naukovih prats "Metodika, mehanizatsiya, avtomatizatsiya ta komp'yuterizatsiya doslidzhen u zemlerobstvi, roslinnitstvi, sadivnitstvi ta ovochivnitstvi. vipusk 9. Kiyiv. – *Tov. PoligrafKonsalting*, 164–171.
8. Mazlumov, A. L. (1970). Selektsiya saharnoy sveklyi/Mazlumov. A. L. *Moskva. Kolos.* 206 p.
9. Balkov, I. Y. (1986). Zakonomernosti nasledovaniya saharistosti i printsipyi otbora vyisokosaharistyih form v selektsii saharnoy sveklyi/Balkov, I. Y. Peretyatko, V. G. Osnovyi povyisheniya saharistosti i tehnologicheskih kachestv saharnoy sveklyi. *Kiev. VNIS.* 70–77.
10. Berezhko, S. T. Kombinatsionnaya sposobnost u tetraploidnoy saharnoy sveklyi i usloviya vneshney sredyi/S.T. Berezhko Geneticheskie issledovaniya saharnoy sveklyi. *Kiev. VNIS.* 75–86.
11. Petrenko, V. P. Ispolzovanie kombinatsionnoy sposobnosti v selektsionnom protsesse/V.P. Petrenko Geneticheskie issledovaniya saharnoy sveklyi. *Kiev. VNIS.* 86–96.

CHAPTER 25

COMPARISON OF CHROMOSOMAL REARRANGEMENTS IN EARLY SEEDLINGS OF CREPIS CAPILLARIS AFTER TREATMENT OF SEEDS BY X-RAYS AND BY THE CHEMICAL MUTAGEN

L. I. WEISFELD

CONTENTS

ABSTRACT

Seeds of *Crepis capillaris* L.(Wall) were worked by means of two methods: 1) exposed to X-rays in doses 2000, 3000, 4000 P or 2) the alkylating agent phosphemid having two groups of ethyleneimine and a pyrimidine bases. In early seedlings we analyzed aberrations of chromosomes in metaphases and of mitotic activity of cells. The seeds were germinated for 24, 27, 31, 36 h. Seedlings of 1–2 mm in length were fixed through 2–12 h after their appearance. Aberrations chromosomes in metaphase plates and mitotic activity of cells were analyzed. Phosphemid induced structural chromosome aberrations of chromatid type. Frequency aberrations on one cell do not exceed the frequency of cells with aberrations using a moderate dose phosphemid. This experiment revealed that in the seedlings of the first arising – 24 h after the beginning of the soaking of seeds frequency aberrations of chromosomes increases as far as their fixing from 3 to 12 h. In seedlings from 27-hourfrequency of chromosome aberrations was slightly higher, than maximal frequency of aberrations is in 24-hour arising (in 12 h after arising). In seedlings, which were selected through 36 h after soaking, number of aberrations of chromosomes with increasing time of fixation increased. Our data allow to conclude that the phase of G1 of dry seeds is heterogeneous relatively mitotical cycle and that a mutagene is saved in cells, in spite of the fact that seeds were washed with running water. Phosphemid suppressed mitotic activity of seedling cells.

Effect of x-rays on dry seeds had opposite direction: frequency of aberrations was maximal in the first terms of fixing, next become below. The number of aberrations per 100 cells exceeded the number of metaphases with aberrations. All of them were aberrations of the chromosome type.

25.1 INTRODUCTION

After discovery of chemical mutagenesis [1–3] in the forties of twentieth century the scientists of many countries began actively to study the cytogenetic manifestations of mutagens in meristem of growing tissues of plants. There was the openly delayed and undelayed action of mutagens and fundamental distinction of radiation and chemical mutagenesis on cytological level (see for example reviews Refs. [4, 5]). Aberrations in ana- and telophases in the plants were studied [3–5].

The much of researches ware carried out on plant of *Crepis capillaris* L. This plant possesses three pairs of the distinguishable chromosomes. Else until discovering of the chemical mutagenesis Navashin [6] was investigating karyology of genus of *Crepis*.

Chemical mutagens have the detained effect on cellular level, that is, cause violations of chromosomes only during the synthesis of DNA. After termination of synthesis of DNA – in the postsynthetic phase of G2 and before his beginning – in phase of G1 the chromosomes are not damaged regardless of presence of the chemical mutagens [4, 5]. A basic question was consisting in determination of mechanism of chemical mutagenesis on cellular level.

Workings in the direction of study of the mechanism of the chemical mutagenesis on cytogenetic level were conducted in the laboratory of N.P. Dubinin with co-workers and by the collective under the direction of B.N. Sidorov and N.N. Sokolov.

Dubinin et al. [8] were studying the cytogenetic effects of ionizing radiation, chemical compounds, in particular of the alkylating agents. This group published a large number of researches, several monographs (see review Ref. [7]). N.P. Dubinin formulated the idea of the mechanism of action of chemical mutagens: mutagens induce potential changes in the chromosomes at all stages of the mitotic cycle, which come to light during a number of cell generations. He called it "chain process" in mutagenesis.

Sidorov and Sokolov had another point of view [9, 10]. In detail their experiments [9, 10] will come into question below. These authors the rearrangements of chromosomes analyzed on the seedlings of *Cr. capillaris*, on the asynchronous populations of cells in regards to phases of mitotic cycle – G1, S, and G2. In our work, which we will describe below, was used a synchronous object are dry seeds of *Cr. capillaris*, whose bioblasts probably are found in the stage of G1.Action of the chemical mutagen, which was named phosphemid, was analyzed by comparison to the action of X-rays. Alkylating agent phosphemidum (synonym phosphemid, phosphazin) is representing di-(etilene imid)-pyrimidyl-2-amidophosphoric acid. This compound is interesting by the fact, that it contains two ethylene imine groups and pyrimidine [11]. We were studying the rearrangements of chromosomes in metaphases of seedlings of *Crepis capillaris*(L.).

25.2   MATERIAL AND METHODOLOGY

In experiments with phosphemid air-dried seeds of *Cr. capillaris* of crop of 1967 were analyzed in 1968: April (8 months of storage after harvest), June (10 months of storage), and July (11 months of storage). In the control and in the experiments used distilled water. The chromosome aberrations were analyzed in seedlings (meristem of tip roots).

In control and in experiments we applied the distilled water, and were analyzing rearrangements of chromosomes in metaphases of meristem cells of root tips.

The100 seeds there were treated in an aqueous solution of phosphemid concentrations: 1×10^{-2} M (22.4 mg was dissolved in 10 mL water), 2×100^{-2} M (22.4 mg was dissolved in 50 mL) or $2 \quad 100^{-3}$ M (2.24 mg dissolved in 50 mL of water). The treatment was carried out at room temperature (19–21 °C) for three hours. Then the seeds were washed with running water for 45 min. The washed seeds were placed in Petri dishes on filter paper moistened with a solution of colchicine (0.01%). Seeds were germinating in a thermostat at 25 °C, but in July 1968 because of hot weather the temperature in the thermostat could reach 27 °C. In parallel control experiments, the seeds were treated with aqua distillate.

After 24, 27, 31 and 36 h after soaking of seeds we choose seedlings, took away plantlets measuring a 1 mm or less and named them "arisings." The term "*arising*" refers to those seedlings that emerge after the beginning of soaking of dry seeds. These seedlings were selected in a Petri dish for further germination and subsequent fixation. In Russian the term "*arising*" is called "*proklev*."

Root tips (0.5–1 cm) were cut off with a razor and placed in the solution: 96% ethanol 3 parts + 1 part of glacial acetic acid. Solution poured out through 3–4 h. Seedlings were washed 45 min in 70% alcohol. These seedlings were kept in 70% alcohol. We were preparing temporary pressure preparations: fixed root tips were stained acetous carmine and crushing in a solution of chloral hydrate between the slide and cover slip. We analyzed chromosome aberrations in metaphase plates of seedlings in the first division after treatment of seeds ($2n$-karyotype). In each seedling were counted up all metaphases. Intact seeds of harvest of 1966, 1067 and 1969 years served as control. The seedlings were fixed at different time intervals from 3 to 24 h.

Mitotic activity in the seedlings was determined on a number of metaphase plates, depending on the number of nuclei in seedlings in the control and experiment. We used the seedlings after different intervals of time after the "arising." In each seedling, we counted between 500 and 1,000 nuclei. Counted up a number of seedlings with metaphase to all watched seedlings at all stages of fixation.

In all the experiments was estimated standard deviation from the mean.

In experiments with an irradiation, seeds exposed to the action of X-rays in doses 3000, 2000 and 4000P. Seedlings took away, as described higher, through 22, 24, 27 h after the soakage of seed, fixed and cooked preparations, as described higher, in the intervals of time through 0, 2, 3, 6, 9 h after arising. In parallel carried out control tests data of that are incorporated.

25.3   RESULTS AND DISCUSSION

Radiation treatment by X-rays in the doses of 2000 and 3000P has evoked the high percent of rearrangements of chromosomes in the early seedlings (Table 25.1).

TABLE 25.1 Rearrangements of Chromosomes in the Early Seedlings of *Crepis Capillaris* after Radiation Treatment by the X-ray of Seeds

Radiation dose, P	Time, hours from a soakage to arising	Time, hours from an arising to fixation	General number of seedlings with metaphases	Number of metaphases Σ	Number of metaphases with rearrangements Σ	Number of metaphases with rearrangements %	Number of rearrangements Σ for 100 metaphases		+m, %
3000	22	0×	5	50	39	78.0	50	100.0	22.0
	24	3	16	334	179	53.6	203	60.8	7.2
		6	13	177	64	36.2	68	38.4	2.2
	27	2	21	184	96	52.2	123	66.8	14.6
		6	11	273	104	38.1	124	45.4	7.3
		9	9	377	101	26.8	106	28.1	1.3
	38	0×	8	96	35	36.4	38	39.6	3.2
2000	24	4	19	134	53	39.7	61	45.5	5.8
	29	3	22	229	58	25.3	61	26.7	1.4
4000	Growth is delayed to 30–39 h								
Control			18	1316	39	2.96	39	2.96	0.0

* – fixed turning up seedlings long less than 1 mm.

In the seedlings selected through 24 and 27 h after a soakage, in the first term of fixation there was the greatest percent of rearrangements of 60.8 and 45.5% accordingly. In the seedlings selected through 24 and 27 h after a soakage, in the first term of fixation there was the greatest percent of rearrangements of 60.8 and 45.5% accordingly.

Through 24 and 27 h after X-rays irradiation of seeds the frequency of rearrangements of chromosomes was beginning from maximum. Their frequency is decreasing with increasing time from arising (see Table 25.1). Through 22 h after irradiation, in seedlings which were fixated in 0 h, was discovered only 5 seedlings with metaphases. At them was discovered the high level of rearrangements of chromosomes: 50 metaphases contained 78% of the rearrangements; through 38 h are arising the 8 seedlings containing 36,4% metaphases with rearrangements. At all variants of arising and of fixation the frequencies of rearrangements the calculating on 100 cells were higher than frequencies of metaphases with rearrangements, that is more than one rearrangement in every cell (see a column of m, % in a table 25.1).

Under exposition of radiation in dose 4000P, the germination of seeds was significantly delayed. In metaphases after radiation of seeds prevailed rearrangements of chromosomal type. Rearrangements of chromatid type were observed, but their frequency statistically significantly did not differ from control (Table 25.2).

TABLE 25.2   Chromosomal and Chromatid Types of Rearrangements in Control Experiment

Number of seedlings	Number of metaphases	Number of rearrangements, %	Chromatid rearrangements, %	Rearrangements of chromosomal type, %
117	1316	2.96	2.58	0.38

Analogical rule of correlation of number of rearrangements and number of metaphases with rearrangements under the influence of the radiation on seedlings was observed in research in Ref. [12]. In this work frequency of rearrangements also decreases later in the course of germinating of seeds after an treatment by X-rays.

Phosphemid on the contrary caused only chromatid rearrangements in all variants of the experience. In moderate concentrations of phosphemid the frequency of rearrangements in one cell was not exceeding of frequency of metaphases with rearrangements. In Fig. 25.1, it is shown that phosphemid in a moderate dose (2×10^{-3} M) causes rearrangements of chromosomes in the seedlings of first arising – through 24 h after the soakage of seed. Their frequency increases with the increase of term of fixation.

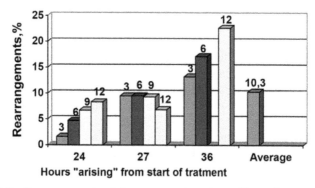

FIGURE 25.1   Rearrangements of chromosomes in early seedlings after treatment of seeds by phosphemid in a concentration 2×10^{-3} M.

On a y-axis – rearrangements, %. On abscissa axis is represented the time after the soakage of seeds: 24, 27 and 36 h. Seedlings are fixed through 3, 6, 9 or 12 h (numerals over columns) after arising. Data are statistically meaningful.

The later arising – through 27 h after soaking characterized by the more frequency of rearrangements than upon latest fixation – 12 h after first arising – 24 h. At later arising – through 36 h again the frequency of rearrangements increased again in comparison with last fixation (27 h of arising) (see a Fig. 25.1). Through 36 h after soaking the rearrangement frequency again increases also since a higher level as compared to a 27-hour arising.

All rearrangements were the chromatid type, i. e. were arising during the synthesis of DNA.

On the basis of these facts we conclude that a mutagen presents in cells during all period of fixations, influencing on chromosomes after their passing to the DNA synthesis, remains in cells through 24, 27 and 36 h after arising, in spite of washing of the seeds by tap water from the phosphemid. Increasing of number of rearrangements through 2, 6, 9 and 12 h in the arising 24 h talks about heterogeneity of phase of G_1 in dry seeds. It is not "delayed effect" of chemical mutagens, typical for asynchronous population of cells, when chemical mutagens do not cause rearrangements of chromosomes at G_2 or G_1 of mitotic cycle. In these experiments the growth of number of rearrangements in seedlings can be result of the saving of mutagen in cells and gradual going over of cell cycle from the C1 stage to the synthesis of DNA. Increasing of number of rearrangements in the arising 36 h can be explained also by the lack of synchronization of passage by cells of cell cycle to the synthesis of DNA in presence a mutagen. Table 25.3 and Fig. 25.2 are showing the mitotic activity (number of the mitoses in a seedlings) at different periods after arising or of fixation.

TABLE 25.3 Mitotic Activity in the 2n Cells of Seedlings of *Crepis Capillaris* After Treatment of Air-Dry Seeds with Phosphemid in a Concentration 2×10^{-3} M

Time from the beginning of treatment, hours	Time from "arising" to fixing, hours	Seedlings, number		
		Total	With mitoses	%
April				
24	3	57	24	42.1
	6	56	32	57.1
	8	54	36	66.7
27	3	47	16	35.5
	5	19	12	65.0
On average		233	120	51.5±3.28
June				
24	3	38	5	13,2

Time from the beginning of treatment, hours	Time from "arising" to fixing, hours	Seedlings, number		
		Total	With mitoses	%
	6	28	16	57.1
	9	19	15	78.9
	12	27	23	85.2
27	3	35	8	22.9
	6	35	17	48.6
	9	23	12	52.2
	12	23	20	87.0
36	3	37	22	59.5
	6	22	10	45.5
	12	20	15	75.0
On average		307	163	53.1±2.85
July				
24	3	43	21	48.8
	4	30	15	50.0
	5	26	22	84.2
	7	17	13	76.5
	9	18	13	72.2
27	3,5	34	13	78.2
	5	44	31	70.5
	6	36	22	61.1
	7	26	24	92.3
	8	22	20	91.0
31	3	34	23	67.6
	4	59	33	55.9
On average		289	360	64.3±2.43

Harvest 1967. Analysis 1968.

On average the mitotic activity was increasing in senescent seeds during 1968 from 51% in April to 64% in July 1968 (see Table 25.3). Mitotic activity within the limits of every arising also increased during a year: April, June, and July (see Fig. 25.2, Table 25.3). Frequency of rearrangements also increases (see Fig. 25.1).

The time after the soakage of seeds was 24, 27 and 36 h. Seedlings are fixed through 3, 4, 5, 6, 7, 8, 9 and 12 h (numerals over columns) after arising. Data are statistically meaningful.

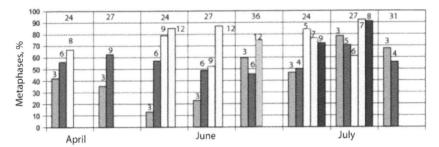

FIGURE 25.2 Mitotic activity in the cells of seedlings of *Crepis capillaris* after treatment of seeds by phosphemid in a concentration 2×10^{-3} M. Harvest 1967. An analysis is in April, June, and July 1968.

Thus, the later the cell enters into mitosis, the longer the synthesis of DNA will be undergoing of the action of mutagen in cells.

By high dose 1×10^{-2} M phosphemid is injuring of the chromosomes considerably stronger on the background of the decline of number of seedlings with metaphases. In 72 explored seedlings were discovered 771 metaphases; from them in 166 (22%) metaphases were discovering rearrangements. In 22 metaphases were discovered plural breakages of chromosomes. Frequency of plural breakages was increasing with the increase of term of fixing. Through 24 h (27 h arising) more than 21% metaphases contained plural breakages of chromosomes.

Besides the large number of breakages of chromosomes the high doses of phosphemid resulted in abnormality of spindle of division. It also affirms that mutagen is interlinked with the proteins of cytoplasm. Rapoport [12], founder of method of the chemical mutagenesis, proved that chemical compounds – ketonic connections, aldehydes, ethylene imine and others are interacting with the proteins of cytoplasm and chromosomes. His experimental data and theoretical conclusions were published in a number of articles, which now are united in the edition of the "Discovery of chemical mutagenesis".

Probably, phosphemid cooperates with proteins by means of two groups of ethylene imine. In addition, the pyrimidin in its structure can be included in structure of chromosomes at the time of the synthesis of DNA. In addition, that phosphemid causes rearrangements of chromosomes, he is repressing mitosis in cells, and id of est. affects proteins of spindle of cell division. It become apparent under the influencing of the high doses of phosphemid, bringing to the plural chromosomal breakages and the conglomeration of chromosomes in the form of "stars" in the center of cell. In the research [13] were determined the number of nuclei labeled by H3-thymidin and of labeled mitosis in seedlings after treatment of dry seeds by H3-thymidin through 24, 30 and 36 h after its introduction. The first labeled nuclei appeared through 10 h after beginning of labeling. These data confirm heterogeneity of phase of G_1 in dry seeds.

In research [8], Sidorov, Sokolov, and Andreev [9] have processed the plantlets of *Cr. capillaris* by ethylene imine. Seedlings were grown in presence a colchicine. During five generations of polyploid cells were discovered the doubled rearrangements of chromosomes. Also into them were educed new rearrangements of chromatid type, which were not redoubled. In the research [9], seedlings were worked in ethylene imine, then washed in tap water, and were cut on very small fragments up to the form of "gruel." Intact seedlings treated by this "gruel," washed in the water off and were placed in Petri dish for sprouting in solution of colchicine. In metaphases of seedlings that treated by "gruel," authors detected new aberrations of chromosome of chromatid type. That proves preservation of mutagen in cages, which were treating in this "gruel." Appearance of new rearrangements in row of generations of cells these authors named secondary mutagenesis.

In that article [9] authors were studying the influence of ethylene imine on dry seeds. Rearrangements only of chromatid type were got. It goes to show that chromosomes in the G_1 are not sensitive to the chemical mutagen. Ionizing radiation, vice versa, damages chromosomes at all stages of cellular cycle.

However, growths of frequency of alterations after treatment of seeds by ethylene imine here are not discovered, unlike our data (see a Table. 25.1; Fig. 25.1). Frequency of rearrangements had as though wavy character (Tables 25.3 and 25.4)

TABLE 25.4　Influence of Ethylene Imine (solution 0.1%) on the Dry Seeds of *Crepis Capillaris* (from Ref. [9])

Time of fixing, hours	Number of cells	Rearrangements	
		Number	%
2	554	36	6.50±1.05
8	672	23	3.42±0.70
12	1018	88	8.64±0.88
16	1000	38	3.80±0,61

There is not growth of number of alterations the depending on time of fixing. Probably, here seedlings were fixed later then in my experiments and length of seedlings was various. In our supervisions, even through 2 h after arising the length of seedlings were varied. Wavelike character of action of ethylene imine in Ref. [9] also talks about heterogeneity of phase of G1 in dry seeds. Maybe, the synthesis of DNA under the mutagen impact is yet more asynchronous.

Although a radiation causes the direct breaking of chromosomes, but after an ionizing radiation also there can be secondary mutagens, if radiation breaks the structure of cells proteins. The phenomenon of preservation of chemical mutagen in the cells of plants has a practical value. The damages of chromosomes are remained in cells are transmitted after dividing and organogenesis of any plants. It is necessary to take into account genetic harm of herbicides of type 2,4-D, of izo-proturon [14], and of pesticides [15] in the same year after their using. It is necessary to take into account chemical contaminating of environment [16].

KEYWORDS

- **Arising**
- **Chemical mutagenesis**
- **Colchicine**
- **Ethylene imine**
- **Rearrangement**
- **Seed**
- **Seedling**
- **Seeds**
- **X-Rays**

REFERENCES

1. Rapoport, J. A. (1946). Ketonic connections and chemical mechanisms of mutations/ Doklaly Academy of Sciences USSR (Press), *54(1)*, 65–68.
2. Auerbach, Ch. & Robson J. M. (1944). Production of mutations by allyl isothiocyanate/ *Nature, 154*, 81.
3. Auerbach, Ch. & Robson J. M. (1946). Chemical Production of Mutations/ *Letters to Editor Nature, 157*, 302.
4. Kihlman, B. A. (1963). Aberrations induced by radiomimetic compounds and their relations to radiation induced aberrations/Radiation-induced chromosome aberrations (S. Wolff, Ed.). Columbia University Press, N.Y., London, 100–122.
5. Loveless, A. (1970). Genetic allied effects of alkylating agents/London. 1966. Translation edited by N.P. Dubinin's. Moscow. Nauka, 255 p.
6. Navashin, M. S. (1985). Problems karyotype and cytogenetic studies in the genus *Crepis*. Moscow. Nauka, 349 p.
7. Dubinina, L. G. (1978). Structural mutations in the experiments with *Crepis capillaris.* Moscow. Nauka, 187 p.
8. Dubinin, N. P. (1966). Some key questions of the modern theory of mutations. *Genetika (Genetics), 7*, 3–20.
9. Sidorov, B. N., Sokolov, N. N. & Andreev, V. A. (1965). Mutagenic effect of ethylene imine in a number of cell generations. *Genetika (Genetics), 1*, 121–122.
10. Andreev, V. S., Sidorov B.N. & Sokolov N.N. (1966). The reasons for long-term mutagenic action of ethylenimine. *Genetika (Genetics), 4*, 28–36.
11. Chernov, V. A. (1964). Cytotoxic substances in chemotherapy of malignant tumors // Moscow. *Medicine,* 320 p.

12. Protopopova, E. M., Shevchenko, V. V. & Generalova, M. V. (1967). Beginning of DNA synthesis in seeds *Crepis capillaris. Genetika (Genetics), 6*, 19–23.
13. Rapoport, J. A. (1993). Discovery of chemical mutagenesis. The chosen works. Moscow. Nauka. 304 p.
14. Protopopova, E. M., Shevchenko, V. V. & Grigorjeva, G. A. (1969). Change of mutagen action of ethylene imine under influence of cellular metabolites. *Doclady Academy of Sciences USSR (Press), 86(2)*, 464–467.
15. Sanjay Kumar. (2010). Effect of 2,4-D and Isoproturon on chromosomal disturbances during mitotic division in root tip cells of *Triticum aestivum* L. *Cytology and Genetics, 44(2)*, 14–21.
16. Pandey, R. M. (2008). Cytotoxic effects of pesticide in somatic cells of *Vicia faba* L. *Cytology and Genetics, 42(6)*, 13–18.
17. Lekjavichjus, R. K. (1983). Chemical mutagenesis and contamination of environment/ Vilnjus. Moksklas, 223 p.

INDEX